PERSEUS IN SICILY:
FROM BLACK HOLE TO CLUSTER OUTSKIRTS

IAU SYMPOSIUM 342

COVER ILLUSTRATION:

(Top) X-ray image of the Per A. obtained by the Chandra X-ray Observatory (NASA/CXC/GSFC/S. A. Walker, et al.) together spectral data by the X-ray Astronomy Satellite "Hitomi" (Hitomi collaboration, JAXA, NASA, ESA, SRON, CSA). (Bottom Left) Innermost radio structure of 3C 84 taken by the RadioAstron (Giovannini et al. 2018) (Bottom Right) VSOP image of the Per A (Asada et al. 2006)

IAU SYMPOSIUM PROCEEDINGS SERIES

INTERNATIONAL ASTRONOMICAL UNION

UNION ASTRONOMIQUE INTERNATIONALE

International Astronomical Union

IAU

PERSEUS IN SICILY: FROM BLACK HOLE TO CLUSTER OUTSKIRTS

PROCEEDINGS OF THE 342nd SYMPOSIUM OF THE INTERNATIONAL ASTRONOMICAL UNION HELD IN NOTO, ITALY 14–18 MAY, 2018

Edited by

KEIICHI ASADA

Academia Sinica, Taiwan

ELISABETE DE GOUVEIA DAL PINO

Universidade de Sao Paulo, Brazil

MARCELLO GIROLETTI

Istituto Nazionale di Astrofisica, Italy

HIROSHI NAGAI

National Astronomical Observatory of Japan

and

RODRIGO NEMMEN

Universidade de Sao Paulo, Brazil

CAMBRIDGE
UNIVERSITY PRESS

CAMBRIDGE UNIVERSITY PRESS
University Printing House, Cambridge CB2 8BS, United Kingdom
1 Liberty Plaza, Floor 20, New York, NY 10006, USA
10 Stamford Road, Oakleigh, Melbourne 3166, Australia

First published 2020

Printed in the UK by Bell & Bain, Glasgow, UK

Typeset in System LaTeX 2ε

A catalogue record for this book is available from the British Library Library of Congress Cataloguing in Publication data

ISBN 9781108471534 hardback
ISSN 1743-9213

Table of Contents

Preface

The Perseus cluster has recently been the stage of ground-breaking discoveries, from the immediate vicinity of the black holes at the centres of its galaxies out to the dynamics of the gas on cluster scales. The importance and timeliness of discussing these and other topics was enhanced by developments in the fields of theory, numerical simulations, and imaging techniques. Multi-wavelength observers and theoreticians, experts from the event horizon out to the Megaparsec scales, were invited to the beautiful Sicilian city of Noto, a location rich in history, culture, nature, and science, near to the INAF 32m radio telescope and the CTA-ASTRI prototype on Mt. Etna.

The symposium has thus seen a broad participation, with science highlights ranging across several wavelengths and spatial scales, and including observations, interpretation, numerical simulations and pure theory. The organisers have been particularly pleased with the diversity of the involved communities, as is really in the spirit of IAU symposia. Talks and speakers came from different astrophysical backgrounds, as well as broad age, gender, and geographical distributions, but shared the common trait of giving excellent presentations.

A recurring quote was that "Perseus is weird", although it eventually remained open to debate whether this weirdness were intrinsic or rather the consequence of an unparalleled level of detail that we have gathered for this system.

On the **finest linear scales**, approaching the event horizon scales, the main novelty has been the **discovery of a wide and collimated radio jet in 3C 84 on the scale of a few hundred gravitational radii**, thanks to new space very-long-baseline-interferometry observations including the RadioAstron satellite. This finding has important implications on the formation of relativistic jets in active galactic nuclei, as discussed in a series of sessions devoted to general relativity magneto-hydrodynamic numerical simulations and to the technical and computational development required by the operations of the Event Horizon Telescope.

The (sub-)parsec scale properties have also been relevant for the connection to the high and very high energy (VHE) **gamma-ray** emission. 3C 84 is reaching **record level emission among just a handful of radio galaxies detected** in gamma rays. The time scales and the spectral properties have been debated in order to constrain the location and the physical properties of the region responsible for the gamma-ray emission in the jet or in the immediate vicinity of the black hole itself. Another Perseus galaxy, IC 310, has been reported to have very short time scale activity at VHE, which can possibly be explained with phenomena occurring on the black hole magnetosphere scale. Alternative acceleration processes involving multi-structures or turbulent magnetic reconnection have been proposed in order to explain current puzzles on the origin of VHE emission, particularly when it is highly variable.

On somewhat larger scales, starting from a few tens of kiloparsecs, amazing images in terms of resolution and dynamic range have been presented, thanks to new observations with Jansky VLA and other instruments. These datasets probe the non-thermal emission in the Perseus and other clusters over a broad range of wavelengths and reveal **a multitude of new structures** extending to hundreds of kpc in size. Their irregular morphology seems to have been influenced both by **the AGN activity and by the sloshing motion of the cluster' gas**. The gas properties have been the subject of X-ray focused talks, with exquisite energy and space resolution by Hitomi and Chandra, respectively. The former, in particular, has revealed a **mostly uniform and low velocity dispersion**.

How the AGN influences the host galaxy and the surrounding environment was also the subject of much debate, with discussions on the relative role of **AGN and supernovae-driven winds**, on the presence of persistent **filaments** both based on recent X-ray and sub-millimetre observations, and on **numerical simulations**. In this context, 3C 84 again appeared as an important prototype.

In terms of future breakthroughs, it has been important to hear about projects such as the X-ray observatory Athena, the Square Kilometre Array radio telescope, and the VHE Cherenkov Telescope Array, which will provide a transformational contribution to the above areas, besides many other topics. They shall eventually reveal whether "Perseus is weird" or if many more system present similar peculiarities. We also had a look at the recent past, with a rich, brilliant, and moving talk about the legacy of Ger de Bruyn for the study of the Perseus system, of galaxy clusters in general, and ultimately the passion for astrophysical research.

This rich science program was interleaved with a busy set of social activities, aimed at creating a good atmosphere among the participants, facilitating interactions about the science themes but also giving a flavour of the scientific, historical, naturalistic, and gastronomic richness and diversity of the host region. At the end of the sessions on Monday, which took place in the historical Tina di Lorenzo theatre, participants assisted to "Lu cuntu di Perseu", an emotional play based on the myth of Perseus. Other activities included a guided tour of the Baroque palaces in historic downtown Noto, the main social excursion to historic Syracuse, and a night of science, music, and food with live jazz music accompanying a buffet dinner by the INAF 32m radio telescope near Noto. At the very end of the symposium, on Friday night, participants had the chance to assist to the opening of the yearly traditional *Infiorata*: a cascade of thousands of flowers, expertly arranged on the road, filling the city with scents and colours. In parallel, the symposium has offered the opportunity to bring astrophysics to the general public: local high schools were involved in the design of a lithograph given to each participant, as well as in the adaption of the play for the international audience. A conference on "The sense of discovery" was offered to the general public, followed by a stargazing session.

We are grateful to the IAU for sponsoring this event and providing generous support to the participation of participants with more limited travel means. We also acknowledge significant funding from the RadioNet consortium, from the Italian National Institute of Astrophysics (INAF), and the local government in Noto.

Editors

Keiichi Asada (Academia Sinica, Institute of Astronomy and Astrophysics)
Elisabete Maria de Gouveia Dal Pino (Instituto de Astronomia, Geofisico e Ciencias Atmosfericas)
Marcello Giroletti (Italian National Institute for Astrophysics)
Hiroshi Nagai (National Astronomical Observatory of Japan)
Rodrigo Nemmen (Astronomy Department – Universidade de Sao Paulo)

Editors

Keiichi Asada
Academia Sinica, Taiwan

Elisabete de Gouveia Dal Pino
University of São Paulo, Brazil

Marcello Giroletti
Italian National Institute for Astrophysics, Italy

Hiroshi Nagai
National Astronomical Observatory of Japan, Japan

Rodrigo Nemmen
University of São Paulo, Brazil

Organising Committee

Scientific Organising Committee

SOC Co-Chairs

Elisabete de Gouveia Dal Pino	(IAG – Universidade de Sao Paulo, Brazil)
Marcello Giroletti	(Institute of Radioastronomy, Italy)
Christine Jones	(USA, Smithsonian Astrophysical Observatory)
Hiroshi Nagai	(National Astronomical Observatory of Japan, Japan)

SOC Members

Keiichi Asada	(Academia Sinica Institute of Astronomy and Astrophysics, Taiwan)
Roger Blandford	(Stanford University, USA)
Geoff Bower	(Academia Sinica Institute of Astronomy and Astrophysics, Taiwan)
Shep Doeleman	(Harvard University, USA)
Andy Fabian	(University of Cambridge, UK)
Paola Grandi	(Institute of Space Astrophysics and Cosmic Physics, Italy)
Chung-Pei Ma	(University of California, USA)
Rodrigo Nemmen	(Astronomy Department – Universidade de Sao Paulo, Brazil)
Lawrence Rudnick	(University of Minnesota, USA)
Aneta Siemiginowska	(Smithsonian Astrophysical Observatory, USA)
Lukasz Stawarz	(Jagiellonian University, Poland)
Feng Yuan	(Shanghai Astronomical Observatory, China)

Local Organising Committee

LOC Co-Chairs

Marcello Giroletti	(INAF Istituto di Radioastronomia, Bologna)
Barbara Neri	(INAF Istituto di Radioastronomia, Bologna)
Carlo Nocita	(INAF Istituto di Radioastronomia, Noto)
Grazia Umana	(INAF OsservatorioAstrofisico di Catania)

LOC Members

Giancarlo Bellassai	(INAF Osservatorio Astrofisico di Catania)
Milena Bufano	(INAF Osservatorio Astrofisico di Catania)
Eugenio Martinetti	(INAF Osservatorio Astrofisico di Catania)
Pier Raffaele Platania	(INAF Istituto di Radioastronomia, Noto)
Gina Santagati	(INAF Osservatorio Astrofisico di Catania)

Participants

1) Akiyama, Kazunori, United States
2) Asada, Keiichi, Taiwan, R.O.C.
3) Bacchini, Fabio, Belgium
4) Balasubramaniam, Karthik A, Poland
5) Barai, Paramita, Brazil
6) Bednarek, Wlodek, Poland
7) Bellassai, Giancarlo, Italy
8) Bhattacharyya, Dipanweeta, India
9) Bogovalov, Sergey, Russia
10) Bourne, Martin, UK
11) Bower, Geoffrey, United States
12) Brentjens, Michiel, Netherlands
13) Britzen, Silke, Germany
14) Bu, Defu, China
15) Bufano, Milena, Italy
16) Cao, Xinwu, China
17) Cassaro, Pietro, Italy
18) Chen, Yi-Hao, United States
19) Chernoglazov, Alexander, Russia
20) Churazov, Eugene, Germany
21) Ciotti, Luca, Italy
22) Clavijo Bohórquez, William E., Brazil
23) Combes, Francoise, France
24) de Gouveia Dal Pino, Elisabete, Brazil
25) Doro, Michele, Italy
26) Duc Thong, Le, Viet Nam
27) Dutta, Jayanta, India
28) Ehlert, Kristian, Germany
29) Fedorova, Elena, Ukraine
30) Forman, William, United States
31) Fukazawa, Yasushi, Japan
32) Gabuzda, Denise, Ireland
33) Gebhardt, Karl, United States
34) Gendron-Marsolais, Marie-Lou, Canada
35) Giovannini, Gabriele, Italy
36) Giroletti, Marcello, Italy
37) Glawion, Dorit, Germany
38) Govoni, Federica, Italy
39) Grandi, Paola, Italy
40) Guainazzi, Matteo, Netherlands
41) Hada, Kazuhiro, Japan
42) Hirano, Ayumi, Japan
43) Hodgson, Jeffrey, South Korea
44) Hull, Charles, Chile
45) Ignesti, Alessandro, Italy
46) Johnston-Hollitt, Melanie, Australia
47) Jones, Christine, United States
48) Kale, Ruta Prabhakar, India
49) Kim, Minsun, South Korea
50) Kim, Dong-Woo, United States
51) Klein, Uli, Germany
52) Knuettel, Sebastian, Ireland
53) Kosmaczewski, Emily, Poland
54) Koulouridis, Elias, France
55) Lakhchaura, Kiran, Indian
56) Lee, Jee Won, South Korea
57) Li, Shuangliang, China
58) Li, Yuan, USA
59) Lim, Jeremy, Australian
60) Lindfors, Elina, Finland
61) Longinotti, Anna Lia, Mexico
62) Lyskova, Natalia, Russian
63) Mahatma, Vijay, United Kingdom
64) Martinetti, Eugenio, Italy
65) Massaro, Francesco, Italy
66) Migliori, Giulia, Italy
67) Millas, Dimitrios, Belgium
68) Mizuno, Yosuke, Germany
69) Morganti, Raffaella, The Netherlands
70) Moscibrodzka, Monika, Netherlands
71) Nagai, Hiroshi, Japan
72) Negrete, Alenka, Mexico
73) Nemmen, Rodrigo, Brazil
74) Neri, Barbara, Italy
75) Nocita, Carlo, Italia
76) Nokhrina, Elena, Russia
77) Oh, Junghwan, South Korea
78) Orienti, Monica, Italy
79) Panessa, Francesca, Italy
80) Park, Jongho, South Korea
81) Pellegrini, Silvia, Italy
82) Perger, Krisztina, Hungary
83) Pezzulli, Edwige, Italy
84) Platania, Pier Raffaele, Italy
85) Popovic, Luka, Serbian
86) Reynolds, Christopher, United Kingdom
87) Rodriguez Ramirez, Juan Carlos, Brazil
88) Roelofs, Freek, Netherlands
89) Sacchi, Andrea, Italy
90) Sahakyan, Narek, Armenia
91) Sanders, Jeremy, Germany
92) Sani, Eleonora, Chile
93) Santagati, Gina, Italy
94) Savolainen, Tuomas, Finland
95) Sijacki, Debora, United Kingdom
96) Sohn, Bong Won, South Korea
97) Tamilan, Mageshwaran, India
98) Tamura, Takayuki, Japan
99) Thimmappa, Rameshan, Poland
100) Torresi, Eleonora, Italy
101) Toscani, Martina, Italy
102) Trippe, Sascha, South Korea
103) Umana, Grazia, Italy
104) VahdatMotlagh, Armin, Iran
105) Vaidya, Bhargav, India
106) Van der Westhuizen, Izak, South Africa
107) Varano, Stefania, Italy
108) Vijayaraghavan, Rukmani, United States
109) Webster, Brendan, United Kingdom
110) Wójtowicz, Anna, Poland
111) Younsi, Ziri, Germany
112) Yuan, Feng, Chinese
113) Yuan, Zhong-Sheng, China
114) Zdziarski, Andrzej, Poland
115) Zhang, Congyao, Germany
116) Zhuravleva, Irina, United States

Perseus in Sicily: From Black Hole to Cluster Outskirts
Proceedings IAU Symposium No. 342, 2018
K. Asada, E. de Gouveia Dal Pino, M. Giroletti,
H. Nagai & R. Nemmen, eds.

doi:10.1017/S1743921318007640

Black hole mass measurements in AGN: Polarization in broad emission lines

Luka Č. Popović[1], Victor L. Afanasiev[2] and Djordje Savić[1]

[1]Astronomical Observatory, 11160 Belgrade, Serbia,
email: lpopovic@aob.rs; djsavic@aob.rs

[2]Special Astrophysical Observatory of the Russian AS,
Nizhnij Arkhyz, Karachaevo-Cherkesia 369167, Russia
email: vafan@sao.ru

Abstract. We present a new method for supermassive black hole (SMBH) mass measurements in Type 1 active galactic nuclei (AGN) using polarization angle across broad lines. This method gives measured masses which are in a good agreement with reverberation estimates. Additionally, we explore the possibilities and limits of this method using the STOKES radiative transfer code taking a dominant Keplerian motion in the broad line region (BLR). We found that this method can be used for the direct SMBH mass estimation in the cases when in addition to the Kepler motion, radial inflows or vertical outflows are present in the BLR. Some advantages of the method are discussed.

Keywords. polarization, spectral line shapes, active galactic nuclei, black holes: masses

1. Introduction

Supermassive Black Holes (SMBHs, $10^6 - 10^9\ M_\odot$) are supposed to reside in the bulges of spiral and elliptical galaxies, and they are in action to shape their cosmic environment, i.e. it seems that there is a connection between the central SMBH and the host galaxy structure. The close connection between the formation and evolution of galaxies and of their central SMBHs involves a variety of physical phenomena of great relevance in modern astrophysics (see e.g. Heckman & Best 2014). The parameters which define a SMBH are mass, spin and electricity, where mass of SMBH is the most important and probably is in the correlation with the galaxy bulge mass. Therefore, one of the most important issue in astrophysics today is to measure masses of SMBHs.

One of the most powerful objects in the Universe are active galactic nuclei (AGNs) which emit huge amount of energy that is created around SMBHs. The emission gas is very close to the central SMBH, and therefore the gas kinematics is directly influenced by the mass of the central black hole. Especially if the gas emits the broad lines from the so called broad line region (BLR). The width of lines emitted from the BLR corresponds to the rotational velocity that can be used for estimation of the black hole masses in AGNs.

In principle there are several methods for the SMBH mass measurements (see Peterson 2014), out of which the reverberation method is the one often used in the case of AGNs. Using this method one can estimate the SMBH mass as:

$$M_{\rm BH} = f\frac{R_{\rm BLR}\sigma_V^2}{G},$$

where $R_{\rm BLR}$ is the photometric dimension of the BLR obtained from the reverberation mapping, and σ_V is the corresponding orbital velocity that can be estimated from the broad line widths. G is the gravitational constant and the virial factor f depends on the BLR inclination and geometry which are unknown and, consequently there is a problem to accurately determine the virial factor (Peterson 2014).

On the other side, the polarization in the broad lines can be very useful for investigation of the nature of the BLR (Smith *et al.* 2002, 2004, 2005; Afanasiev *et al.* 2014; Afanasiev & Popović 2015, Afanasiev *et al.* 2018 etc.). The polarization in the broad lines, especially the polarization angle (φ) that shows horizontal S shape across the broad line profile, which indicates a dominant Keplerian-like motion in the BLR, and dominant equatorial scattering as a polarization mechanism in the broad lines (see Afanasiev & Popović 2015).

In this contribution we shortly describe the method and give some overview of the numerical tests of the validity (limitations) of the method which have been performed by STOKES code (see Savić *et al.* 2018)

2. Polarization method for SMBH measurements

The method is described in more details in Afanasiev & Popović (2015). We used the fact that in the case of equatorial scattering, the polarization angle (φ_i) is connected with the distance (R_i) of an emitting cloud (see Fig. 1 in Afanasiev & Popović 2015) as

$$\tan \varphi_i \sim R_i,$$

that in the case of a Keplerian like motion can be connected with the velocity ($V_i = (\lambda_i - \lambda_0)/\lambda_0$, where λ_0 is the transition wavelength) of the emitting cloud (i.e. λ_i is the wavelength emitted by the cloud) as

$$V_i = \sqrt{\frac{GM_{\rm BH}}{R_i}},$$

where $M_{\rm BH}$ is the mass of SMBH.

It is not difficult to obtain that $R_i = R_{sc} \cdot \tan \varphi_i$, where $R_{\rm sc}$ is the the distance between the SMBH and the scattering region (supposed to be in the inner part of the torus, see Fig. 1 in Afanasiev & Popović 2015). Considering the above relations one can obtain the relationship between the velocity (V_i) and $\tan \varphi$ as:

$$\log\left(\frac{V_i}{c}\right) = a - 0.5 \cdot \log(\tan(\Delta\varphi_i)),$$

where c is the speed of light, and constant a depends on the SMBH mass ($M_{\rm BH}$) as

$$a = 0.5 \log\left(\frac{GM_{\rm BH}\cos^2(\theta)}{c^2 R_{\rm sc}}\right),$$

that can be used for the SMBH mass measurements.

We show in Afanasiev & Popović (2015) that the method gives SMBH mass estimates which can be compared with the reverberation mapping results. The method seems to be very perspective, since, as it can be seen in Fig. 1 (left panel), very often we can observe a horizontal S-shaped polarization angle in the case of Type 1 AGNs (AGNs with the broad emission lines, see Afanasiev *et al.* 2018).

3. Modeling and method validity and limitations

Following the method described by Afanasiev & Popović (2015), we theoretically model the broad line polarization due to equatorial scattering. We assume that the unpolarized lines are emitted from the disk-like BLR with dominant Keplerian motion and scattered by free electrons at the inner part of the dusty torus. We used 3D Monte Carlo radiative

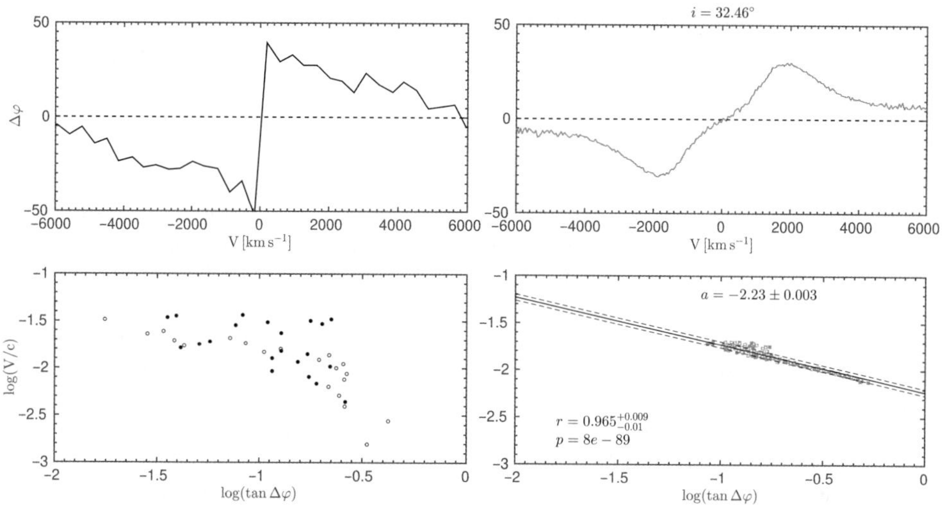

Figure 1. The observed (left) and modeled (right) polarization angle changes across the Hα line profile (panels up) for NGC 4151 and $\log(\frac{V}{c})$ vs. $\tan(\Delta\varphi)$ (panels down) relationship (in more detail see Savić *et al.* 2018).

transfer code STOKES (see, e.g. Goosmann & Gaskell 2007, Marin *et al.* 2015, Marin 2018, Rojas Lobos *et al.* 2018). The size of the BLR was obtained from the reverberation mapping with the outer edge set due to dust sublimation. Model parameters of the BLR and the scattering region (SR) are described in more details in Savić *et al.* (2018), and here will not be repeated. The models are showing that the mass estimates with the polarization method are in the frame of 10% error-bars in the case that the SR distance which is twice the size of the outer limit of the BLR. The largest number of photons is scattered only once from the inner edge of the SR. The number of multiple scattering events is negligible and the assumption of a single scattering approximation is valid. In Fig. 1 (right panel) we show the modeled polarization angle across the broad Hα profile for NGC 4151, and as one can see from Fig 1, the modeled profile is very similar to the observed one. Also the relationship between velocity and $\tan\varphi$ across the line profile can be reproduced (Fig 1. bottom panels). This indicates that the single element equatorial approximation is valid.

Additionally, we considered the possible inflows/outflows in the BLR (see Savić *et al.* 2018) and we have shown that the method for SMBH mass measurements can be used when the contribution of the inflows/outflow is much smaller in comparison with Keplerian motion. Therefore, this method could potentially be used for highly ionized lines such as C III] and C IV that are observed in the optical domain for high-redshifted quasars.

For high inclinations, the polar scattering becomes dominant and we have Type-2 objects for which the method is no longer valid. For face-on AGNs, the optical polarization is usually much lower than 1% (Smith *et al.* 2002, Marin 2014) and the amount of interstellar polarization can dominate the equatorial scattering induced polarization in the innermost part of the AGNs. The variability of Type-1 AGNs must also be taken into account. The flux variations of the continuum and broad lines can be up to a factor of 10 or greater between minimum and maximum activity state (Shapovalova *et al.* 2008, 2010). When observed in the minimum activity (up to Type-2), the absence or the weak flux in the BLRs could not be detected and the method cannot be used.

4. Discussion and conclusions

In this paper we shortly describe the polarization method for the SMBH mass measurement using the shape of polarization angle across the broad line profiles (Afanasiev *et al.* 2014, Afanasiev & Popović 2015, Savić *et al.* 2018). We found that our method gives very good results in comparison with the reverberation one.

As conclusions we can list the advantages and disadvantages of the method. First, let us mention the advantages, which are: i) the method does not need assumption of virialization (as in the case of the reverberation one), since the horizontal S-like shape of polarization angle across the broad line indicates presence of the Keplerian-like motion (see Fig. 1); ii) the method is not time consuming, i.e. from one epoch observation of broad line, one can obtain the SMBH mass; iii) the method can be applied to all broad lines from the UV to the optical, i.e. one can measure the SMBH mass of different redshifted AGNs in the same (consistent) way.

On the other hand, there are some disadvantages of the method which are: i) the polarization in the broad line of AGNs is not so big, therefore the effect can be hidden by the other (additional) effects of polarization or depolarization; ii) in the case of very strong outflow and inflows, the effect cannot be clearly seen, and could not be used for the SMBH mass determination and iii) there can be a problem with the estimation of distance of the equatorial scattering region (R_{sc}).

Finally, let us conclude that the method is very useful and it represents a new method for the SMBH mass determination that can be used also in the case of high redshifted AGNs.

Acknowledgement

This work was supported by the Russian Foundation for Basic Research (project N15-02-02101) and the Ministry of Education, Science and Technological Development (Republic of Serbia) through the project Astrophysical Spectroscopy of Extragalactic Objects (176001).

References

Afanasiev, V. L., & Popović, L. Č. 2015, *ApJ*, 800L, 35

Afanasiev, V. L., Popović, L. Č., & Shapovalova, A. I. 2018, *MNRAS* sent

Afanasiev, V. L., Popović, L. Č., Shapovalova, A. I., Borisov, N. V., & Ilić, D. 2014, *MNRAS*, 440, 519

Goosmann, R. W. & Gaskell, C. M. 2007, *A&A*, 465, 129

Heckman T. M. & Best P. N. 2014 *ARAA* 52 589

Marin, F. 2014, *MNRAS*, 441, 551

Marin, F. 2018, ArXiv e-prints [arXiv:1805.09098]

Marin, F., Goosmann, R. W., & Gaskell, C. M. 2015, *A&A*, 577, A66

Peterson, B. M. 2014, *SSRev*, 183, 253

Rojas Lobos, P. A., Goosmann, R. W., Marin, F., & Savić, D. 2018, *A&A*, 611, A39

Savić, Dj., Goosmann, R.; Popović, L. Č., Marin, F., & Afanasiev, V. L. 2018, *A&A*, accepted (arXiv:1801.06097)

Shapovalova, A. I., Popović, L. Č., Burenkov, A. N., *et al.* 2010, *A&A*, 509, A106

Shapovalova, A. I., Popović, L. Č., Collin, S., *et al.* 2008, *A&A*, 486, 99

Smith, J. E., Young, S., Robinson, A., *et al.* 2002, *MNRAS*, 335, 773

Smith, J. E., Robinson, A., Alexander, D. M., *et al.* 2004, *MNRAS*, 350, 140

Smith, J. E., Robinson, A., Young, S., *et al.* 2005, *MNRAS*, 359, 846

Perseus in Sicily: From Black Hole to Cluster Outskirts
Proceedings IAU Symposium No. 342, 2018
K. Asada, E. de Gouveia Dal Pino, M. Giroletti,
H. Nagai & R. Nemmen, eds.

doi:10.1017/S1743921318005732

Fully analytical solutions for Bondi accretion in galaxies with a central Black Hole

Luca Ciotti and Silvia Pellegrini

Department of Physics and Astronomy, University of Bologna,
via Gobetti 93/2, 40129 Bologna, Italy
email: luca.ciotti@unibo.it

Abstract. The fully analytical solution for isothermal Bondi accretion on a black hole (MBH) at the center of JJ two-component Jaffe (1983) galaxy models is presented. In JJ models the stellar and total mass density distributions are described by the Jaffe profile, with different scale-lengths and masses, and to which a central MBH is added; all the relevant stellar dynamical properties can also be derived analytically. In these new accretion solutions the hydrodynamical and stellar dynamical properties are linked by imposing that the gas temperature is proportional to the virial temperature of the stellar component. The formulae that are provided allow to evaluate all flow properties, and are then useful for estimates of the accretion radius and the mass flow rate when modeling accretion on MBHs at the center of galaxies.

Keywords. galaxies: elliptical and lenticular, cD, galaxies: ISM, galaxies: nuclei, hydrodynamics, stellar dynamics

1. Introduction

Observational and numerical investigations of accretion on massive black holes (hereafter MBH) at the center of galaxies often lack the resolution to follow gas transport down to the parsec scale. In these cases, the *classical* Bondi (1952) solution on to an isolated central point mass is then commonly adopted, for estimates of the accretion radius (i.e., the sonic radius), and the mass accretion rate (see Ciotti & Pellegrini 2017, hereafter CP17, and references therein). However, two major problems affect the direct application of the classical Bondi solution, namely the facts that 1) the boundary values of density and temperature of the accreting gas are assigned at infinity, and 2) in a galaxy, the gas experiences the gravitational effects of the galaxy itself (stars plus dark matter), and the MBH gravity becomes dominant only in the very central regions. The solution commonly adopted is to use values of the gas density and temperature "sufficiently near" the MBH. It is therefore important to quantify the systematic effects, on the estimates obtained from the classical Bondi solution for the accretion radius and the mass accretion rate, due to measurements taken at finite distance from the MBH, and under the effects of the galaxy potential well. A first step analysis of this problem was carried out in Korol *et al.* (2016, hereafter KCP16) where the Bondi problem was generalized to the case of mass accretion at the center of galaxies, including also the effect of electron scattering on the accreting gas. CP17 showed that the whole accretion solution can be given in an analytical way for the *isothermal* accretion in Jaffe (1983) galaxies with a central MBH. Ciotti & Pellegrini (2018, hereafter CP18, in preparation), extend the study to JJ two-component galaxy models (Ciotti & Ziaee Lorzad 2018, hereafter CZ18), where the *stellar* and *total* mass density distributions are both described by the Jaffe profile, with different scale-lengths and masses, and a MBH is added at the center. In particular, JJ models offer the *unique* opportunity to have a quite realistic family of galaxy models with a central MBH, allowing both for the fully analytical solution of the Bondi (isothermal) accretion problem, *and* the fully analytical solution of the Jeans equations.

2. The models

2.1. *The Bondi solution*

In the Bondi problem, the gas is perfect, has a spatially infinite distribution, and is accreting on to a MBH, of mass $M_{\rm BH}$. The gas density ρ and pressure p are linked by the polytropic relation

$$p = \frac{k_{\rm B}\rho T}{<\mu> m_{\rm p}} = p_\infty \tilde{\rho}^\gamma, \quad \tilde{\rho} \equiv \frac{\rho}{\rho_\infty}, \tag{2.1}$$

where $\gamma = 1$ in the isothermal case, $m_{\rm p}$ is the proton mass, and p_∞ and ρ_∞ are respectively the gas pressure and density at infinity. The sound speed is $c_{\rm s} = \sqrt{\gamma p/\rho}$, and in the isothermal case $T(r) = T_\infty$ and so $c_{\rm s}(r) = c_\infty$. With the introduction of the Bondi radius $r_{\rm B}$ and of the Mach number $\mathcal{M}$, given respectively by

$$r_{\rm B} \equiv \frac{GM_{\rm BH}}{c_\infty^2}, \quad \mathcal{M}(r) = \frac{v(r)}{c_{\rm s}(r)}, \tag{2.2}$$

the time-independent continuity equation becomes

$$x^2 \tilde{\rho}(x)\mathcal{M}(x) = \frac{\dot{M}_{\rm t}}{4\pi r_{\rm B}^2 \rho_\infty c_\infty} \equiv \lambda, \quad x \equiv \frac{r}{r_{\rm B}}, \tag{2.3}$$

where $\dot{M}_{\rm t}$ is the mass accretion rate on the MBH at the center of the galaxy, and λ the accretion parameter.

For a Jaffe galaxy of total mass $M_{\rm g}$ and scale-length $r_{\rm g}$, the gravitational potential and the two parameters determining the accretion solution are (CP17):

$$\Phi_{\rm g} = \frac{GM_{\rm g}}{r_{\rm g}} \ln \frac{r}{r + r_{\rm g}}, \quad \mathcal{R} \equiv \frac{M_{\rm g}}{M_{\rm BH}}, \quad \xi \equiv \frac{r_{\rm g}}{r_{\rm B}}. \tag{2.4}$$

In the Bondi problem for isothermal accretion reduces to the solution of

$$\frac{\mathcal{M}^2}{2} - \ln \mathcal{M} = f(x) - {\rm e}^\lambda, \quad f = \frac{\chi}{x} - \frac{\mathcal{R}}{\xi} \ln \frac{x}{x + \xi} + 2 \ln x, \tag{2.5}$$

where $\chi \equiv 1 - L/L_{\rm Edd}$ measures the effect of electron scattering radiation pressure due to the accretion luminosity L. Solutions exist only for $\lambda \leqslant \lambda_{\rm t}$, the *critical* accretion parameter, a value determined by the position of the minimum $x_{\rm min}$ for the function f. For the Jaffe galaxy the position of the only minimum of f (corresponding to the sonic radius of the critical solution) can be calculated analytically, with $x_{\rm min} = x_{\rm min}(\chi, \mathcal{R}, \xi)$, so that quite surprisingly one can evaluate $\lambda_{\rm t}$ analytically. In the peculiar case of $\chi = 0$ (and/or $M_{\rm BH} = 0$), a solution of the accretion problem is possible only for $\mathcal{R} \geqslant 2\xi$. Moreover, the radial trend of the Mach number can also be calculated analytically in terms of the so-called W Lambert-Euler function.

CP18 apply the previous results to JJ models, that are characterized by a *total* Jaffe density distribution $\rho_{\rm g}$ (stars plus dark matter) of total mass $M_{\rm g}$ and scale-length $r_{\rm g}$, and a stellar Jaffe distribution of stellar mass M_*, and scale radius r_*. Remarkably, almost all the stellar dynamical properties of JJ models with a central MBH can be expressed by analytical functions (CZ18), so that they are a family of two-component galaxy models with a central MBH allowing for a full analytical treatment of Bondi accretion and of stellar dynamics (CP18).

2.2. *Structure and dynamics of the JJ models*

The stellar and total density distributions of JJ galaxies are given respectively by

$$\rho_*(r) = \frac{\rho_n}{s^2(1+s)^2}, \quad \rho_g(r) = \frac{\rho_n \mathcal{R}_g \xi_g}{s^2(\xi_g+s)^2}, \quad s \equiv \frac{r}{r_*}, \tag{2.6}$$

where we define

$$\rho_n \equiv \frac{M_\star}{4\pi r_*^3}, \quad \Psi_n \equiv \frac{GM_\star}{r_*}, \quad \mu \equiv \frac{M_{BH}}{M_\star}, \quad \mathcal{R}_g \equiv \frac{M_g}{M_\star}, \quad \xi_g \equiv \frac{r_g}{r_*}. \tag{2.7}$$

Typically, $\mu \approx 10^{-3}$. When $\xi_g \geqslant 1$, the density distribution of the dark halo $\rho_{DM} = \rho_g - \rho_*$ is nowhere negative, provided that $\mathcal{R}_g = \alpha\xi_g$ with $\alpha \geqslant 1$. A galaxy with $\alpha = 1$ is called a *minimum halo* model, and the associated halo is well approximated by the NFW (Navarro *et al.* 1997) profile over a very large radial range (CZ18). For JJ models the Osipkov-Merrit anisotropic Jeans equation, and the projected values of the velocity dispersion at the galaxy center, can be expressed analytically. In particular, the virial temperature of the stellar component can be written as $T_V = <\mu> m_p\sigma_V^2/3$, where the virial velocity dispersion of stars is given by $\sigma_V^2 = \Psi_n\alpha\mathcal{F}_g(\xi_g)$, and $\mathcal{F}_g(\xi_g)$ is a simple analytical function, with $\mathcal{F}_g(1) = 1/2$ and $\mathcal{F}_g(\infty) = 1$.

2.3. *Linking stellar dynamics to fluidodynamics*

In CP18 the idea is to self-consistently "close" the accretion solution, determining a fiducial value for the gas temperature as a function of the galaxy model hosting accretion. For assigned values of $\xi_g \geqslant 1$, $\mathcal{R}_g$ (or α), and μ, we fix $T_\infty = \beta T_V$, with $\beta > 0$, so that $c_\infty = \sigma_V\sqrt{\beta/3}$. All the accretion parameters can therefore be obtained in terms of the galaxy properties as

$$\mathcal{R} = \frac{\alpha\xi_g}{\mu}, \quad \xi = \frac{\mathcal{R}\beta\mathcal{F}_g}{3}, \quad \frac{r_B}{r_*} = \frac{3\mu}{\alpha\beta\mathcal{F}_g}, \quad \frac{r_{min}}{r_*} = x_{min}(\chi, \mathcal{R}, \xi)\,\frac{r_B}{r_*}, \tag{2.8}$$

where r_{min} is the position of the sonic radius. Then the critical accretion parameter λ_t, and the Mach number profile, can be computed analytically using the results of CP17. It turns out that for JJ models with $\chi = 0$ (and/or $M_{BH} = 0$) the Bondi solution exists only for $\beta \leqslant \beta_c \equiv 3/(2\mathcal{F}_g)$ with $3/2 \leqslant \beta_c \leqslant 3$, and that β_c determines the behavior of the solution also in presence of a MBH. The first three panels in Fig. 1 show some representative cases of the quantities in eq. (2.8). One of the most relevant features is the considerable jump of r_{min} from very external to very internal galactic regions, even for a slight increase of the gas temperature. Finally, CP18 evaluate the departure of the estimated mass accretion rate $\dot{M}_e(r) \equiv 4\pi r_B^2\lambda_{cr}\rho(r)c_\infty$ obtained from the classical Bondi solution, from the true value $\dot{M}_t$, as a function of the distance from the center:

$$\frac{\dot{M}_e(r)}{\dot{M}_t} = \frac{\lambda_{cr}\tilde{\rho}(x)}{\lambda_t} = \frac{\lambda_{cr}}{x^2\mathcal{M}(x)}, \tag{2.9}$$

where $\mathcal{M}(x)$ is the solution of eq. (2.5). Here $\rho(r)$ is taken along the solution for accretion within the potential of the galaxy and used as "proxy" for the true value ρ_∞, and $\lambda_{cr} = e^{3/2}/4$ is the isothermal critical accretion parameter of the Bondi solution on an isolated MBH. Note how the bias increases from values much lower than unity in the outer galactic regions to very large values near the center; this allows to directly estimate the so-called "boost factor" (see CP18 for a thorough discussion).

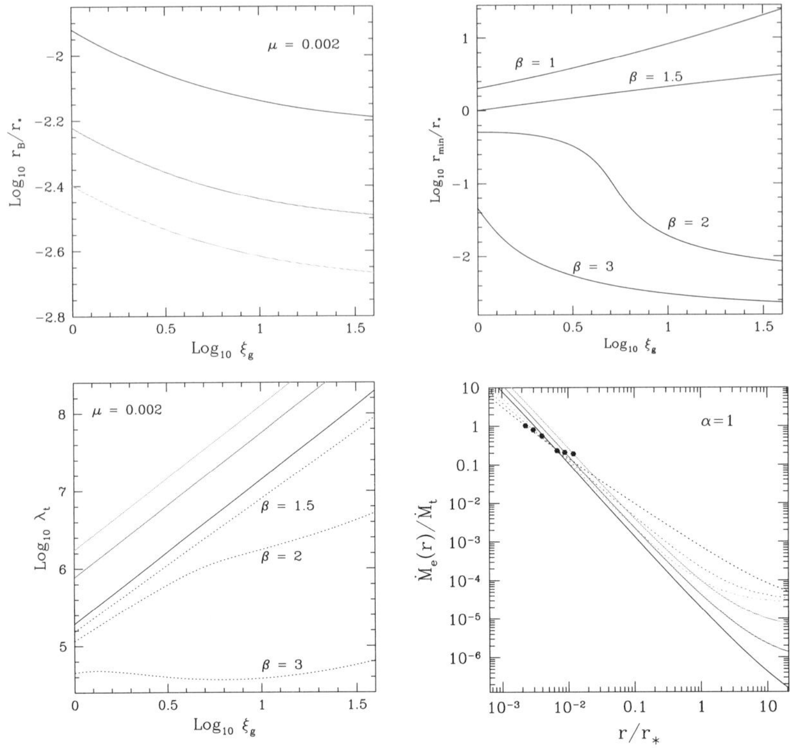

Figure 1. Relevant scale-lengths of the isothermal accretion solution in JJ models, as a function of $\xi_g = r_g/r_*$. A MBH-to-galaxy stellar mass ratio $\mu = 2 \times 10^{-3}$ is assumed, and $T_\infty = \beta T_V$. Top left: the ratio r_B/r_* with $\beta = 1$ and $\alpha = 1$ (black), 2 (blue), 3 (red). Top right: ratio r_{min}/r_* for $\alpha = 1$ and $\beta = 1, 3/2, 2, 3$: for large values of $\mathcal{R}$ and $\beta < \beta_c$ the ratio is almost independent of α but strongly dependent on gas temperature; for $\beta > \beta_c$, the sonic radius collapses near the center. Bottom left: the critical accretion parameter λ_t as a function of ξ_g, for $\alpha = 1, 2, 3$, $\chi = 1$, and $\beta = 1$ (solid lines). The dotted curves refer to $\alpha = 1$ and three different values of β. Bottom right: ratio between the estimate of the accretion rate $\dot{M}_e$ and the true accretion rate $\dot{M}_t$, as a function of r/r_*, in the minimun halo case ($\alpha = 1$), and $\xi_g = 1$ (red), $\xi_g = 3$ (blue), and $\xi_g = 20$ (black). The dotted lines correspond to $\beta = 3$, and the solid dots mark the position of the Bondi radius.

References

Bondi, H., 1952, *MNRAS* 112, 195

Ciotti, L., & Ziaee Lorzad, A. 2018, *MNRAS* 473, 5476 (CZ18)

Ciotti, L., & Pellegrini, S. 2017, *ApJ* 848, 29 (CP17)

Jaffe, W., 1983, *MNRAS* 202, 995

Korol, V., Ciotti, L., & Pellegrini, S. 2016, *MNRAS* 460, 1188 (KCP16)

Navarro, J. F., Frenk, C. S., & White, S. D. M. 1997, *ApJ* 490, 493

Perseus in Sicily: From Black Hole to Cluster Outskirts
Proceedings IAU Symposium No. 342, 2018
K. Asada, E. de Gouveia Dal Pino, M. Giroletti,
H. Nagai & R. Nemmen, eds.

doi:10.1017/S1743921318007263

Modelling the polarised emission from black holes on event horizon-scales

Ziri Younsi[1,2], Oliver Porth[1], Yosuke Mizuno[1], Christian M. Fromm[1,3] and Hector Olivares[1]

[1]Institut für Theoretische Physik, Max-von-Laue-Straße 1,
D-60438 Frankfurt am Main, Germany
email: younsi@itp.uni-frankfurt.de

[2]Mullard Space Science Laboratory, University College London,
Holmbury St. Mary, Dorking, Surrey RH5 6NT, UK
email: z.younsi@ucl.ac.uk

[3]Max-Planck-Institut für Radioastronomie, Auf dem Hügel 69, D-53121 Bonn, Germany

Abstract. Upcoming VLBI observations will resolve nearby supermassive black holes, most notably Sagittarius A* and M87, on event horizon-scales. Recent observations of Sagittarius A* with the Event Horizon Telescope have revealed horizon-scale structure. Accordingly, the detection and measurement of the back hole "shadow" is expected to enable the existence of astrophysical black holes to be verified directly. Although the theoretical description of the shadow is straightforward, its observational appearance is largely determined by the properties of the surrounding accretion flow, which is highly turbulent. We introduce a new polarised general-relativistic radiative transfer code, BHOSS, which accurately solves the equations of polarised radiative transfer in arbitrary strong-gravity environments, providing physically-realistic images of astrophysical black holes on event horizon-scales, as well as also providing insight into the fundamental properties and nature of the surrounding accretion flow environment.

Keywords. gravitation, methods: numerical, radiative transfer, relativity, polarisation

1. Introduction

It is widely believed that all galaxies host a supermassive black hole (SMBH) at their center. Advances in very-long-baseline-interferometry (VLBI) have enabled the Event Horizon Telescope Collaboration (EHTC) to image (with event horizon-scale resolution) the nearby SMBHs Sagittarius A* (Sgr A*) and M87, with the first observational results expected soon (e.g., Doeleman *et al.* 2008, Goddi *et al.* 2017). When observing an astrophysical black hole, theorists anticipate seeing a "shadow", the silhouette of the unstable photon region surrounding the event horizon (e.g., Cunningham & Bardeen 1973, Grenzebach *et al.* 2014, Younsi *et al.* 2016). Gravity near the event horizon is so strong that spacetime is curved and light rays no longer travel along straight lines. Moreover, the surrounding environment is hot, magnetised and turbulent, serving to obscure and modify the shadow's observed properties. As such, both the equations of motion for light rays and the radiative transfer equation must be solved in full general relativity.

We present results from our new general-relativistic radiative transfer (GRRT) code, BHOSS (Younsi *et al.* 2018, in prep.), which solves the equations of polarised radiative transfer for arbitrary spacetime metrics (e.g., Younsi *et al.* 2016, Mizuno *et al.* 2018), and is fully coupled to many three-dimensional general-relativistic magnetohydrodynamical (GRMHD) codes, including BHAC (Porth *et al.* 2017) and HARM (Gammie *et al.* 2003).

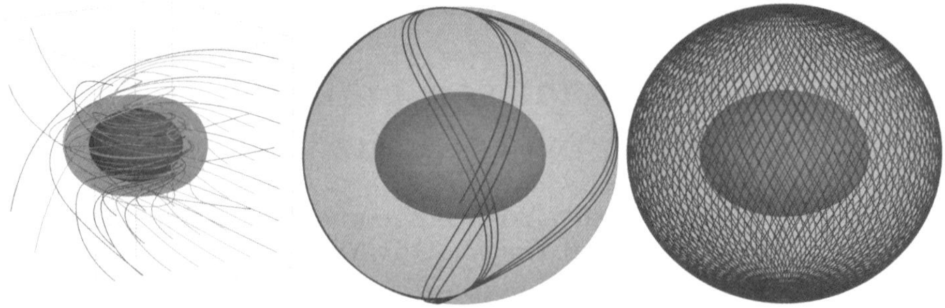

Figure 1. *Photon geodesics around an extremal Kerr black hole.* **Left panel:** 7×7 pixel observer grid of rays with event horizon in red and ergo-region in blue. **Middle panel:** 6π longitudinal oscillations of a spherical photon orbit (event horizon in green). **Right panel:** 100π longitudinal oscillations of the spherical photon orbit. The orbit is stable for long integration times.

2. General-relativistic ray-tracing and radiative transfer

Ray-tracing. In order to construct an image of a black hole, the paths of light rays (photons) comprising an image are ray-traced by solving the geodesic equations of motion:

$$\frac{\mathrm{d}x^{\alpha}}{\mathrm{d}\lambda} = k^{\alpha}, \tag{2.1}$$

$$\frac{\mathrm{d}k^{\alpha}}{\mathrm{d}\lambda} = -\Gamma^{\alpha}{}_{\mu\nu}k^{\mu}k^{\nu}, \tag{2.2}$$

where x^{α} and k^{α} are, respectively, the photon position and 4-momentum, $\Gamma^{\alpha}{}_{\mu\nu}$ denote the Christoffel symbols, λ is the affine parameter and Greek indices range from 0–3 throughout. Appropriate initial conditions may be determined using a pre-defined observer "camera" (see, e.g., Younsi *et al.* 2016 and left panel of Fig. 1). Tensor index raising and lowering operations are performed with the metric tensor, $g_{\mu\nu}$, e.g., $k_{\mu} = g_{\mu\nu}k^{\nu}$. Geodesics in any spacetime may be calculated by specifying the appropriate $g_{\mu\nu}$.

Numerical errors in geodesic integration grow over time, and are minimised in `BHOSS` by using several different high-order integration schemes. A robust test of stability is the long-term integration of spherical photon orbits (e.g., Teo 2003, Chan *et al.* 2017), which are notoriously challenging to maintain for long times but readily maintained in `BHOSS` (see middle and right panels of Fig. 1).

Polarised radiative transfer. Along each ray (image pixel) the intensity and polarisation properties are calculated from the equations of polarised radiative transfer. The observer's polarisation basis can be represented by parallel and perpendicular 4-vectors of this basis, $f^{\alpha}_{(\parallel)}$ and $f^{\alpha}_{(\perp)}$, which are orthogonal both to each other and to k^{α}, and solved as:

$$\frac{\mathrm{d}f^{\alpha}}{\mathrm{d}\lambda} = -\Gamma^{\alpha}{}_{\mu\nu}k^{\mu}f^{\nu}. \tag{2.3}$$

Defining the Lorentz-invariant Stokes vector as $\boldsymbol{\mathcal{S}} := (\mathcal{I}, \mathcal{Q}, \mathcal{U}, \mathcal{V})^{\mathrm{T}}$, where $\boldsymbol{\mathcal{S}} \equiv \mathbf{S}/\nu^3$ and $\mathbf{S} := (I, Q, U, V)^{\mathrm{T}}$, the full polarised GRRT equation may be written as:

$$\frac{\mathrm{d}}{\mathrm{d}\lambda}\begin{pmatrix}\mathcal{I}\\ \mathcal{Q}\\ \mathcal{U}\\ \mathcal{V}\end{pmatrix} = -k_{\mu}u^{\mu}\left\{R(\chi)\begin{pmatrix}\bar{\varepsilon}_I\\ \bar{\varepsilon}_Q\\ 0\\ \bar{\varepsilon}_V\end{pmatrix} - \left[R(\chi)\begin{pmatrix}\bar{\alpha}_I & \bar{\alpha}_Q & 0 & \bar{\alpha}_V\\ \bar{\alpha}_Q & \bar{\alpha}_I & \bar{\rho}_V & 0\\ 0 & -\bar{\rho}_V & \bar{\alpha}_I & \bar{\rho}_Q\\ \bar{\alpha}_V & 0 & -\bar{\rho}_Q & \bar{\alpha}_I\end{pmatrix}R(-\chi)\right]\begin{pmatrix}\mathcal{I}\\ \mathcal{Q}\\ \mathcal{U}\\ \mathcal{V}\end{pmatrix}\right\}, \tag{2.4}$$

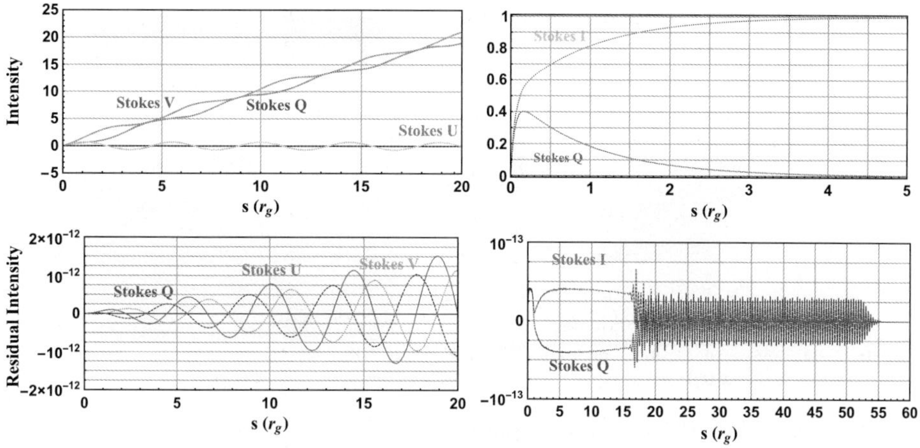

Figure 2. *Polarised GRRT tests I.* **Left column:** pure polarised emission in Stokes (Q, U, V), $I = 0$ (*top panel*) and residuals compared to analytic solution (*bottom panel*). **Right column:** pure emission and absorption in Stokes (I, Q), $U = V = 0$ (*top panel*) and residuals compared to analytic solution (*bottom panel*). Analytic solutions are plotted as dashed lines in upper panels.

where $\overline{\varepsilon}_i$ are the invariant emissivities, $\overline{\alpha}_i$ the invariant absoprtivities and $\overline{\rho}_i$ are the invariant Faraday rotation and conversion coefficients. Invariant (barred) quantities are related to their standard (fluid rest frame) counterparts as $\overline{\varepsilon}_i = \varepsilon_i/\nu^2$, $\overline{\alpha}_i = \nu\alpha_i$ and $\overline{\rho}_i = \nu\rho_i$, where ν is the frequency (see Younsi *et al.* 2012). The 4-velocity of the fluid, e.g., an accretion flow around a black hole, is denoted by u^μ and $R(\chi)$ is a rotation matrix where $\chi \in [-\pi, \pi]$ is the angle between the polarisation basis and the plasma magnetic field, also termed the electric vector position angle (EVPA). Note that $\varepsilon_U = \alpha_U = \rho_U = 0$ is fixed by rotating to a frame aligning the magnetic field with U. Equation (2.4) is integrated along each photon geodesic, returning the Stokes parameters, $\mathbf{S}$.

3. Preliminary tests and future work

We present preliminary tests of the above polarised GRRT framework in BHOSS. Several polarised GRRT codes exist in the literature (Broderick & Blandford 2003a, 2003b, Schnittman & Krolik 2010, Shcherbakov & Huang 2011, Dexter 2016, Mościbrodzka & Gammie 2018, Pihajoki *et al.* 2018) which contain several standard tests. Figure 2 shows two ray transport tests (see Dexter 2016): (i) transport of pure polarised emission with Faraday rotation and conversion, where $\epsilon_Q = \epsilon_U = \epsilon_V = \rho_Q = \rho_V = 1$, $\rho_U = 0$, and (ii) pure emission and absorption in Stokes I and Q, with $\varepsilon_I = \alpha_I = 10$, $\varepsilon_Q = \alpha_Q = 9$. The numerical integration performs well and, as expected, Stokes I is zero to machine precision in test (i) and Stokes U and V are similarly zero in test (ii).

In Fig. 3 we present more detailed polarised GRRT imaging tests. The left panel shows the EVPA map of a thin Keplerian accretion disk around a Schwarzschild black hole, and is in good agreement with that of Fig. 3 in Dovčiak *et al.* 2008. The right panel presents the image of a Novikov-Thorne accretion disk with an electron scattering atmosphere (see Schnittman & Krolik 2010 for further details). The obtained image is in good qualitative agreement with previous results obtained by Schnittman & Krolik 2010 and Dexter 2016.

We have presented a numerical GRRT code capable of integrating the equations of polarised GRRT to high accuracy in arbitrary spacetime geometries, demonstrating excellent agreement with previous results and analytic solutions in the literature. Since BHOSS already performs GRRT in time-dependent optically-thin and optically-thick GRMHD backgrounds, and for arbitrary spacetime geometries, we will next perform polarised

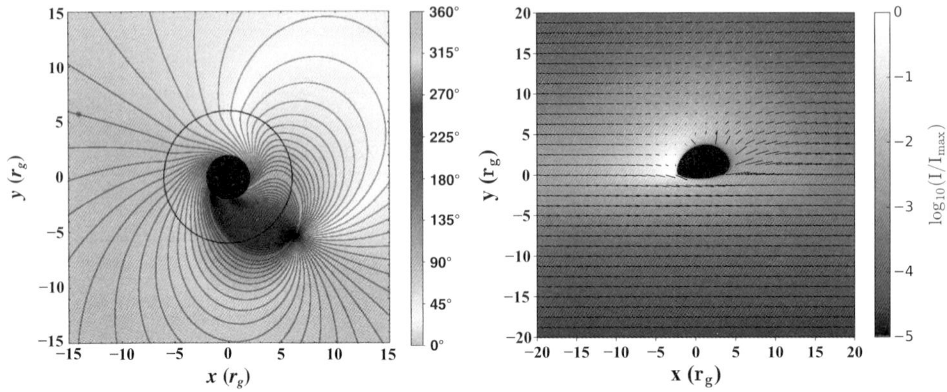

Figure 3. *Polarised GRRT tests II.* **Left panel:** EVPA contour map ($\widetilde{\chi} \equiv \chi + \pi$) of a Keplerian accretion disk around a Schwarzschild BH as viewed at 30°. The black ring denotes the innermost stable circular orbit (ISCO), wherein the disk material is in free-fall. The red dot corresponds to the $\widetilde{\chi} = 0°$ contour. **Right panel:** Novikov-Thorne accretion disk with electron scattering atmosphere around an extremal Kerr black hole. Colour denotes the total intensity integrated over the range 0.1–10 keV, $\dot{M} = 0.1\ M_{\rm Edd}$, $M = 10\ M_{\odot}$ and viewed at 75°. Orientation of black ticks indicates the polarisation direction, and the length of each tick denotes the polarisation degree ($\sim 5\%$ at bottom of panel). Significant depolarisation and enhancement can be seen.

GRRT calculations on GRMHD simulation data, determining time-variable polarised emission from accreting black holes and providing meaningful theoretical predictions to be compared with upcoming and future observing campaigns.

Acknowledgements

We thank Luciano Rezzolla, Hung-Yi Pu and Thomas Bronzwaer for helpful input. Support comes from the ERC Synergy Grant "BlackHoleCam - Imaging the Event Horizon of Black Holes" (Grant 610058). ZY is supported by a Leverhulme Trust Early Career Fellowship. This research has made use of NASA's Astrophysics Data System.

References

Doeleman, S. S., Weintroub, J., Rogers, A. E. E., Plambeck, R., *et al.* 2008, *Nature*, 78–80, 455
Goddi, C., Falcke, H., Kramer, M., *et al.* 2017, *Int. J. Mod. Phys. D*, 26, 1730001-239
Cunningham, C. T. and Bardeen, J. M. 1973, *Astrophys. J.*, 237–264, 183
Grenzebach, A., Perlick, V. and Lämmerzahl, C. 2014, *Phys. Rev. D*, 12, 89
Younsi, Z., Zhidenko, A., Rezzolla, L., Konoplya, R. & Mizuno, Y. 2016, *Phys. Rev. D*, 8, 94
Mizuno, Y., Younsi, Z., Fromm, C. M., Porth, O., De Laurentis, M., Olivares, H., Falcke, H., Kramer, M. & Rezzolla, L. 2018, *Nature Astronomy*, 2, 585–590
Porth, O., Olivares, H., Mizuno, Y., Younsi, Z., Rezzolla, L. *et al.* 2017, *CA&C*, 1, 4
Gammie, C. F., McKinney, J. C. & Tóth, G. 2003, *Astrophys. J.*, 458, 589
Teo, E. 2003 *Gen. Rel. and Grav.*, 35, 11
Chan, C.-k., Medeiros, L., Ozel, F. & Psaltis, D. 2017, ArXiv e-prints:1706.07062
Younsi, Z., Wu, K. and Fuerst, S. V. 2012, *Astron. Astrophys.*, A13, 545
Broderick, A., & Blandford, R. 2003a, *MNRAS*, 342, 1280; 2003b, *Ap&SS*, 288, 161
Shcherbakov, R. V. & Huang, L. 2011, *MNRAS*, 410, 1052
Dexter, J. 2016, *MNRAS*, 462, 115
Mościbrodzka, M. & Gammie, C. F. 2018, *MNRAS*, 475, 43–54
Pihajoki, P., Mannerkoski, M., Nättilä, J. & Johansson, P. H. 2018, *Astrophys. J.*, 863, 8
Schnittman, J. D. & Krolik, J. H. 2010, *Astrophys. J.*, 712, 908–924
Dovčiak, M., Muleri, F., Goosmann, R. W., Karas, V. & Matt, G. 2008, *MNRAS*, 391, 32–38

Perseus in Sicily: From Black Hole to Cluster Outskirts
Proceedings IAU Symposium No. 342, 2018
K. Asada, E. de Gouveia Dal Pino, M. Giroletti,
H. Nagai & R. Nemmen, eds.

doi:10.1017/S1743921318007688

Particle acceleration and the origin of the very high energy emission around black holes and relativistic jets

Elisabete M. de Gouveia Dal Pino[1], Grzegorz Kowal[2], Luis Kadowaki[1], Tania E. Medina-Torrejón[1], Yosuke Mizuno[3] and Chandra Singh[4]

[1]Instituto de Astronomia, Geofísica e Ciências Atmosféricas (IAG-USP), Universidade de São Paulo, R. do Matão, 1226 05508-090 São Paulo, SP Brasil
email: dalpino@iag.usp.br

[2]EACH, Universidade de São Paulo

[3]Physics Institute, University of Tel Aviv

[4]Physics Institute, University of Frankfurt

Abstract. Particle acceleration induced by fast magnetic reconnection may help to solve current puzzles related to the interpretation of the very high energy (VHE) and neutrino emissions from AGNs and compact sources in general. Our general relativistic-MHD simulations of accretion disk-corona systems reveal the growth of turbulence driven by MHD instabilities that lead to the development of fast magnetic reconnection in the corona. In addition, our simulations of relativistic MHD jets reveal the formation of several sites of fast reconnection induced by current-driven kink turbulence. The injection of thousands of test particles in these regions causes acceleration up to energies of several PeVs, thus demonstrating the ability of this process to accelerate particles and produce VHE and neutrino emission, specially in blazars. Finally, we discuss how reconnection can also explain the observed VHE luminosity-black hole mass correlation, involving hundreds of non-blazar sources like Perseus A, and black hole binaries.

Keywords. acceleration of particles, black hole physics, galaxies: active, MHD

---

1. Introduction

Since de Gouveia Dal Pino & Lazarian (2005) proposed for the first time that particles could be accelerated in a first-order Fermi process within fast reconnection sites (current sheets), considerable progress has been attained in this field mainly boosted by numerical modeling employing both PIC (e.g. Drake *et al.* 2006, Zenitani & Hoshino 2008, Lyubarsky *et al.* 2008, Clausen-Brown *et al.* 2012, Cerutti *et al.* 2014, Sironi *et al.* 2014, Guo *et al.* 2016) and test particle-plus-MHD techniques (e.g., Kowal, de Gouveia Dal Pino & Lazarian 2011, 2012, del Valle, de Gouveia Dal Pino & Kowal 2016). These studies have successfully tested the theoretical predictions identifying an exponential growth of the particles energy with time while trapped within current sheets (e.g., Kowal *et al.* 2011, 2012), and a power-law energy spectrum for the accelerated particles with spectral indices ~ -1 to -2 (e.g. Kowal *et al.* 2012, del Valle *et al.* 2016), and a power-law dependence of the acceleration rate with the particle energy (with power-law indices $\sim 0.3 - 0.6$ for a non-relativistic background (del Valle *et al.* 2016). These findings are general and in principle applicable to any system, specially in turbulent magnetically dominated regions.† This mechanism is now regarded as important also beyond the solar system (where it has been extensively investigated). In particular, it may explain current puzzles related to very high energy phenomena observed in compact sources like AGNs, black hole X-ray binary systems, and GRBs. In this talk, we discuss this process in the framework of these sources, i.e., in the surrounds of black holes (BHs) and relativistic jets (see also de Gouveia Dal Pino *et al.* 2015, 2016 for recent reviews).

2. Magnetic Reconnection Acceleration in Relativistic Jets

AGNs with highly beamed jets pointing to the line of sight, namely blazars, are the most common sources of $\gamma-$rays. This emission is generally attributed to diffusive shock particle acceleration along the jet which is strongly Doppler boosted producing apparently very high uxes. However, in some cases (e.g., PKS2155-304 source; Aharonian *et al.* 2007), observed very high variability, of the order of minutes in the TeV range, implies extremely compact and fast acceleration/emission regions ($< R_S/c$) with Lorentz factors much larger than the typical bulk values ($\Gamma \sim 1-10$) expected for such sources, in order to avoid electron-positron pair creation. So far, the only model able to explain this high variability at TeV emission seems to be magnetic reconnection involving misaligned current sheets inside the jet (e.g. Giannios *et al.* 2009; see also Kushwaha *et al.* 2017). Fast reconnection acceleration has been also invoked to explain the transition from magnetically to kinetically dominated flow and the prompt gamma-ray emission in GRBs (Zhang & Yan 2011).

To probe this process, we have recently performed 3D relativistic MHD simulations of rotating Poyinting flux dominated tower jets with initial helical fields (Singh, Mizuno & de Gouveia Dal Pino 2016). Considering models with a ratio between the magnetic and the rest mass energy of the flow $\sigma = 1$, and different density ratios between the jet and the environment, we induced precession perturbations that quickly developed current-driven kink (CDK) modes. Figure 1 (left) shows an example of a jet with density larger than that of the environment. The results evidence the propagation of a helical kinked structure along the jet that causes substantial dissipation of magnetic into kinetic energy. We also identify regions of maximum current density (labeled with black points) that trace filamentary current-sheets, where turbulent fast magnetic reconnection driven by the CDK instability takes place with rates $\sim 0.05 V_A$ (Singh *et al.* 2016; see also Kadowaki *et al.*, in prep.). The right diagram of Figure 1 depicts the kinetic energy evolution of a distribution of 10,000 particles that were accelerated *in situ* within the reconnection regions of the relativistic jet depicted in the left panel. Starting with energies $sim1$ MeV, the particles undergo an exponential acceleration once they are trapped in the reconnection sheets (as in Kowal *et al.* 2012, del Valle *et al.* 2017), until a saturation level, when then the accelerated particles Larmor radius become larger than the acceleration regions. The initial background magnetic field in this simulation had a maximum value $B = 0.13$ G at the jet axis, but we have also considered simulations with values up to 100 times larger. In this case, when the particles enter the exponential regime, they are accelerated up to energies $\sim 10^{19}$ eV at sub-pc distances within an AGN jet (Medina-Torrejon, de Gouveia Dal Pino, Kowal, Mizuno, Singh, Kadowaki, in prep.). The implications of these results are very important for relativistic jets in general, as they indicate that these accelerated protons can produce $\gamma-$rays in very compact regions and also neutrinos, as recently detected in the blazar TXS 0506+056 (IceCube collaboration *et al.* 2018).

† The presence of turbulence makes magnetic reconnection naturally very fast, with reconnection velocities being a substantial fraction of the local Alfven speed (Lazarian & Vishniac 1999, Kowal *et al.* 2009).

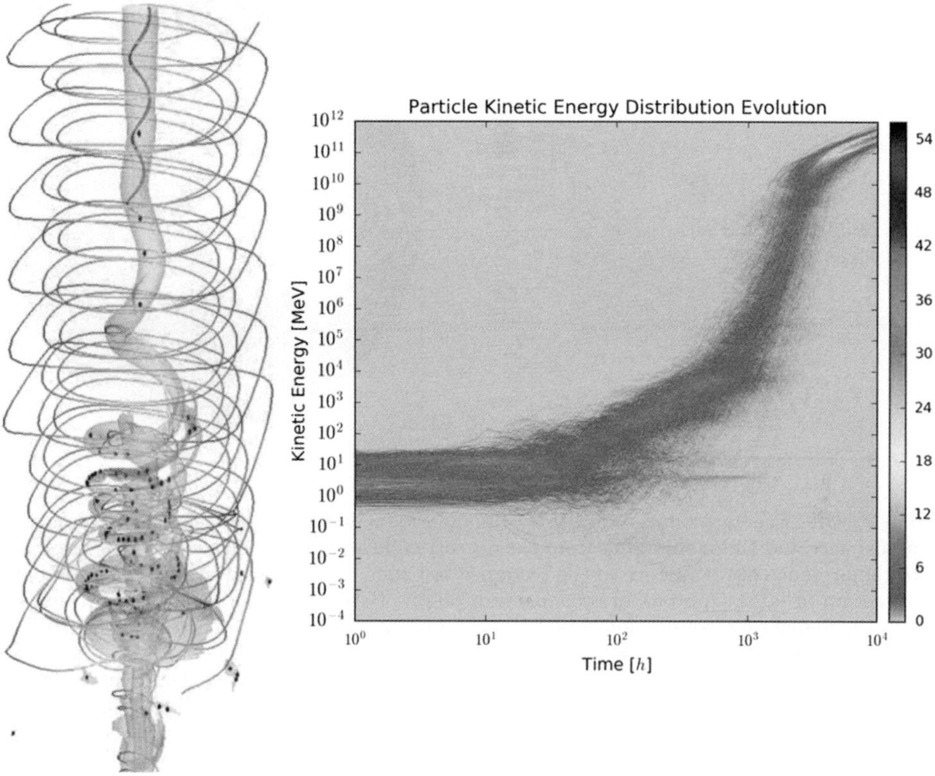

Figure 1. Current-driven kink instability and fast magnetic reconnection in a relativistic MHD jet. The left diagram depicts 3D density isosurfaces (green) with superposed magnetic field lines (black) (extracted from Singh *et al.* 2016). The black points identify the sites of fast reconnection which were detected employing the technique described in Kadowaki, de Gouveia Dal Pino & Stone (2018; see also Kadowaki *et al.*, in prep.). Right diagram shows the time evolution of the kinetic energy of 10,000 test particles (protons) that were injected with an initial Maxwelian velocity distribution in the relativistic jet of the left diagram (with initial maximum background magnetic field scaled to $B = 0.13G$). Once particles are trapped within reconnection regions they are exponentially accelerated up to energies $\sim 10^{17}$ eV in this model. Simulations performed with magnetic field one-hundred times larger, allow for particle acceleration up to 10^{19} eV (Medina-Torrejon, de Gouveia Dal Pino, Kowal, Mizuno, Singh, Kadowaki, in prep.).

3. Magnetic reconnection acceleration around black holes

Recently, a few non-blazar sources which belong to the branch of low luminosity AGNs (or simply LLAGNs), have been also detected at TeV energies by ground based gamma-ray observatories (e.g., the radio galaxies Per A, M87, Cen A, and IC 310; Sol *et al.* 2013; see also E. Torresi these Procs.). The angular resolution and sensitivity of these detectors is still so poor that it is hard to establish if it comes from the jet or from the core. These very high energy (VHE) detections were surprising because, besides being highly underluminous, the viewing angle of the jets of these sources is of several degrees, therefore allowing for only moderate Doppler boosting. These characteristics make it hard to explain the VHE of these sources adopting the same standard scenario as in blazars. Furthermore, observations of short time scale variability in the γ−ray emission of these sources indicate that it is produced in a very compact region that could be, perhaps, the core. These findings led to the search for alternative particle acceleration

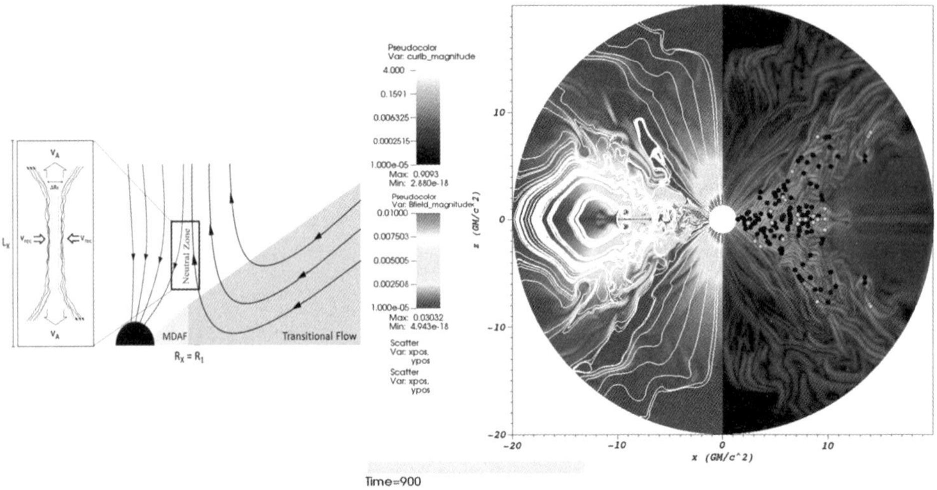

Figure 2. Left: Schematic representation of magnetic reconnection between the lines of the BH magnetosphere and those emerging from the accretion disk into the corona. Reconnection is fast due to the presence of turbulence in the reconnection zone (as indicated in the detail) and allows for extraction of large amounts of reconnection power. Particle acceleration may also naturally occur in the magnetic reconnection zone by Fermi process (adapted from Singh *et al.* 2015). This representation is suitable for a magnetized ADAF type accretion disk, but we obtain similar results when considering a geometrically thin accretion disk (de Gouveia Dal Pino & Lazarian 2005, Kadowaki *et al.* 2015). Right: 2D general relativistic MHD (GR-MHD) simulation of a torus accretion flow around a BH evidencing the development of turbulent reconnection induced by the MRI instability in the magnetic field and current-density distributions. The black and white dots in the current-density plot identify the sites of fast reconnection using the technique described in Kadowaki, de Gouveia Dal Pino & Stone (2018).

scenarios involving the production of the VHE in the core region of these sources (e.g. de Gouveia Dal Pino *et al.* 2010; 2016 and references therein).

Another striking evidence was found by Kadowaki, de Gouveia Dal Pino & Singh (2015) who plotted in the same diagram the $\gamma-$ray luminosity versus the BH mass for more than 230 sources including non-blazar LLAGNs, black hole X-ray binaries (BHBs), blazars and GRBs, spanning over ~ 10 orders of magnitude in mass and power. This diagram reveals two distinct branches, or correlation trends. One that is followed by blazars and GRBs, and another one by LLAGNs and BHBs (see Kadowaki *et al.* 2015, Singh, de Gouveia Dal Pino & Kadowaki 2015). The association between blazars and GRBs is already expected as their non-thermal radiation is commonly attributed to Doppler boosted emission by accelerated particles along the relativistic jets (as discussed in Section 2). The lack of correlation of these sources with non-blazars and BHBs in this diagram suggests that another location and/or population of relativistic particles may be producing the $\gamma-$ray emission. de Gouveia Dal Pino & Lazarian (2005) proposed an emission model based on particle acceleration by magnetic reconnection in the core region of BH sources. This model was revisited more recently (Kadowaki *et al.* 2015; Singh *et al.* 2015) and a sketch of it is depicted in Figure 2 (left). There is direct evidence of this process from GR-MHD simulations of accretion disks around BHs, as indicated by Figure 2 (right). Fast magnetic reconnection between the lines of the BH magnetosphere and those arising from the accretion flow (with rates around 0.1 V_A; see Kadowaki, de Gouveia Dal Pino & Stone in prep.) allows for the release of magnetic energy that heats and accelerates the plasma. The magnetic reconnection power calculated analytically from this core model matches quite well with the observed $\gamma-$ray luminosity of the non-blazar LLAGNs and BHBs plotted in the Luminosity-BH mass diagram described above (see Kadowaki *et al.* 2015; Singh *et al.* 2015)! This suggests that magnetic reconnection in the core region of these sources can explain the observed γ−ray emission. Detailed calculations of the observed spectral energy distribution (SED) of a number of BHBs (Khiali, de Gouveia Dal Pino, & del Valle 2015) and LLAGNs, like Per A (Ramirez-Rodriguez *et al.* these Procs.) using this model, have demonstrated that TeV emission in these sources (due to proton-proton and proton-photon interactions) can be effectively reproduced by accelerated cosmic rays by reconnection, and the emission is *not* entirely absorbed by electron-positron pairs, as suspected before (Ramirez-Rodriguez, de Gouveia Dal Pino & Alves-Batista, in prep.). Besides, the same process leads to neutrino emission too that can account for the observed extragalactic diffuse emission by the IceCube (see Khiali & de Gouveia Dal Pino 2016).

4. Summary and Conclusions

In this talk, we have stressed the importance of turbulent magnetic reconnection and particle acceleration by reconnection in the surrounds of BHs and relativistic jets based on arguments sustained by theory and numerical simulations. Our main conclusions can be summarized as follows:

• Magnetic reconnection is important in accretion and jet systems for dissipation of magnetic energy, conversion into kinetic energy, and particle acceleration.

• In magnetized plasmas particles can be accelerated by fast magnetic reconnection (driven by turbulence) in a Fermi process and develop a power-law spectrum $N(E) \sim E^{-2,-1}$.

• Magnetic reconnection acceleration can be very efficient in blazar jets and produce cosmic rays with energies up to $10^{17,19}$ eV. These could be responsible for γ−ray flares and neutrino emission in the magnetically dominated regions of these jets. This could be also the dominating mechanism responsible for the recent observed γ−ray and neutrino emission in the TXS 0506+056 blazar and deserves further testing.

• The magnetic reconnection power can also explain the γ−ray emission from BHBs and non-blazar LLAGNs as coming from the *core* of these sources. This power matches well with the observed correlation of γ−ray luminosity versus BH mass for these sources that spans 10 orders of magnitude. Accelerated CRs by magnetic reconnection in the surrounds of these sources can produce TeV γ−rays that are not entirely absorbed by pair production, as well as neutrinos (see also Ramirez-Rodriguez *et al.* these Procs.).

Acknowledgments

We acknowledge support from the Brazilian agencies FAPESP (2013/10559-5 grant) and CNPq (306598/2009-4 grant). The simulations presented in this lecture have made use of the computing facilities of the GAPAE group (IAG-USP) and the Laboratory of Astroinformatics IAG/USP, NAT/Unicsul (FAPESP grant 2009/54006-4).

References

Aharonian, F., Akhperjanian, A. G., Bazer-Bachi, A. R., *et al.* 2007, *ApJ Letts.*, 664, L71
Clausen-Brown, E. & Lyutikov, M. 2012, *MNRAS*, 426, 1374
Cerutti, B., Uzdensky, D. A. & Begelman, M. C. 2014, *ApJ*, 746, 148
de Gouveia Dal Pino, E. M., & Lazarian, A. 2005, *A&A*, 441, 845
de Gouveia Dal Pino, E. M., Piovezan, P. P., & Kadowaki, L. H. S. 2010, *A&A*, 518, A5
de Gouveia Dal Pino, E. M., & Kowal, G. 2015, in Magnetic Fields in Diffuse Media, Astrophysics and Space Science Library, Eds. A. Lazarian. E. de Gouveia Dal Pino, C. Melioli, 407, 373
de Gouveia Dal Pino, E., del Valle, M. V., Kadowaki, L., *et al.* 2016, Frontier Research in Astrophysics II, id.56, 56
del Valle, M. V., de Gouveia Dal Pino, E.M., & Kowal, G. 2016, *MNRAS*

Drake, J. F., Swisdak, M., Che, H., & Shay, M. A. 2006, *Nature*, 443, 553
Giannios, D., Uzdensky, D. A., & Begelman, M. C., 2009, *MNRAS*, 228, 395 L29-L33
Guo, F., Li, H., Daughton, W. & Liu, Y. H. 2016, *ApJ*, Physics of Plasmas, 23, 055708
IceCube, T., Fermi-LAT, MAGIC, *et al.* 2018, Science, arXiv:1807.08816
Kadowaki, L. H. S., de Gouveia Dal Pino, E. M., & Singh, C. B. 2015, *ApJ*, 802, 113
Kadowaki, L. H. S., de Gouveia Dal Pino, E. M., & Stone, J. 2018, *ApJ* (in press), 2018, arXiv:180308557K
Khiali, B., de Gouveia Dal Pino, E. M., & del Valle, M. V. 2015, *MNRAS*, 449, 34
Khiali, B. & de Gouveia Dal Pino, E. M. 2016, *MNRAS*, 455, 838
Kowal, G., Lazarian, A., Vishniac, E. T., & Otmianowska-Mazur, K., 2009, *ApJ*, 700, 63
Kowal, G., de Gouveia Dal Pino, E. M., & Lazarian, A. 2011, *ApJ*, 735, 102
Kowal, G., de Gouveia Dal Pino, E. M., & Lazarian, A. 2012, *PRL*, 108, 241102
Kushwaha, P., Sinha, A., Misra, R., Singh, K. P., & de Gouveia Dal Pino, E. M. 2017, *ApJ*, 849, 138
Lazarian, A. & Vishniac, E. T. 1999, *ApJ*, 517, 700
Lyubarsky, Y. & Liverts, M. 2008, *ApJ*, 682, 1436
Singh, C. B., de Gouveia Dal Pino, E. M., & Kadowaki, L. H. S. 2015, *ApJ Letts.*, 799, L20
Singh, C. B., Mizuno, Y., & de Gouveia Dal Pino, E. M. 2016, *ApJ*, 824, 48
Sironi, L., & Spitkovsky, A. 2014, *ApJ Letts.*, 783, L21
Sol, H., Zech, A., Boisson, C. *et al.* 2013, *Astroparticle Physics*, 43, 215
Zhang, B., & Yan, H. 2011, *ApJ*, 726, 90
Zenitani, S. & Hoshino, M. 2008, *AJ*, 677, 530

Perseus in Sicily: From Black Hole to Cluster Outskirts
Proceedings IAU Symposium No. 342, 2018
K. Asada, E. de Gouveia Dal Pino, M. Giroletti,
H. Nagai & R. Nemmen, eds.

doi:10.1017/S1743921318007834

Numerical methods for General Relativistic particles

Fabio Bacchini[1], **Bart Ripperda**[2,3], **Alexander Y. Chen**[2] **and Lorenzo Sironi**[4]

[1]CmPA, Department of Mathematics, KU Leuven, Celestijnenlaan 200B, B-3001, Leuven, Belgium
email: fabio.bacchini@kuleuven.be

[2]Department of Astrophysical Sciences, Princeton University, 4 Ivy Lane, Princeton, NJ 08544

[3]Center for Computational Astrophysics, Flatiron Institute, 162 Fifth Avenue, New York, NY 10010, USA

[4]Department of Astronomy, Columbia University, 550 W 120th St, New York, NY 10027, USA

Abstract. We present recent developments on numerical algorithms for computing photon and particle trajectories in the surrounding of compact objects. Strong gravity around neutron stars or black holes causes relativistic effects on the motion of massive particles and distorts light rays due to gravitational lensing. Efficient numerical methods are required for solving the equations of motion and compute i) the black hole shadow obtained by tracing light rays from the object to a distant observer, and ii) obtain information on the dynamics of the plasma at the microscopic scale. Here, we present generalized algorithms capable of simulating ensembles of photons or massive particles in any spacetime, with the option of including external forces. The coupling of these tools with GRMHD simulations is the key point for obtaining insight on the complex dynamics of accretion disks and jets and for comparing simulations with upcoming observational results from the Event Horizon Telescope.

Keywords. acceleration of particles, accretion disks, black hole physics, gravitation, relativity, methods: numerical

1. Introduction and overview

Particle-based numerical methods are a powerful tool in plasma physics. From collisionless magnetic reconnection (Sironi & Spitkovsky 2014) to turbulent processes (Roytershteyn *et al.* 2015), numerical methods such as Particle-in-Cell (PiC) have produced a variety of new physical results. In the astrophysical context, relativistic particle methods are used to model processes at the microscopic scale characterized by extremely high-energy, while the global scale is modeled with magnetohydrodynamics (MHD) approaches.

Current studies of such phenomena, when carried out with particles methods, are generally limited to locally flat regions of spacetime. Numerical schemes that take into account the underlying curvature of spacetime have not been actively developed (but see e.g. Levinson & Cerutti 2018). Therefore, the microscopic dynamics of plasmas around compact objects such as black holes and neutron stars, where gravity plays an important role, is not well understood.

The physics of plasmas around black holes has received much attention in recent studies, thanks to ambitious projects that aim at the direct imaging of black holes for the first time (Falcke 2017). In this context, global simulations of black hole magnetospheres have been carried out with GRMHD approaches. Furthermore, from simulation data, synthetic radiation maps have been obtained, allowing for the reconstruction of the observed image as seen from a distant telescope. These images allow for direct comparison of observational data with theoretical predictions.

Obtaining such images involves a complex numerical approach that combines plasma physics on the global scale, optics, and microscopic processes. Here, advanced particle methods for general relativity play multiple roles: first, particle simulations can be used to obtain energy distributions from which the synthetic radiation maps are used. Current approaches generally assume Maxwellian or κ-distributions (Davelaar *et al.* 2018). Second, once the radiation maps are available, rays of light must be traced from the physical point in space to the observer's position, in order to obtain an image that mirrors realistic observations. This step involves so called "ray tracing" techniques, which essentially consist of simulating bundles of photons (i.e. massless particles) traveling on geodesics.

Here, we aim both at improving existing particle methods used for simulating geodesic motion (for ray tracing), and at extending to the inclusion of the Lorentz force. This step is necessary in order to obtain particle methods that can ultimately be applied to the dynamics of plasmas around compact objects, in order to obtain realistic energy distributions and therefore more physically correct radiation maps used in the imaging process.

2. General relativistic particles in curved spacetime

The motion of particles subject to strong gravity is governed by the equation of motion

$$\frac{d^2x^\mu}{d\tau^2} + \Gamma^\mu_{\lambda\sigma}\frac{dx^\lambda}{d\tau}\frac{dx^\sigma}{d\tau} = qF^{\mu\nu}u_\nu, \tag{2.1}$$

where $\mu = 0, 1, 2, 3$, for the four-position x^μ and an affine parameter τ. The contravariant four-momentum is $g^{\mu\nu}u_\nu := dx^\mu/d\tau$. $\Gamma^\mu_{\lambda\sigma}$ is the Christoffel symbol for the metric $g_{\mu\nu}$. In presence of electromagnetic fields, the force term on the right-hand side is given by the particle charge, q, and the Maxwell tensor, $F_{\mu\nu} = \partial_\mu A_\nu - \partial_\nu A_\mu$, where A_μ is the electromagnetic four-potential.

The equations above are typically written in the 3+1 language In this formulation, any metric of signature $(-, +, +, +)$ is given in the form

$$g_{\mu\nu} = \begin{pmatrix} -\alpha^2 + \beta_k\beta^k & \beta_i \\ \beta_j & \gamma_{ij} \end{pmatrix}, \tag{2.2}$$

where α is the lapse function, β^i is the shift three-vector, and γ_{ij} is the spatial part of $g_{\mu\nu}$. Latin indices assume values from 1 to 3. With the definitions above, the equation of motion (2.1) can be rewritten in terms of first-order evolution equations in the variables x^i and $u_i = g_{i\mu}u^\mu$, such that

$$\frac{dx^i}{dt} = \gamma^{ij}\frac{u_j}{u^0} - \beta^i, \tag{2.3}$$

$$\frac{du_i}{dt} = -\alpha u^0\partial_i\alpha + u_k\partial_i\beta^k - \frac{u_ju_k}{2u^0}\partial_i\gamma^{jk} + q\left(F_{i0} + F_{ij}\frac{u^j}{u^0}\right), \tag{2.4}$$

where

$$u^0 = \left(\gamma^{jk}u_ju_k + \epsilon\right)^{1/2}/\alpha, \tag{2.5}$$

with $\epsilon = 0$ for photons and $\epsilon = 1$ for massive particles. In this formalism, the coordinate time $t := x^0$ is used as affine parameter, and $u^0 := dt/d\tau$.

The system of equations (2.3–2.4) is suitable for numerical integration. It can be shown that the definition of u^0 above enforces the conservation of the norm of the four-velocity, $u^\mu u_\mu = -\epsilon$. Additionally, integrating in coordinate time t rather than in proper time τ makes it easier to embed the motion of test particles in the time evolution of global electromagnetic fields.

In this work, we consider stationary metrics with no dependence on coordinate time t, hence the metric $g_{\mu\nu}$ depends on x^i only. As a consequence, in all cases the energy, $E = -u_0$ is conserved. This physical property is often not respected during simulations, due to numerical errors affecting any discretization scheme. However, in some cases the lack of exact energy conservation can be detrimental for the accuracy of the results. Here, we present a new implicit numerical scheme that conserves energy to machine precision during the computation. The scheme is constructed by enforcing that the underlying Hamiltonian (i.e. the energy) is constant in time.

For simplicity, consider the case of pure geodesic motion, $A_\mu = 0$. The Hamiltonian for stationary metrics is defined as

$$H(x^a, u_b) = \alpha(\gamma^{jk} u_j u_k + \epsilon)^{1/2} - \beta^j u_j. \tag{2.6}$$

It is straightforward to derive the equations of motion by differentiating H, such that

$$\frac{dx^i}{dt} = \frac{\partial H(x^a, u_b)}{\partial u_i} = \frac{\alpha \gamma^{ij} u_j}{(\gamma^{jk} u_j u_k + \epsilon)^{1/2}} - \beta^i, \tag{2.7}$$

$$\frac{du_i}{dt} = -\frac{\partial H(x^a, u_b)}{\partial x^i} = -(\gamma^{jk} u_j u_k + \epsilon)^{1/2} \partial_i \alpha - \frac{1}{2} \frac{\alpha u_j u_k}{(\gamma^{jk} u_j u_k + \epsilon)^{1/2}} \partial_i \gamma^{jk} + u_j \partial_i \beta^j, \tag{2.8}$$

which are precisely equations (2.3)–(2.4) above. One can immediately verify that the corresponding discretized system

$$\frac{\Delta x^i}{\Delta t} = \frac{\Delta^i_u H(x^a, u_b)}{\Delta u_i}, \tag{2.9}$$

$$\frac{\Delta u_i}{\Delta t} = -\frac{\Delta^i_x H(x^a, u_b)}{\Delta x^i}, \tag{2.10}$$

satisfies the condition

$$\frac{\Delta H(x^a, u_b)}{\Delta t} = \frac{\Delta^x_i H(x^a, u_b)}{\Delta x^i} \frac{\Delta x^i}{\Delta t} + \frac{\Delta^i_u H(x^a, u_b)}{\Delta u_i} \frac{\Delta u_i}{\Delta t} = 0, \tag{2.11}$$

and therefore it conserves energy. Here, Δ indicates a total discrete derivative, such that $\Delta H(x^a, u_b) = H(x^{a,n+1}, u_b^{n+1}) - H(x^{a,n}, u_b^n)$. The operators Δ^i_x and Δ^u_i, instead, correspond to discrete partial derivatives, namely with respect to x^i and u_i. The exact form of the discrete operators can be found in full in Bacchini *et al.* (2018).

A remarkable feature of this scheme is the absolute freedom in the definition of the Hamiltonian H. This implies that for systems characterized by a Hamiltonian different from that of equation (2.6), the algorithm retains its energy (or in general, first integrals) conservation properties. Thus, the extension to more complicated physical situations becomes straightforward, provided that the corresponding Hamiltonian formulation is available. For this reason, it is immediate to include the Lorentz force, by employing the corresponding Hamiltonian,

$$H(x^a, \pi_b) = \alpha \sqrt{1 + \gamma^{ij} \left(\pi_i - \frac{q}{m} A_i\right) \left(\pi_j - \frac{q}{m} A_j\right)} - \beta^k \left(\pi_k - \frac{q}{m} A_k\right) - q A_0, \tag{2.12}$$

(a)

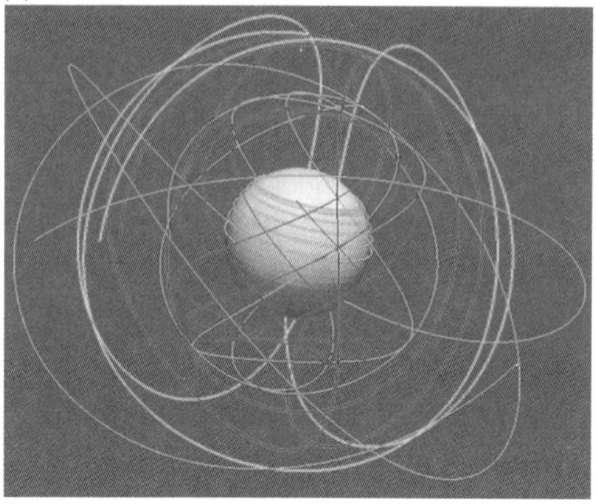

Unstable spherical photon orbits around an extremal Kerr black hole.

(b)

Simulation of the observed shadow of an extremal Kerr black hole distorting the view of four-color background.

Figure 1. Simulations of photon trajectories with the new Hamiltonian method.

where the conjugate three-momentum $\pi_i = u_i + qA_i$ appears in place of u_i as an independent variable. The solution procedure remains unchanged with respect to the case of pure geodesic motion, hence making it possible to simulate charged particles in general relativistic contexts while retaining exact energy conservation.

3. Results and outlook

The numerical solution of equations (2.9)–(2.10) yields geodesic trajectories characterised by conservation of energy to machine precision. The advantage implied by this feature is two-fold. First, the motion of particles around metric singularities is particularly affected by numerical energy dissipation that characterizes the numerical schemes typically employed (e.g. explicit Runge-Kutta or implicit midpoint rule). The new "Hamiltonian" scheme, eliminates such errors and therefore does not require the extreme reduction of the time step demanded by other schemes. Second, since there is no secular growth of error in the energy, bound orbits confined within certain regions of space around a compact object can be simulated indefinitely without the spurious escape of the particle observed when using explicit methods.

The features above are exemplified in Figure 1a, where several spherical photon orbits are shown around an extremal Kerr black hole. Such orbits are highly unstable, and cannot be easily simulated with standard methods unless the time step is greatly reduced. The Hamiltonian scheme, by contrast, inherently stabilizes the motion on the correct path. As a side effect, all other constants of the motion are exactly conserved (energy, angular momentum, and Carter constant).

For ray tracing simulations, the features above are especially attractive because they allow for the fast computation of so called "black hole shadows" via integration of a large number of geodesic paths. One example of such a calculation is shown in Figure 1b. Here, an extremal Kerr black hole is placed between a distant observer and a four-color background. The distortion of spacetime caused by the presence of the black hole bends the light rays close to the event horizon, creating the asymmetrical shadow. Such a picture can be directly compared to upcoming observations of black holes, with the aim of confirming our current theoretical models. Hence, it is crucial that the numerical solution is as accurate as possible, which is ensured by numerical methods such as the Hamiltonian scheme here presented.

Finally, as mentioned above, the scheme is directly applicable to the motion of charged particles. This is a very fundamental step for obtaining accurate simulations of plasmas around compact object from the microscopic perspective. The results of such particle simulations can be used as a basis for the calculation of energy distributions from which synthetic radiation maps are drawn. This lays the ground for improving our understanding of the dynamics of energetic outflows and flares that are currently being measured (Genzel *et al.* 2003). Additionally, when the full particle feedback to the electromagnetic fields (in the PiC fashion) is taken into account, previously unreachable energy scales can actually be explored. Since the development of e.g. kinetic waves and instabilities is expected to be quantitatively different under these conditions, numerical schemes such as those here presented are the ideal tool to enrich our knowledge of these complicated physical processes.

References

Bacchini, F., Ripperda, B., Chen, A., & Sironi, L. 2018, *The Astrophysical Journal Supplement Series*, 237, 6

Davelaar, J., Mościbrodzka, M., Bronzwaer, T., & Falcke, H. 2018, *Astronomy & Astrophysics*, 612, A34

Falcke, H. 2017, *Journal of Physics: Conference Series*, 942, 012001

Genzel, R., Schödel, R., Ott, T., *et al.* 2003, *Nature*, 425, 934

Levinson, A. & Cerutti, B. 2018, arXiv:1803.04427

Roytershteyn, V., Karimabadi, H., Omelchenko, Y., & Germaschewski, K. 2015, *AGU Fall Meeting Abstracts*, SH11E-2417

Sironi, L. & Spitkovsky, A. 2014, *The Astrophysical Journal Letters*, 783, L21

Perseus in Sicily: From Black Hole to Cluster Outskirts
Proceedings IAU Symposium No. 342, 2018
K. Asada, E. de Gouveia Dal Pino, M. Giroletti,
H. Nagai & R. Nemmen, eds.

doi:10.1017/S1743921318007676

On the prospects of imaging Sagittarius A* from space

Freek Roelofs[1], **Heino Falcke**[1,2], **Christiaan Brinkerink**[1], **Monika Moscibrodzka**[1], **Leonid I. Gurvits**[3,4], **Manuel Martin-Neira**[5], **Volodymyr Kudriashov**[1], **Marc Klein-Wolt**[1], **Remo Tilanus**[1,6], **Michael Kramer**[2] **and Luciano Rezzolla**[7]

[1]Department of Astrophysics, Institute for Mathematics, Astrophysics and Particle Physics, Radboud University, PO Box 9010, 6500 GL Nijmegen, The Netherlands
email: f.roelofs@astro.ru.nl

[2]Max-Planck-Institut für Radioastronomie, Auf dem Hügel 69, D-53121 Bonn, Germany

[3]Joint Institute for VLBI ERIC, P.O. Box 2, 7990 AA Dwingeloo, The Netherlands

[4]Department of Astrodynamics and Space Missions, Delft University of Technology, 2629 HS Delft, The Netherlands

[5]European Space Research and Technology Centre (ESTEC), The European Space Agency, Keplerlaan 1, 2201 AZ Noordwijk, The Netherlands

[6]Leiden Observatory, Leiden University, P.O. Box 9513, 2300 RA Leiden, The Netherlands

[7]Institut für Theoretische Physik, Goethe-Universität, Max-von-Laue-Str. 1, 60438 Frankfurt, Germany

Abstract. Very Long Baseline Interferometry (VLBI) at sub-millimeter waves has the potential to image the shadow of the black hole in the Galactic Center, Sagittarius A* (Sgr A*), and thereby test basic predictions of the theory of general relativity. We investigate the imaging prospects of a new Space VLBI mission concept. The setup consists of two satellites in polar or equatorial circular Medium-Earth Orbits with slightly different radii, resulting in a dense spiral-shaped uv-coverage with long baselines, allowing for extremely high-resolution and high-fidelity imaging of radio sources. We simulate observations of a general relativistic magnetohydrodynamics model of Sgr A* for this configuration with noise calculated from model system parameters. After gridding the *uv*-plane and averaging visibilities accumulated over multiple months of integration, images of Sgr A* with a resolution of up to 4 μas could be reconstructed, allowing for stronger tests of general relativity and accretion models than with ground-based VLBI.

Keywords. Galaxy: center, techniques: interferometric, techniques: high angular resolution

1. Introduction

Sagittarius A* (Sgr A*) is a strong radio source in the Galactic Center. Infrared monitoring of stellar orbits in the Galactic Center has led to the conclusion that at the position of this radio source, there is a supermassive black hole with a mass of of $4.3 \pm 0.4 \times 10^6 M_\odot$ (Ghez *et al.* 2008; Gillessen *et al.* 2009). Ray-tracing simulations of synchrotron emission originating in an accretion flow led to the prediction of the appearance of a "black hole shadow" at millimeter wavelengths (Falcke *et al.* 2000).

The Event Horizon Telescope (EHT) is a mm VLBI array aiming to image a black hole shadow. As the largest black hole in the sky in angular size (53 μas), Sgr A* is the most promising candidate. The other main target, with a comparable shadow size, is M87. In April 2017, the EHT conducted observations that may lead to the first black

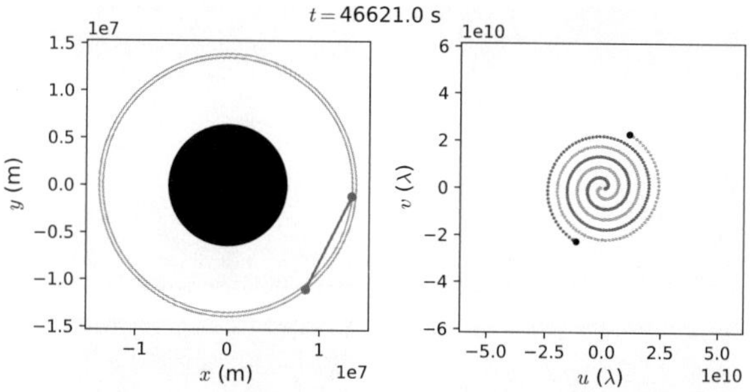

Figure 1. Example of satellite positions (left) and *uv*-coverage (right) for circular MEOs. Left, the satellites and baseline are shown in red, the orbits in blue, and the Earth in black. Right, the red and blue points show the *uv*-track for the two baseline directions. The black dots represent current *uv*-coordinates. From the initial satellite positions, the distance between them increases until the Earth occults the ISL. As the inner satellite catches up again, the spiral is traversed inwards.

hole shadow images. These could inform us about the nature of the accretion flow (e.g. Dexter *et al.* 2010; Mościbrodzka *et al.* 2014; Broderick *et al.* 2016), and be used to test general relativity (e.g. Psaltis *et al.* 2015; Goddi *et al.* 2017).

Several observational constraints make setting strong limits on accretion flow parameters and the validity of general relativity a challenging task for the EHT. The array is sparse, making it challenging to produce a robust image. The angular resolution of the EHT is about 23 μas, set by the observing frequency and the size of the Earth. Mizuno *et al.* (2018) show that this resolution is likely not high enough to distinguish between different theories of gravity, for which a resolution of $\sim 5-10$ μas would be required. At mm and lower wavelengths, the atmosphere introduces strong phase corruptions, making it difficult to push for higher frequencies. For Sgr A*, interstellar scattering distorts the image significantly at 230 GHz. Also, rapid variability poses challenges for imaging this source (Lu *et al.* 2016; Johnson *et al.* 2017; Bouman *et al.* 2017).

2. Two-satellite Space VLBI concept

The effects mentioned above could be mitigated with a new Space VLBI (SVLBI) concept. An initial design study has been performed for the purpose of the Event Horizon Imager (Martin-Neira *et al.* 2017; Kudriashov 2017). This concept involves two satellites in polar or equatorial Medium Earth Orbits (MEOs) at slightly different radii, observing at frequencies up to ~ 690 GHz. This setup results in a dense spiral-shaped *uv*-coverage with long baselines (Fig. 1). Because of the absence of the atmosphere, SVLBI allows for observations at higher frequencies, resulting in higher angular resolutions. At high frequencies, the emission originates closer to the black hole event horizon, producing a sharper outline of the black hole shadow due to strong lensing effects (Fig. 2). Variability of the geometry is expected to be mitigated as the observed image is dominated by lensing effects. Finally, the effect of scattering is mitigated at higher frequencies (Fig. 2).

Several technical challenges need to be overcome before the concept could turn into an actual mission. In the initial design study of Martin-Neira *et al.* (2017) and Kudriashov (2017), correlation takes place on the fly as the satellites share their local oscillator signals over an intersatellite link (ISL). In order to set a delay window that is not prohibitively large, the satellite positions and velocities should be known to within tens of wavelengths. GNSS satellites in combination with ISL ranging measurements may be able to provide sufficient precision, but more research into the feasibility is needed. Also, sufficiently stable clocks should be installed for the interferometer to operate in the submm-regime.

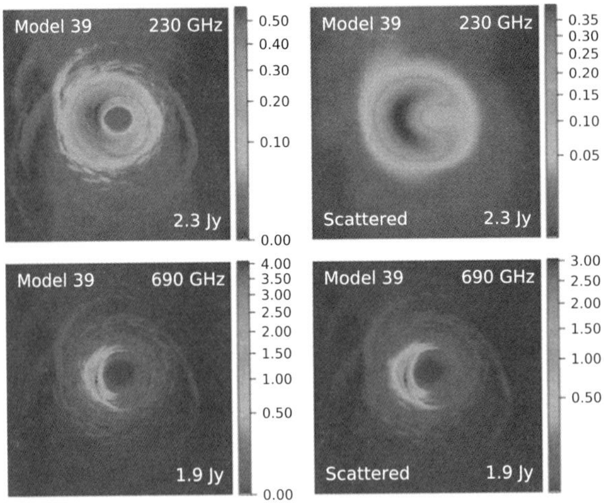

Figure 2. Time-averaged GRMHD source models of Sgr A* from Mościbrodzka *et al.* (2014). Images with note 'scattered' were convolved with the scattering kernel from Bower *et al.* (2006). The field of view is 210 μas. Colors indicate brightness/pixel in mJy.

3. Simulated SVLBI observations

In order to assess what image quality may be obtained with the EHI, we perform simulated observations of a time-averaged ray-traced general relativistic magnetohydrodynamics (GRMHD) source model of Sgr A* at 690 GHz from Mościbrodzka *et al.* (2014) (Fig. 2), using the `eht-imaging` library (Chael *et al.* 2016). We assume orbital radii of 13,892 and 13,913 km, which gives a maximum baseline length of $5.7 \times 10^{10}\lambda$, corresponding to a 3.6 μas angular resolution. We blur the source model with the scattering kernel appropriate for Sgr A* (Bower *et al.* 2006; Johnson and Gwinn 2015). Gaussian thermal noise on the complex visibilities was calculated from model system parameters as

$$\sigma = \frac{1}{0.88}\sqrt{\frac{\mathrm{SEFD}_1\mathrm{SEFD}_2}{2\Delta\nu t_{\mathrm{int}}}}, \quad \text{where SEFD} = \frac{2k_{\mathrm{B}}T_{\mathrm{sys}}}{A_{\mathrm{eff}}},$$

$\Delta\nu$ is the observing bandwidth, t_{int} is the integration time per measurement, k_{B} is Boltzmann's constant, T_{sys} is the system temperature, and A_{eff} is the effective area of the antenna (Thompson *et al.* 2017). The factor 1/0.88 results from 2-bit quantization losses. Here, we adopt $\Delta\nu = 2.4$ GHz, $t_{\mathrm{int}} = 180$ s, $T_{\mathrm{sys}} = 513$ K, resulting in $\sigma = 0.42$ Jy for 3-meter dishes and $\sigma = 0.11$ Jy for 6-meter dishes with efficiencies of 0.58.

For our source models, this noise figure allows for a signal-to-noise ratio (S/N) larger than 1 only on the short baselines. In order to get high S/N values on long baselines as well, we apply two strategies. We grid the *uv*-plane, averaging all visibilities in a single cell. Furthermore, we integrate for multiple iterations ($\sim$ months) of the spiral. For these strategies to work in the low-S/N regime, the orbits should be determined precisely enough in post-processing to get coherent phases. Alternatively, the system noise should be reduced significantly so that traditional fringe fitting can be performed. A complication arising with long integrations is the variable character of the source. Lu *et al.* (2016) showed that for the EHT, an image of the average structure may be obtained by observing multiple days and averaging visibilities at equal *uv*-coordinates. Our strategy of gridding and averaging in the *uv*-plane thus helps mitigating this variability.

Figures 3 and 4 show the resulting S/N *uv*-maps and reconstructed images. Especially with high S/N, the black hole shadow can be reconstructed with an order of magnitude higher resolution than the EHT can achieve. In Roelofs *et al.* (2019), we show that

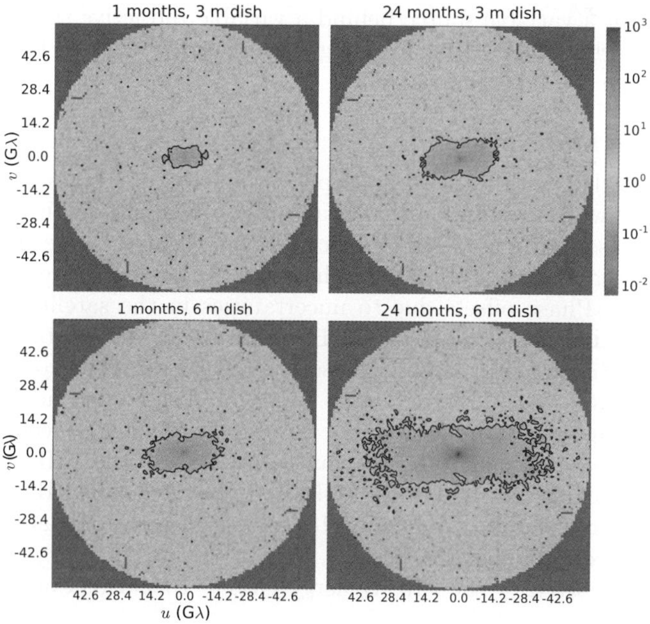

Figure 3. S/N map of the gridded visibilities of model 39 (scattered) at 690 GHz after integrating for 1 (left) and 24 (right) months, with a reflector diameter of 3 (upper panels) and 6 (lower panels) meters. Contours indicate the points with an S/N of 3.

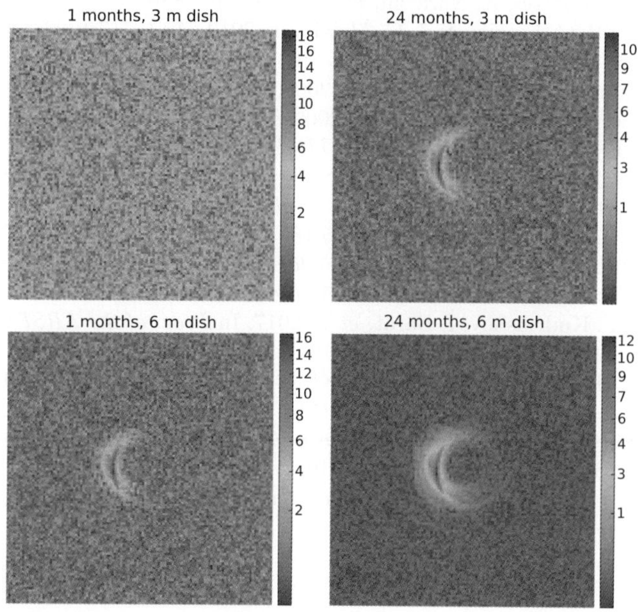

Figure 4. FFT of the gridded visibilities of model 39 (scattered) at 690 GHz after integrating for 1, 6, and 24 months (left to right), with a reflector diameter of 3 (upper panels) and 6 (lower panels) meters. The field of view is 210 μas for all images. Colors indicate brightness/pixel in mJy.

this is also true for a variable source behind a variable refractive scattering screen, and that the selected orbits allow for sharp images when projection effects due to the source declination of $-29°$ are taken into account.

4. Conclusions and outlook

We have shown that with the EHI concept, an image of the black hole shadow of Sgr A* could be made with a resolution that is an order of magnitude higher than the EHT, as many limitations of ground-based VLBI could be mitigated. More engineering studies are necessary in order to assess whether the system described here could fly within a reasonable budget. Phase offsets due to uncertainties in the satellite orbits should be included in the simulations when a detailed system model is available. The necessity of using a third satellite to form closure phases, which are immune to these corruptions, should be investigated. Also, the extent to which black hole and accretion parameters can be recovered and general relativity can be tested with this concept as compared to the EHT should be quantified. We also plan to investigate the EHI imaging prospects for other sources such as M87 and other AGN, and polarized GRMHD models (e.g. Gold *et al.* 2017; Mościbrodzka and Gammie 2018), using different orbital configurations for filling the *uv*-plane on different timescales.

This work is supported by the ERC Synergy Grant "BlackHoleCam" (Grant 610058).

References

Bouman, K. L., Johnson, M. D., Dalca, A. V., *et al.* 2017, arXiv:1711.01357
Bower, G. C., W. M. Goss, H. Falcke, *et al.* 2006, *ApJL*, 648:L127–L130
Broderick, A. E., V. L. Fish, M. D. Johnson, *et al.* 2016, *ApJ*, 820:137
Chael, A. A., M. D. Johnson, R. Narayan, *et al.* 2016, *ApJ*, 829:11
Dexter, J., E. Agol, P. C. Fragile, & J. C. McKinney 2010, *ApJ*, 717:1092–1104
Falcke, H., F. Melia, & E. Agol 2000, *ApJL*, 528:L13–L16
Ghez, A. M., S. Salim, N. N. Weinberg, *et al.* 2008, *ApJ*, 689:1044–1062
Gillessen, S., F. Eisenhauer, S. Trippe, *et al.* 2009 *ApJL*, 692:1075–1109
Goddi, C., H. Falcke, M. Kramer, *et al.* 2017, *IJMPD*, 26:1730001–239
Gold, R., J. C. McKinney, M. D. Johnson, & S. S. Doeleman 2017. *ApJ*, 837:180
Johnson, M. D., K. L. Bouman, L. Blackburn, *et al.* 2017, *ApJ*, 850:172
Johnson, M. D. & C. R. Gwinn 2015. *ApJ*, 805:180
Kudriashov, V. 2017. *International Research Fellow Final Report, ESA-TECE-TN-007251*
Lu, R.-S., F. Roelofs, V. L. Fish, *et al.* 2016, *ApJ*, 817:173
Martin-Neira, M., V. Kudriashov, I. Barat, *et al.* 2017, In *Proc. of the ARSI Workshop*, ESTEC
Mizuno, Y., Z. Younsi, C. M. Fromm, *et al.* 2018, *ArXiv e-prints*
Mościbrodzka, M., H. Falcke, H. Shiokawa, & C. F. Gammie 2014, *A&A*, 570:A7
Mościbrodzka, M. & C. F. Gammie 2018, *MNRAS*, 475:43–54
Psaltis, D., F. Özel, C.-K. Chan, & D. P. Marrone 2015, *ApJ*, 814:115
Roelofs, F., H. Falcke, C. Brinkerink, *et al.* 2019, *A&A*, 625:A124
Thompson, A. R., J. M. Moran, & G. W. Swenson, Jr. 2017, *Interferometry and Synthesis in Radio Astronomy, 3rd Edition*

Perseus in Sicily: From Black Hole to Cluster Outskirts
Proceedings IAU Symposium No. 342, 2018
K. Asada, E. de Gouveia Dal Pino, M. Giroletti,
H. Nagai & R. Nemmen, eds.

doi:10.1017/S1743921318007792

The Hot Universe with XRISM and Athena

M. Guainazzi[1] **and M. S. Tashiro**[2,3]

[1]ESTEC/ESA, Keplerlaan 1, 2201AZ Noordwijk, The Netherlands
email: `Matteo.Guainazzi@sciops.esa.int`

[2]ISAS/JAXA, 3-1-1 Yoshinodai, Chuo-ku, Sagamihara, Kanazgawa 252-5210, Japan
email: `tashiro@astro.isas.jaxa.jp`

[3]Graduate School of Science and Engineering, Saitama University, 255 Shimo-Okubo, Sakawa, Saitama, Saitama 338-8570, Japan
email: `tashiro@phy.saitama-u.ac.jp`

Abstract. X-ray spectroscopy is key to address the theme of "The Hot Universe", the still poorly understood astrophysical processes driving the cosmological evolution of the baryonic hot gas traceable through its electromagnetic radiation. Two future X-ray observatories: the JAXA-led XRISM (due to launch in the early 2020s), and the ESA Cosmic Vision L-class mission *Athena* (early 2030s) will provide breakthroughs in our understanding of how and when large-scale hot gas structures formed in the Universe, and in tracking their evolution from the formation epoch to the present day.

Keywords. space vehicles: instruments, telescopes, X-rays: general

---

1. The Hot Universe

The last few decades have witnessed a momentous improvement in our understanding of the origin and evolution of the Universe. According to the standard ΛCDM cosmology, the Big Bang was followed by a short (a fraction of a second) phase of strongly accelerated expansion ("inflation"), followed by a phase of slower but continuing expansion. Density fluctuations in the early Universe led to the formation of the first structures. These seeds grew over cosmic time via hierarchical collapse. While the evolution of large scale structures is primarily driven by the cosmological parameters and by the interaction with the gravitationally dominant dark matter, still poorly understood astrophysical processes drive the evolution of the **"Hot Universe"**, the baryonic hot gas traceable through its electromagnetic radiation. Accretion onto the dark matter potential wells heated gas to million degree temperatures. Galaxy clusters (the largest gravitationally bound structures in the Universe) lay at the nodes of the cosmic web (Schaye *et al.* 2015). While about 80% of their total mass is in the form of dark matter, the remainder is dominated by diffuse, hot, metal-enriched X-ray emitting plasma (the "Intergalactic Cluster Medium, ICM), with stars constituting only 15% of visible baryons. Furthermore, it is now ascertained that a Super-Massive Black Hole (SMBH) sits at the center of most, if not any, galaxies. SMBHs can inject sufficient energy into the ICM to substantially affect its structure, dynamical state and chemical abundance (Fabian 2012). This "feedback" process can be augmented by SuperNova (SN) winds, coupling the ICM evolution with the life cycle of stars in galaxies. The evolution of the ICM can be therefore shaped by processes beyond the pure gravitational collapse. By tracing clusters and groups from the local Universe back to their formation epochs (at z∼2–3), observational cosmology can be put on a solid experimental grounds because a full understanding of the entire range of astrophysical process at play is required to get a consistent picture of the evolution of cosmic structures. Last, but not least, the census of baryons is largely incomplete, $\simeq$40% at z<2 still eluding detection despite intense observational efforts (Nicastro *et al.* 2017; see, however, Nicastro *et al.* 2018).

Major astrophysical questions remain therefore to answer (Nandra *et al.* 2013):

• How do baryons in groups and clusters accrete and dynamically evolve in the dark matter halos?

• What drives the chemical and thermodynamic evolution of the Universe's largest structures?

• What is the interplay of galaxies, super-massive black holes, and intergalactic gas evolution in groups and clusters?

• Where are the missing baryons at low redshift, and what are their physical states?

X-ray is the only band of the electromagnetic spectrum allowing to probe physical properties of this plasma such as temperature, densities, abundances, velocity field, and ionisation state. All these quantities can be reliably measured only with high-resolution X-ray spectroscopic observations able to characterize recombination transitions of highly ionized (He- and H-like) ions of a wide range of metals. Such measurements allow us to fully characterize the physical state of the plasma as well as the relative share of the energy going into thermal and non-thermal processes in the ICM, the latter generating turbulence and kpc-scale bulk motions. These processes have still to be observed at the relevant spatial scales (corresponding to a few arc-seconds in the local Universe). Furthermore, comparing abundances in the local and high-redshift Universe will trace the evolution of the chemical enrichment and the impact of SN feedback on the growth of large-scale structure in the Universe. Comparing abundances in galaxies, galaxy groups and clusters and in the intergalactic medium is needed to model the life-cycle of matter. All these measurements require a significant improvement in the performance of X-ray spectroscopic instruments over several parameter spaces: energy resolution, spatial resolution, and collecting area.

2. The *Hitomi* heritage

However, high-resolution X-ray spectroscopy is just in its infancy. Only with the advent of modern X-ray observatories such as *Chandra* and XMM-Newton at the beginning of the century, X-ray detectors with a resolving power higher than 100 became routinely operational (Paerels & Kahn 2003). These missions, still enormously successful, carry grating systems able to disperse the incoming X-ray photons onto energy-sensitive detectors. In the soft X-ray band (E$\leqslant$2 keV) resolving powers of $\sim$3–400, and $\sim$1000 were achieved by the grating systems on board XMM-Newton (den Herder *et al.* 2001) and *Chandra* (Brinkman *et al.* 1987; Canizares *et al.* 2005), respectively. In the energy range around $\simeq$6 keV, where the all-important K-shell transitions of iron occur, the *Chandra* high-energy gratings achieve a resolving power of $\simeq$170, albeit with a tiny effective area ($\simeq$30 cm^2).

A huge step forward was expected with the JAXA-led X-ray observatory *Hitomi*, whose successful launch on 2016 February 17 led to hope that a new era in observational X-ray spectroscopy had started. *Hitomi* carried the *Soft X-ray Spectrometer* (SXS), a pixellated micro-calorimeter detector with an unprecedented $\leqslant$5 eV energy resolution over the 0.3–12 keV energy band (Kelley *et al.* 2016). Such a performance corresponds to an improvement in resolving power larger then one order of magnitude at 6 keV with respect to the *Chandra* high-energy transmission gratings, together with an one-order-of magnitude larger effective area at the same energies. Regrettably, the spacecraft was lost after only six weeks of operations due to a chain of anomalies of the attitude control system coupled to human errors (JAXA 2016). However, careful planning during the commissioning phase led to *Hitomi* observing for almost one week the core of the Perseus

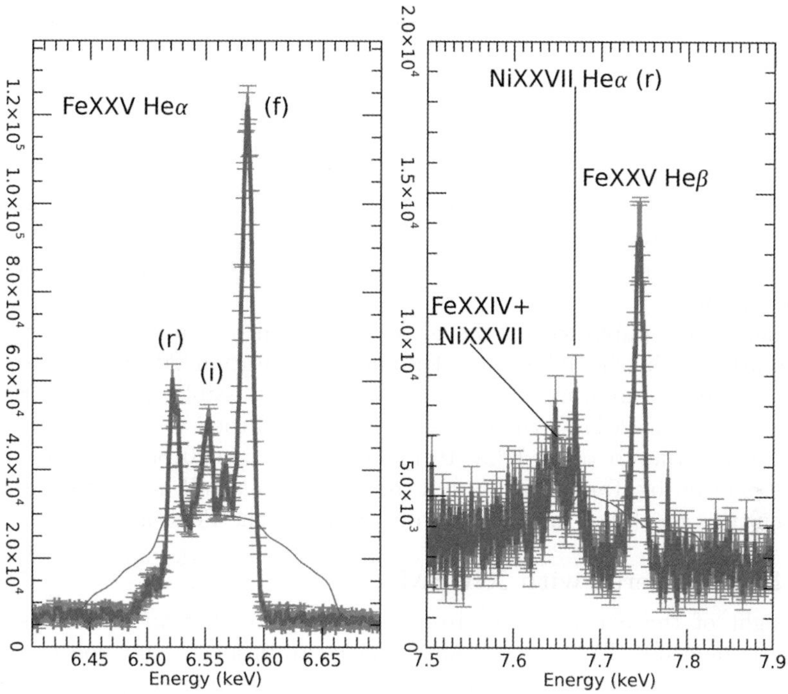

Figure 1. *Hitomi*/SXS spectrum of the Perseus Cluster in the 6.4–6.7 keV (*left panel*) and 7.5–7.9 keV energy range (*right panel*), respectively. The main emission lines discussed in the manuscript are labeled. The error bars correspond to 1σ Poissonian uncertainties. The *red lines* represent the same spectrum at CCD-resolution. Technical details: the spectrum was extracted from the combined calibrated event lists of Obs.# 100040020, -30, -40, and -50 ($\simeq 2.43\times10^5$ s net exposure time), downloaded from the *Hitomi* science archive. It corresponds to a circular extraction region around the cluster core with a 150" radius. It was generated with HEASOFT version 6.22, and the calibration files version 10, released on 15 February 2018.

galaxy cluster, one of the X-ray brightest in the local Universe ($z = 0.0179$). While the SXS was still not formally commissioned, it was operated with the closed "gate valve" with a $\sim$300 μm beryllium window along the optical path as contamination prevention (leading to the total suppression of photons at energies $\leqslant$2 keV), and its data were largely self-calibrated (Tsujimoto *et al.* 2018), its spectra provided a transformational view of the properties of the ICM in Perseus (Fig. 1).

The first surprise came from the analysis of the dynamical state of the ICM. Based primarily on the Fe XXV multiplet (see the *left panel* of Fig. 1), the SXS measured a line-of-sight velocity dispersion of the turbulent gas of 164 ± 10 km s^{-1}, in a region 30–60 kpc from the central nucleus, with only a marginally larger velocity in the core (187 ± 13 km s^{-1}) (Hitomi Collaboration 2016). This level of precision is unprecedented in X-ray astronomy. The measured velocities correspond to a turbulent pressure $\simeq$4% ($<$8%) of the thermal pressure. This is puzzling, because a powerful radio-loud Active Galactic Nucleus (AGN; NGC1275) is present at the cluster core, injecting an amount of energy in the ICM sufficient to evacuate bubbles filled with relativistic plasma (Böhringer *et al.* 1993; Fabian *et al.* 2000): one of the most spectacular examples of "radio-mode" AGN feedback, postulated to prevent radiative cooling of the ICM core gas. The *Hitomi* result is important not only for its implications on the astrophysics of the ICM but also because, if confirmed on a larger sample of clusters at higher redshift, would imply that the deviations from the hydrostatic equilibrium in the ICM are small, and therefore that X-ray derived galaxy cluster masses using this assumption are reliable probes of cosmological parameters (Allen *et al.* 2011).

Another important result came from the study of the chemistry of the ICM in the Perseus cluster. The unprecedented combination of resolving power and effective area allowed the *Hitomi*/SXS to measure the relative elemental abundance of iron-peak elements such as chromium, manganese, and nickel (cf. the *right panel* of Fig. 1). *Hitomi* demonstrated that these abundances are consistent with solar, disproving prior claims of significant over-abundance based on CCD-resolution data (Hitomi Collaboration 2017a). This constrains the progenitor of type Ia SN to be a combination of near- and sub-Chandrasekhar.

Additional papers published on the SXS observation of the Perseus cluster discuss: resonant scattering in the ICM core (Hitomi Collaboration 2018a), tight constraints on the X-ray decay signature of sterile neutrinos, a possible dark matter candidate (Hitomi Collaboration 2017b); the bulk velocity field (Hitomi Collaboration 2018b) and the temperature structure of the ICM (Hitomi Collaboration 2018c). Readers are referred to the contribution by Tamura in these Proceedings for a full description of the whole range of scientific results of the *Hitomi* observations of the Perseus Cluster.

3. The Hot Universe with XRISM

In the light of the extraordinary results obtained by *Hitomi* during its, alas, too short!, operational life, JAXA and NASA decided to propose a mission to recover one of its fundamental scientific objectives: "Resolving astrophysical problems by precise high-resolution X-ray spectroscopy". This is the **X-Ray Imaging and Spectroscopy Mission** (XRISM, a.k.a XARM) (Tashiro *et al.* 2018). The European and Canadian Space Agencies, as well as European institutes participate in the mission development together with a wide range of scientific institutions in Japan and the United States. The mission is designed to fulfill the main theme: "Revealing material circulation and energy transfer in cosmic plasma and elucidating evolution of cosmic structure and objects" with a payload that closely replicates the soft X-ray telescopes and instruments on *Hitomi*:

• a micro-calorimeter (*Resolve*) with a requirement energy resolution $\leqslant$7 eV in the 0.3-12 keV energy range over a 3'×3' field-of-view covered by an array of 35 sky sensitive pixels;

• an array of CCD detector (*Xtend*) with a large field-of-view (larger than 30'×30'), and an energy resolution $\leqslant$200 eV at 6 keV at the beginning of the operational life;

• a large-area, light weight soft X-ray telescope with a $\simeq$1.7' Half Energy Width (HEW) or better, and an area comparable to that of the soft X-ray telescopes on-board *Hitomi*.

This configures a payload whose X-ray spectroscopic performance largely exceeds any currently operational mission (Fig. 2).

One of the XRISM Science Objective is directly related to the "Hot Universe": "Structure, formation and evolution of cluster of galaxies". A micro-calorimeter like *Resolve* will be able to address themes that have been only initiated by *Hitomi*. Elucidating how the gravitational energy released by cluster mergers is converted into thermal energies, and how the energy is distributed between particle and collective gas motions is one of the main goals of XRISM. These objectives will be pursued through measurements of the velocity field structure in the central regions of cool clusters to examine local heating sources (AGN jets; magneto-hydrodynamic interaction between the ICM and member galaxies); measurements of temperature and collective motions of gas stripped from galaxy group accreting onto a cluster to investigate if the infalling galaxies contribute to the ICM heating; and measurements of the turbulent velocity in relaxed and disturbed galaxies that will allow us to evaluate how gravitational energy is distributed among thermal energy, kinetic motions of the ICM, and relativistic particles. Not less

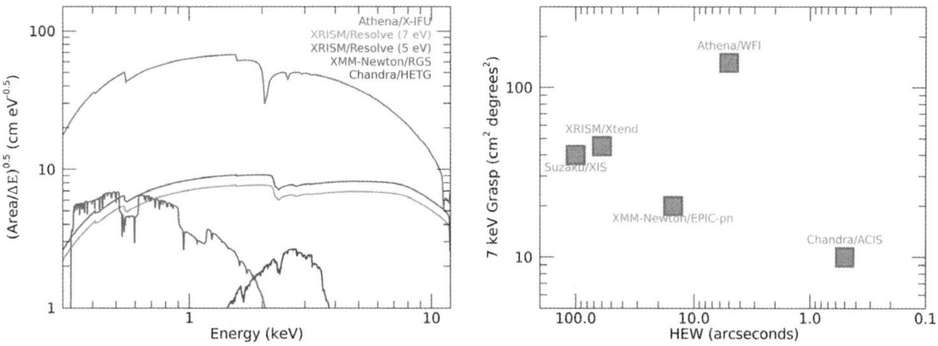

Figure 2. *Left panel:* Weak-line X-ray spectroscopy figure-of-merit for selected operational and future X-ray observatories. The figure of merits is the square root of the ratio between the effective area and the energy resolution. For the *Resolve* instrument on-board XRISM two values are shown, based on the energy resolution requirements (7 eV) and the proven flight resolution of the *Hitomi* SXS ($\leqslant$5 eV). *Right panel*: 7 keV Grasp versus HEW for selected operational and future X-ray observatories. The 1 keV grasp, where the *Athena*/SPO area is optimized, is $\simeq$2800 cm^2 degrees^2 for the *Athena*/WFI, $\simeq$400 cm^2 degrees^2 for the EPIC-pn, and $\simeq$50 cm^2 degrees^2 for the *Chandra*/ACIS.

importantly in a cosmological context, extending the sample of measurements of the ICM turbulent pressure will allow us to correct the hydrostatic bias potentially affecting the X-ray cluster mass functions and therefore remove systematics in the determination of cosmological parameters. XRISM will continue the investigation on the metallicity of the gas trapped in the filament of the cosmic web as a probe of the contribution of different SN explosion types and progenitor populations to the cosmic nucleosynthesis. XRISM is due to launch in the early 2020s (Tashiro *et al.* 2018).

4. The Hot Universe with Athena

The next step in this challenge is *Athena*, the second L-class mission of the ESA "Cosmic Vision" program. *Athena* (Nandra *et al.* 2013) is a large area observatory, aiming at addressing the science themes of the "Hot and Energetic Universe". *Athena* aims at tracing the chemical and physical evolution of large-scale cosmic structures from the epoch of their formation (z$\sim$2–3) to the present Universe; and to study the evolution of accreting black holes in the Universe and of the processes through which they affect the cosmological evolution of the galaxy where they reside, by performing a full census of AGN up to the epoch of reionization. However, besides these basic core science themes, *Athena* will be an observatory fully open to the international astronomical community, with fast ($\leqslant$4 hours) and efficient ($\simeq$50%) response to Targets of Opportunity occurring in a random position in the sky. The large majority of its observing time will be allocated on the basis of proposals evaluated in a peer review process. This high-level scientific objectives will be achieved through an innovative payload:

- a single telescope based on Silicon Pore Optics (SPO) technology developed in Europe (Collon *et al.* 2016), with a 12 m focal length, 5" HEW angular resolution at energies lower than 7 keV, and an effective area $\geqslant$1.4 m^2 (0.25 m^2) at 1 (6) keV;
- the Wide Field Imager (WFI; Meidenger *et al.* 2016), an active pixel sensor Si detector with a wide field-of-view (40'$\times$40'), and spectral-imaging capability with a CCD-like energy resolution ($\leqslant$150 eV at 6 keV);
- the X-ray Integral Field Unit (X-IFU; Barret *et al.* 2018) a cryogenic imaging spectrometer with a 2.5 eV energy resolution in the energy range between 0.2 and 12 keV, 5' diameter effective field-of-view and a $\simeq$5" pixel size.

This payload configuration guarantees that the *Athena* performance will exceed any existing or planned X-ray observatory in the decade of the 20s by more than one order-of-magnitude in several parameter spaces simultaneously, such as the photon collecting area or the truly unprecedented combination of spatial resolution, energy resolution and field-of-view enabled by the X-IFU (Fig. 2).

4.1. *Formation and evolution of galaxy group and clusters*

Athena will probe for the first time the physical properties of the ICM in galaxy clusters and group at the epoch of their formation ($z \simeq 2$). Measurement of X-ray surface brightness and temperature with the X-IFU and WFI will allow to derive the entropy profile, and compare it with the expectations of a pure gravitational collapse as opposed to external effects such as feedback by SN winds and/or AGN. These accurate measurements will permit tracing the evolution of scaling relations between the total mass of the cluster with X-ray observables such as the temperature, X-ray luminosity, entropy up to $z \simeq 2$ for a wide range of masses (Pointecouteau *et al.* 2013), thus ideally complementing the full census of $z \leqslant 1.5$ massive clusters to be achieved by eROSITA (Merloni *et al.* 2012). In the local Universe, the X-IFU will map the non-thermal component of the ICM energy budget through spatially-resolved spectroscopy of the line spectral broadening due to turbulence, and of the line shift due to bulk motion, determining for the first time the large scale properties of the ICM for nearby clusters by resolving the relevant spatial scales. Maps of emission measure, temperature and metallicity in the outskirts of local clusters will unveil the processes occurring where new material is accreting into the dark matter potential and energy is transferred to the ICM via merging events (Ettori *et al.* 2013).

4.2. *Chemical history of hot baryons*

High-resolution spectroscopy of the ICM X-ray line-rich spectrum is the only way to probe the metal abundances of the gas and its cosmological evolution. In the local Universe, *Athena* will map the distribution of the most abundant elements out to the outskirts of galaxy clusters. The peripheral abundance will be used to estimate the contribution of AGN feedback in expelling pre-enriched gas. *Athena* will measure the metallicity of a wide range of elements up to $z \simeq 2$, with accuracies ranging from a few percent (Fe, Si) to 10–20% (O, Ca) (Pointecouteau *et al.* 2013). Because the ICM is an almost perfectly isolated system in the dark matter potential well, its abundances are the resultant of the metal synthesis by different types of SNe over cosmic time (a measurement pioneered by *Hitomi* on the Perseus Cluster: Hitomi Collaboration 2017a). Ratios of iron-peak elements such as chromium and nickel against iron trace SN Ia, those of intermediate-Z elements from O to Si trace core-collapsed SN, lower Z elements are mostly sensitive to AGB stars (Werner *et al.* 2008). The *Athena* X-IFU will measure the whole range of these "metals", and provide a full characterization of the element synthesis history (Ettori *et al.* 2013).

4.3. *AGN feedback in galaxy groups and clusters*

The *Athena* X-IFU will measure velocities of the hot ICM gas with a precision of 10-20 km s^{-1}, and temperatures and metallicity with a precision of a few percent on spatial scales of $\simeq$5" in nearby cluster cores. This implies determining the kinematics of the hot gas in galaxies, groups and clusters on scales small enough to be able to resolve the regions where powerful AGN jets impact, and potential affect its dynamical state. This will allow us to assess how the energy carried by relativistic particles is dissipated and distributed to the ICM, and how the balance between heating and cooling is maintained in regions where the most massive galaxies are formed. In other terms, while we have now a good understanding of the macro-physics of the radio mode of AGN feedback in nearby clusters (Hlavacek-Larrondo *et al.* 2015), *Athena* will investigate for the first time the micro-physics of this process thanks to X-IFU spatially-resolved spectroscopy of the hot gas. By estimating the energy stored in cavities, ripples and motions, *Athena* will estimate the energy input of AGN for the first time. On a larger (Mpc) scale, velocity measurements of shocked expanding hot shells around radio lobes will allow an estimate of the integrated jet power, an elusive quantity preventing the whole power of strong shocks from being estimated (Croston *et al.* 2013). Furthermore, the WFI large field-of-view (coupled with the smooth vignetting curve of the SPO, Willingale *et al.* 2013) will enable population studies of AGN-induced disturbances and bow shocks surrounding radio lobes, permitting to characterize morphological disturbances of the surface brightness distribution in tens of clusters at $z \leqslant 0.1$, and to correlate their mechanical energy with the properties of the environment and the AGN jet power (Croston *et al.* 2013).

4.4. *Missing baryons and WHIM*

About one half of the baryons in the Universe are still elusive. Cosmological simulations shows that the bulk of them should be locked in a hot medium, the Warm-Hot Intergalactic Medium (WHIM), with temperature ranging between 10^5 to 10^7 K. However, they have escaped detection so far (see, however, Nicastro *et al.* 2018 for a different view). This cosmic "hide-and-seek" will end with *Athena* that will be able to detect the WHIM, and characterize its physical properties, both in absorption against a bright background source (Gamma-Ray Bursts, GRB; blazars, etc.), or through emissions along tenuous and diffuse filaments. During its nominal 4 year operation life, *Athena* is expected to observe about 90 systems in absorption, follow-up $\simeq$7 GRB in emission, and observe about 100 shallow and 1 deep side-line in emission. Depending of the speed of the spacecraft re-pointing, this will allow to probe filament with Oxygen column densities down to 10^{15-16} cm^{-2} (Kaastra *et al.* 2013).

5. Summary

After the global disarray for the early demise of *Hitomi* operations, the future of X-ray astronomy looks bright again thanks to a plethora of new missions and observatories due launch in the next few years. In this context, the enhanced high-resolution capabilities of the XRISM/*Resolve*, enhanced in the *Athena*/X-IFU by its sharp angular resolution and better energy resolution, and combined to deep wide-field spectral imaging with the XRISM/*Extend* and the *Athena*/WFI will yield a giant leap in our understanding of the formation and evolution of large-scale baryonic structure in the "Hot Universe". Furthermore, besides their nominal core science goals XRISM and *Athena* are observatories open to the astronomical community. Their observational programs will be primarily driven by the collective scientific wisdom of the community active at the time of their operations. Thanks to the unique capabilities of XRISM and *Athena* - largely exceeding those of any existing or planned spectroscopic X-ray missions - we should "expect the unexpected" - as the far too short *Hitomi* history clearly shows.

Acknowledgments

This paper is written on behalf of the XRISM Science Management Office and of the Athena Science Study Team, whose comments on an early version of the manuscript are gratefully acknowledged. We thank Dr. Didier Barret for a careful revision of the manuscript.

References

Allen, S. W., Evrard, A. E., & Mantz, A. B. 2011, *ARA&A*, 49, 409
Barret, D., *et al.* 2016, *SPIE*, 9905, 2
Barret, D., *et al.* 2018, *SPIE*, in press (arXiv:1807.06092)
Böhringer, H., Voges, W., Fabian, A. C., Edge, A. C., & Neumann, D. M. 1993, *MNRAS*, 264, 25
Brinkman, A. C., *et al.* 1987, *ApL&C*, 26, 73
Canizares, C. R., *et al.* 2005, *PASP*, 117, 1144
Collon, M., *et al.* 2016, *SPIE*, 9905, 28
Croston, J. H., *et al.* 2013, arxiv:1306.2323
den Herder, J. W., *et al.* 2001, *A&A*, 365, 7
Ettori, S., *et al.* 2013, arXiv:1306.2322
Fabian, A. C. *et al.* 2000, *MNRAS*, 318, 65
Fabian, A. C. 2012, *ARA&A*, 50, 455
Hitomi Collaboration 2016, *Nature*, 535, 117
Hitomi Collaboration 2017a, *Nature*, 551, 478
Hitomi Collaboration 2017b, *ApJ*, 837, 15
Hitomi Collaboration 2018a, *PASJ*, 70, 10
Hitomi Collaboration 2018b, *PASJ*, 70, 9
Hitomi Collaboration 2018c, *PASJ*, 70, 11
Hlavacek-Larrondo, J., *et al.* 2015 *ApJ*, 805, 35
Kaastra, J., *et al.* 2013, arXiv:1306.2324
Kelley, R. L., *et al.* 2016 *SPIE*, 9905, id. 99050V
JAXA 2016 *Hitomi Experience Report: Investigation of Anomalies Affecting the X-ray Astronomy Satellite 'Hitomi' (ASTRO-H)*
Meidinger, N., *et al.* 2016, *SPIE*, 9905, 2
Merloni, A., *et al.* 2012, arXiv:1209.3114
Nandra, K., *et al.* 2013, arXiv:1306.2307
Nicastro, F., Krongold, T., Mathur, S., & Elvis, M. 2017, *AN*, 338, 218
Nicastro, F. *et al.* 2018, arXiv:1806.08395
Paerels, F. B. S., & Kahn S. M. 2003, *ARA&A*, 41, 291
Pointecouteau, E., *et al.* 2013, arXiv:1306.2319
Schaye, J., *et al.* 2015, *MNRAS*, 446, 521
Tashiro, M., *et al.* 2018, *SPIE*, in press
Tsujimoto, M., *et al.* 2018, *JATIS*, 4, 1205
Werner, N., Finoguenov, A., Kaastra, J. S., Simionescu, A., Dietrich, J. P., Vink, J., & Böhringer, H. 2018, *SSRv*, 134, 337
Willingale, R., *et al.* 2013, arXiv:1307.1709

Perseus in Sicily: From Black Hole to Cluster Outskirts
Proceedings IAU Symposium No. 342, 2018
K. Asada, E. de Gouveia Dal Pino, M. Giroletti,
H. Nagai & R. Nemmen, eds.

doi:10.1017/S1743921318003861

Low frequency observations of radio relics and halos

Ruta Kale

National Centre for Radio Astrophysics,
Tata Institute of Fundamental Research,
S. P. Pune University Campus,
Ganeshkhind, Pune 411007, India
email: ruta@ncra.tifr.res.in

Abstract. Diffuse radio emission from galaxy clusters in the form of radio halos and relics are tracers of the shocks and turbulence in the intra-cluster medium. The imprints of the physical processes that govern their origin and evolution can be found in their radio morphologies and spectra. The role of mildly relativistic population of electrons may be crucial for the acceleration mechanisms to work efficiently. Low frequency observations with telescopes that allow imaging of extended sources over a broad range of low frequencies (< 2 GHz) offer the best tools to study these sources. I will review the Giant Metrewave Radio Telescope (GMRT) observations in the past few years that have led to: i) statistical studies of large samples of galaxy clusters, ii) opening of the discovery space in low mass clusters and iii) tracing the spectra of seed relativistic electrons using the Upgraded GMRT.

Keywords. acceleration of particles, radiative mechanisms: non-thermal, shock waves, turbulence, galaxies:clusters:general

1. Introduction

Clusters of galaxies are gravitationally bound systems of masses $\sim 10^{14} - 10^{15}$ $M_\odot$ that are composed of dark matter, galaxies and the intra-cluster medium (ICM). The ICM is the most massive baryonic component forming $10 - 15\%$ of the total mass and mainly contains thermal plasma that emits in X-rays via thermal Bremsstrahlung mechanism. It also contains non-thermal components such as the magnetic fields and relativistic particles but these elude detection in most observing bands. The detection of diffuse radio emission of synchrotron origin from the ICM provides direct evidence for the presence of relativistic electrons ($\sim$GeV) and magnetic fields ($\sim 0.1-$ a few μG) in galaxy clusters. These sources are typically classified into radio halos and radio relics based on their morphology and location relative to the X-ray emitting thermal ICM (see Feretti *et al.* 2012; Brunetti & Jones 2014, for reviews). Due to their steep spectra ($\alpha > 1.0$, $S_\nu \propto \nu^{-\alpha}$) these sources are typically studied at low radio frequencies ($\leqslant 2$GHz). The short radiative lifetime (~ 0.1 Gyr) and long diffusion times ($\geqslant$Gyr to reach Mpc distance) of relativistic electrons in the ICM requires that radio halos and relics have mechanisms of in-situ re-acceleration associated with their origin (e. g. Jaffe 1977).

Radio relics are elongated or arc-like, polarized radio sources that are found at the peripheries of clusters. These occur as single or sometimes in pairs around galaxy clusters (Fig. 1, top). Radio relics are proposed to trace shocks at the cluster outskirts where particles are accelerated (e.g. Enßlin *et al.* 1998). The polarization indicating aligned magnetic fields, spectral indices showing steepening from outer to the inner edges of the relics (e. g. Giacintucci *et al.* 2008; Bonafede *et al.* 2009; van Weeren *et al.* 2010;

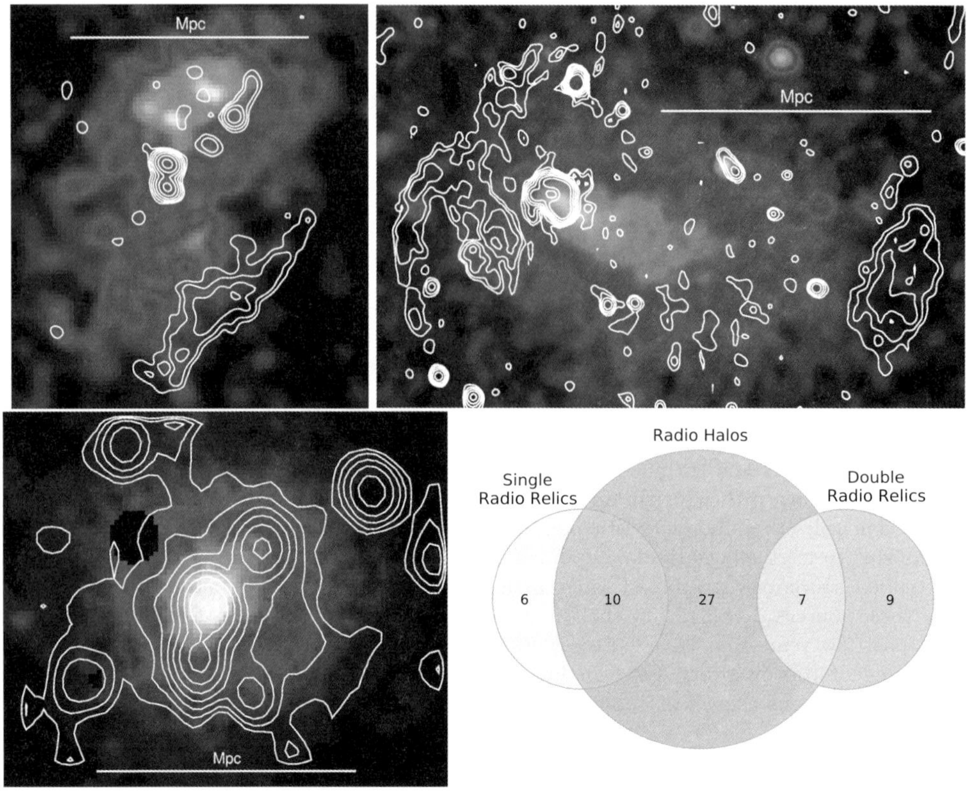

Figure 1. The single radio relic in the cluster PLCK G200.9-28.2 (top left), the double radio relic in Abell 3376 (top right) and the radio halo in PLCKG171.9-40.7 (bottom left) are shown in white radio contours overlaid on the respective X-ray images of the host clusters in colour. The relics are elongated arc-like sources at the edges of the clusters and a radio halo is a centrally located Mpc-scale extended radio source. Bottom right:- The Venn diagram represents the currently known clusters that are host to one or more kinds of these emission. There are significant number of clusters that are host to one or more kinds of such radio emission. The sample is taken from Yuan *et al.* (2015).

Kale *et al.* 2012) and the co-spatiality with X-ray detected shocks (e. g. Akamatsu *et al.* 2012; Ogrean & Brüggen 2013) provide support for relics as tracers of shocks.

Radio halos are centrally located in the cluster, Mpc sized and unpolarized (Fig. 1, bottom left). The origin of radio halos had been proposed to be in the secondary electrons generated by the hadronic collisions in the ICM (e. g. Dennison 1980; Blasi & Colafrancesco 1999). The stringent upper limits on the gamma rays associated with this process (e. g. Ackermann *et al.* 2010; Arlen *et al.* 2012) have led to a scenario where a primary mechanism such as turbulent reacceleration (e. g. Petrosian 2001; Brunetti *et al.* 2001) may be playing a crucial role (e. g. Brunetti *et al.* 2017). An empirical scaling relation between the cluster mass (or X-ray luminosity) and the radio power of radio halos is known (e. g. Cassano *et al.* 2013).

The connection between cluster mergers and occurrence of radio halos and relics has been found observationally (Buote 2001; Cassano *et al.* 2010; Kale *et al.* 2015b). Cluster mergers are a natural origin for the shocks and turbulence that are proposed to play a role in the generation of such sources. Indeed a significant fraction of clusters that host radio halos and relics show presence of both types of sources (Fig. 1, bottom right).

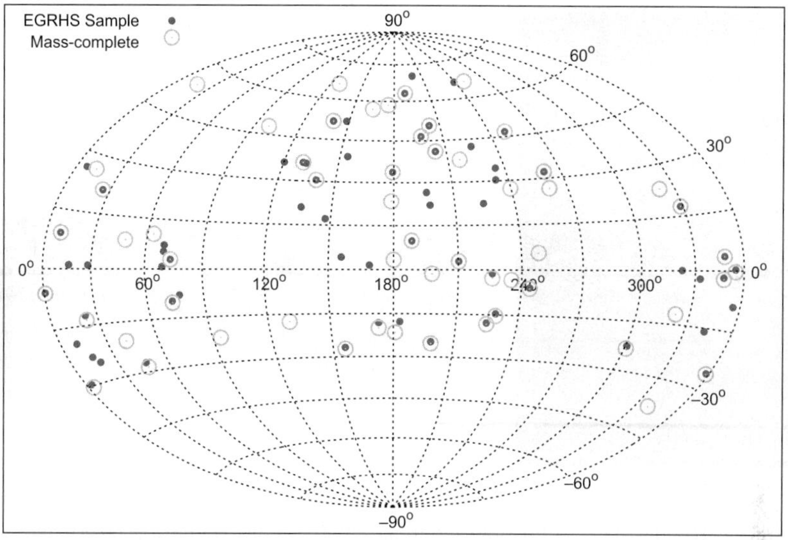

Figure 2. The largest uniform surveys of galaxy clusters aimed towards searching for diffuse radio emission from the intra-cluster medium are shown on the celestial sphere. The clusters surveyed in the Extended GMRT Radio Halo Sample (Venturi *et al.* (2007, 2008); Kale *et al.* (2013, 2015b)) are marked with red filled points. The clusters that form the mass-complete extension (Cuciti *et al.* 2015) of the EGRHS sample are shown in green open points.

However it is still a matter of investigation as to why some merging clusters host radio halos and relics while others do not.

For the reacceleration mechanisms to work, a seed population of relativisitic electrons is needed as the efficiencies of acceleration are low (e. g. Markevitch *et al.* 2005; Kang & Ryu 2011; Pinzke *et al.* 2017). These seeds may be due to the radio galaxies in clusters and the secondary electrons. Low frequency observations are crucial in order to trace the seed population as it is expected to be aged synchrotron plasma with steep spectra.

In this review I will focus on how low frequency observations have helped to get insights into the occurrence statistics of radio halos and relics, led to the discoveries of such diffuse sources in low mass clusters and are helping to trace the seed populations of relativistic electrons.

2. Radio surveys of galaxy clusters

Typically the search for diffuse radio emission has been done in all sky surveys such as the NRAO VLA Sky Survey (NVSS), WENSS and more recently in the GLEAM survey. Although this approach can result in discoveries, it is not useful to quantify the non-detections. For that purpose tailored surveys such as the GMRT Radio Halo Survey (Venturi *et al.* 2007, 2008) are crucial. A short summary of the recent and ongoing surveys is provided below.

The Extended GMRT Radio Halo Survey

The EGRHS is an extension of the GRHS and together have surveyed a total of 64 clusters selected from the REFLEX and eBCS catalogues of clusters (Kale *et al.* 2013, 2015b). These were the clusters in the redshift range 0.2 – 0.4 and brighter than 5×10^{44} erg s^{-1} in X-ray luminosity and were above the declination of $-31°$ (Fig. 2). Radio halos were found in 22% and relics in 5% of the sample. Upper limits using the method of injection of models (Venturi *et al.* 2008) were reported on a total of 31 clusters. Using X-ray morphology estimators, strong evidence for the connection between cluster merger

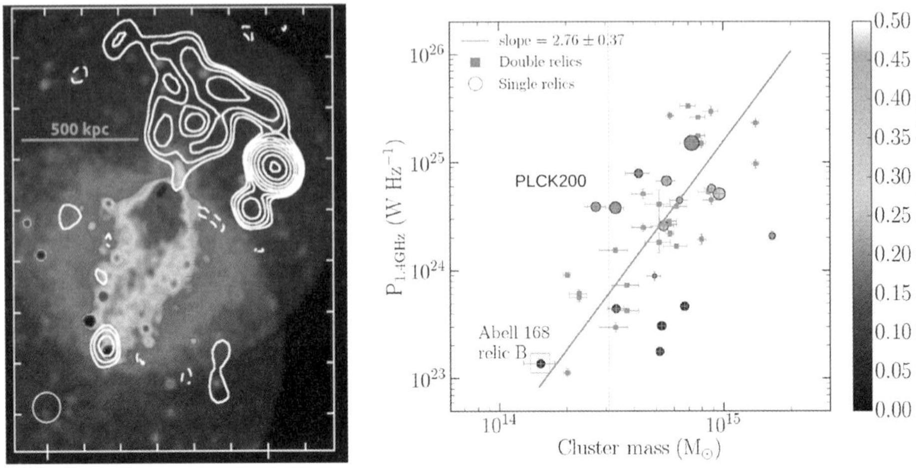

Figure 3. Left:- The single relic cluster with lowest host cluster mass discovered from the GLEAM survey. Right:- The radio power versus cluster mass plane for radio relics. The double radio relics (grey squares) and the single relics (circles scaled according the relic size and coloured according to the redshift shown in the color bar) are shown. The vertical line is plotted to highlight the two lowest mass clusters with single radio relics labelled as "PLCK200" and "Abell168 relic B". The best fit line for the scaling relation plotted as the grey solid line. Images in both the panels are adapted from Dwarakanath *et al.* (2018).

and occurrence of radio halos was found. This is the first largest uniformly surveyed sample of galaxy clusters and formed the basis for further exploration of radio power - cluster mass scaling and a study of the Brightest Cluster Galaxies (e. g. Cassano *et al.* 2013; Kale *et al.* 2015).

Mass complete sample: Ongoing survey

The detections of a large number of galaxy clusters by the Planck satellite using the Sunyaev-Zel'dovich effect resulted in a sample of clusters with well constrained total masses (Planck Collaboration *et al.* 2014). A sample of galaxy clusters with mass-completeness $\sim 80\%$ in the redshift range 0.08 – 0.33 with masses, $> 6 \times 10^{14}$ $M_\odot$ was constructed and explored occurrence rates of radio halos (Cuciti *et al.* 2015). The sample is shown in Fig. 2. The fraction of radio halos in high and low mass sub-samples divided at 8×10^{14} $M_\odot$ was found to be $60 - 80\%$ and $20 - 30\%$, respectively. Two underluminous radio halos have been found in this sample (Cuciti *et al.* 2018) and the analysis of the remaining sample is ongoing and will provide the statistics of occurrence in the coming years.

3. Low mass, merging galaxy clusters

The EGRHS and the mass-complete samples have focused mainly on the most massive clusters. The low mass clusters that show merging signatures are promising sites to constrain the properties of shocks and turbulence. A radio relic was recently used to trace a low mass cluster (de Gasperin *et al.* 2017a).

We carried out a search for diffuse radio emission of extent > 500 kpc towards the newly detected clusters in the Early SZ Planck catalogue (Planck Collaboration *et al.* 2015). This resulted in the discovery of a radio halo in the cluster PLCKG171.9-40.7 (Giacintucci *et al.* 2013) and a single radio relic in the cluster PLCK G200.9-28.2 (Kale *et al.* 2017). The single relic cluster with the mass of 2.7×10^{14} $M_\odot$ was the lowest mass host to a single relic known at that time. We found an even lower mass cluster with a single relic using Murchison Widefield Array (MWA, Tingay *et al.* (2013))

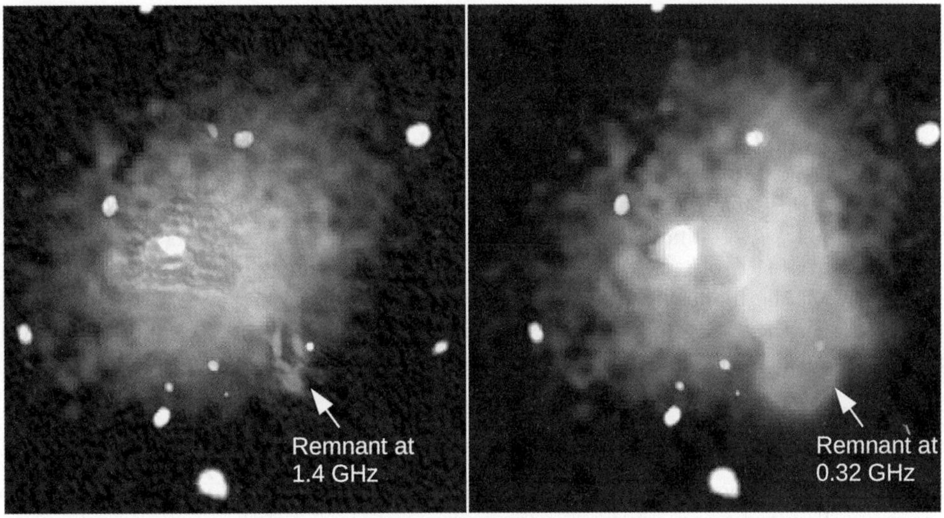

Figure 4. The cluster Abell 4038 in optical (green), X-ray (red) and radio bands (blue) is shown in both the panels. The 1.4 GHz image is shown in the left and the 0.32 GHz image is shown in the right panel. The low frequencies trace the steep spectrum emission that is much more extended compared to the higher frequencies.

which has the best existing short baseline coverage at low frequencies (< 200 MHz) in the southern hemisphere. The cross matching of the galaxy cluster catalogues with the 200 MHz GaLactic and Extra-galactic All-sky MWA survey (GLEAM, Hurley-Walker *et al.* (2017)) led to the discovery of a radio relic in the cluster Abell 168 (Dwarakanath *et al.* 2018). The confirmation of the relic was done with the GMRT (325 and 610 MHz) and Very Large Array (1 – 2 GHz). The single arc-like relic at the periphery of this cluster has a smaller ring-like radio source in its wake. Based on the curved spectrum of this source we termed this as a remnant relic that may have been responsible for providing seed relativistic electrons in the outer relic. This is a unique case where the reaccelerated plasma at the shock and the possible seed plasma are detected.

4. Broadband studies with the Upgraded GMRT: characterising the seeds

The remnants of radio galaxies in galaxy clusters can be important sources of seed relativistic electrons. Galaxy clusters have been found to be host to diffuse sources that are filamentary or ring-like with sizes of 100 – 200 kpc and with steep spectra (Slee *et al.* 2001). A number of such sources have been recently discovered using deep radio observations (e. g. Shimwell *et al.* 2015, 2016). Such remnants can be revived by adiabatic compression due to shocks (Enßlin & Gopal-Krishna 2001) or by gentle reacceleration (van Weeren *et al.* 2017; de Gasperin *et al.* 2017b). In order to characterise the seed population, their spectra need to be studied. The spectra can often be curved and can vary across the extent of the source (e. g. 1 – 2 GHz study of A2256 relics Owen *et al.* 2014).

The Upgraded GMRT (uGMRT) is now operational and provides observations with near seamless coverage between 0.12 – 1.4 GHz with instantaneous bandwidths of up to 400 MHz (Gupta *et al.* 2017). The observing bands have been released on shared risk basis starting mid-2016. The uv-coverage has improved due to the enhanced bandwidth and extended sources can be imaged up to sizes twice of what were possible to be imaged with the legacy GMRT (Deo & Kale 2017). The radio relic in Abell 4038 is a prototype of steep spectrum relics which require low frequency observations to reveal their extent (Fig. 4). Our GMRT study led us to the conclusion at it may be an adiabatically compressed remnant of the radio galaxy hosted by the Brightest Cluster Galaxy (Kale & Dwarakanath 2012). We have mapped the spectrum across the extent of the radio relic in the cluster Abell 4038 (Kale & Dwarakanath 2012) with the uGMRT in the band 300 – 500 MHz and 1050 – 1450 MHz and found variations of the spectral curvature across the extent of the relic (Kale *et al.* 2018). Images with rms 70μJy beam^{-1} at 0.4 GHz and 30μJy beam^{-1} 1.26 GHz were made and used to study the variation of the spectral index and its curvature across the relic. Between 0.3 – 1.4 GHz, the spectral curvatures range between 0.5 – 1.6, with the highest curvatures in a curved region of the relic skirting an arc-like feature seen in the X-ray image. The variations thus are connected to the local physical conditions and possibly indicate a change in magnetic field across the source as well.

5. Conclusions

Radio halos and relics are direct probes of the non-thermal components in the ICM. Sensitive low frequency observations possible with the current and upcoming interferometers are well suited to reveal the spectra and morphologies of these sources to constrain the theoretical models. In recent years the role of reacceleration by turbulence and shocks in the formation of radio halos and relics has gained considerable support. The availability of seed relativistic electrons so that the reacceleration mechanisms are efficient may be one of the crucial factor dividing merging clusters with and without radio halos and relics. The uniform surveys of galaxy clusters, namely the EGRHS and the ongoing survey mass-complete sample are providing the occurrence statistics of radio halos and relics. The unexplored regime of low mass merging clusters is becoming accessible due to the sensitive low frequency telescopes such as the MWA (see M. Johnston-Hollitt, this volume). The seed relativistic electron population is likely going to be steep spectrum and thus their characterisation relies on sensitive measurements in sub-GHz frequency ranges. The uGMRT is operational and we have carried out the first study of the spectrum of a remnant radio galaxy in the cluster A4038. The prospects to study radio halos and relics with the uGMRT are promising.

Acknowledgements

RK would like to thank the organisers for the invitation to give a talk at this excellent conference. RK acknowledges the support from the DST INSPIRE Faculty Award by the Government of India. We thank the staff of the GMRT that made these observations possible. GMRT is run by the National Centre for Radio Astrophysics of the Tata Institute of Fundamental Research. This research has made use of the NASA/IPAC Extragalactic Database (NED) which is operated by the Jet Propulsion Laboratory, California Institute of Technology, under contract with the National Aeronautics and Space Administration. This research made use of Astropy, a community-developed core Python package for Astronomy (Astropy Collaboration, 2018). The scientific results reported in this article are based in part on data obtained from the Chandra Data Archive. The National Radio Astronomy Observatory is a facility of the National Science Foundation operated under cooperative agreement by Associated Universities, Inc. Based on observations obtained with XMM-Newton, an ESA science mission with instruments and contributions directly funded by ESA Member States and NASA

References

Ackermann, M., Ajello, M., Allafort, A., *et al.* 2010, *ApJ*, 717, L71
Akamatsu, H., Takizawa, M., Nakazawa, K., *et al.* 2012, *PASJ*, 64, 67
Arlen, T., Aune, T., Beilicke, M., *et al.* 2012, *ApJ*, 757, 123

Blasi, P., & Colafrancesco, S. 1999, Astroparticle Physics, 12, 169
Bonafede, A., Giovannini, G., Feretti, L., Govoni, F., & Murgia, M. 2009, *A&A*, 494, 429
Brunetti, G., & Jones, T. W. 2014, International Journal of Modern Physics D, 23, 30007
Brunetti, G., Setti, G., Feretti, L., & Giovannini, G. 2001, *MNRAS*, 320, 365
Brunetti, G., Zimmer, S., & Zandanel, F. 2017, *MNRAS*, 472, 1506
Buote, D. A. 2001, *ApJ*, 553, L15
Cassano, R., Ettori, S., Giacintucci, S., *et al.* 2010, *ApJ*, 721, L82
Cassano, R., Ettori, S., Brunetti, G., *et al.* 2013, *ApJ*, 777, 141
Cuciti, V., Brunetti, G., van Weeren, R., *et al.* 2018, *A&A*, 609, A61
Cuciti, V., Cassano, R., Brunetti, G., *et al.* 2015, *A&A*, 580, A97
de Gasperin, F., Intema, H. T., Ridl, J., *et al.* 2017a, *A&A*, 597, A15
de Gasperin, F., Intema, H. T., Shimwell, T. W., *et al.* 2017b, Science Advances, 3, e1701634
Dennison, B. 1980, *ApJ*, 239, L93
Deo, D. K., & Kale, R. 2017, Experimental Astronomy, 44, 165
Dwarakanath, K. S., Parekh, V., Kale, R., & George, L. T. 2018, *MNRAS*, 477, 957
Enßlin, T. A., Biermann, P. L., Klein, U., & Kohle, S. 1998, *A&A*, 332, 395
Enßlin, T. A., & Gopal-Krishna. 2001, *A&A*, 366, 26
Feretti, L., Giovannini, G., Govoni, F., & Murgia, M. 2012, *AApR*, 20, 54
Giacintucci, S., Kale, R., Wik, D. R., Venturi, T., & Markevitch, M. 2013, *APJ*, 766, 18
Giacintucci, S., Venturi, T., Macario, G., *et al.* 2008, *A&A*, 486, 347
Gupta, Y., Ajithkumar, B., Kale, H. S., *et al.* 2017, Current Science, 113, 707
Hurley-Walker, N., Callingham, J. R., Hancock, P. J., *et al.* 2017, *MNRAS*, 464, 1146
Jaffe, W. J. 1977, *ApJ*, 212, 1
Kale, R., & Dwarakanath, K. S. 2012, *ApJ*, 744, 46
Kale, R., Dwarakanath, K. S., Bagchi, J., & Paul, S. 2012, *MNRAS*, 426, 1204
Kale, R., Parekh, V., & Dwarakanath, K. S. 2018, *MNRAS*, arXiv:1808.04057
Kale, R., Venturi, T., Cassano, R., *et al.* 2015a, *A&A*, 581, A23
Kale, R., Venturi, T., Giacintucci, S., *et al.* 2013, *A&A*, 557, A99
Kale, R., Wik, D. R., Giacintucci, S., *et al.* 2017, *MNRAS*, 472, 940
Kale, R., Venturi, T., Giacintucci, S., *et al.* 2015b, *A&A*, 579, A92
Kang, H., & Ryu, D. 2011, *ApJ*, 734, 18
Markevitch, M., Govoni, F., Brunetti, G., & Jerius, D. 2005, *ApJ*, 627, 733
Ogrean, G. A., & Brüggen, M. 2013, *MNRAS*, 433, 1701
Owen, F. N., Rudnick, L., Eilek, J., *et al.* 2014, *ApJ*, 794, 24
Petrosian, V. 2001, *ApJ*, 557, 560
Pinzke, A., Oh, S. P., & Pfrommer, C. 2017, *MNRAS*, 465, 4800
Planck Collaboration, Ade, P. A. R., Aghanim, N., *et al.* 2014, *A&A*, 571, A29
—. 2015, *A&A*, 581, A14
Shimwell, T. W., Markevitch, M., Brown, S., *et al.* 2015, *MNRAS*, 449, 1486
Shimwell, T. W., Luckin, J., Brüggen, M., *et al.* 2016, *MNRAS*, 459, 277
Slee, O. B., Roy, A. L., Murgia, M., Andernach, H., & Ehle, M. 2001, *AJ*, 122, 1172
Tingay, S. J., Goeke, R., Bowman, J. D., *et al.* 2013, *PASA*, 30, e007
van Weeren, R. J., Röttgering, H. J. A., Brüggen, M., & Hoeft, M. 2010, Science, 330, 347
van Weeren, R. J., Andrade-Santos, F., Dawson, W. A., *et al.* 2017, Nature Astronomy, 1, 0005
Venturi, T., Giacintucci, S., Brunetti, G., *et al.* 2007, *A&A*, 463, 937
Venturi, T., Giacintucci, S., Dallacasa, D., *et al.* 2008, *A&A*, 484, 327
Yuan, Z. S., Han, J. L., & Wen, Z. L. 2015, *ApJ*, 813, 77

Perseus in Sicily: From Black Hole to Cluster Outskirts
Proceedings IAU Symposium No. 342, 2018
K. Asada, E. de Gouveia Dal Pino, M. Giroletti, H. Nagai & R. Nemmen, eds.

doi:10.1017/S1743921318003897

Probing the non-thermal emission in the Perseus cluster with the JVLA

M. Gendron-Marsolais[1], J. Hlavacek-Larrondo[1], R. J. van Weeren[2], T. Clarke[3], A. C. Fabian[4], H. T. Intema[2], G. B. Taylor[5], K. M. Blundell[6] and J. S. Sanders[7]

[1]Département de Physique, Université de Montréal, Montréal, QC H3C 3J7, Canada
email: marie-lou@astro.umontreal.ca

[2]Leiden Observatory, Leiden University, Niels Bohrweg 2, NL-2333CA, Leiden, The Netherlands

[3]Naval Research Laboratory, Code 7213, 4555 Overlook Ave. SW, Washington, DC 20375, USA

[4]Institute of Astronomy, University of Cambridge, Madingley Road, Cambridge CB3 0HA

[5]Department of Physics and Astronomy, University of New Mexico, Albuquerque, NM 87131, USA

[6]University of Oxford, Astrophysics, Keble Road, Oxford OX1 3RH, UK

[7]Max-Planck-Institut für extraterrestrische Physik, 85748 Garching, Germany

Abstract. We present deep low radio frequency (230-470 MHz) observations from the Karl G. Jansky Very Large Array of the Perseus cluster, probing the non-thermal emission from the old particle population of the AGN outflows. Our observations of this nearby relaxed cool core cluster have revealed a multitude of new structures associated with the mini-halo, extending to hundreds of kpc in size. Its irregular morphology seems to have been influenced both by the AGN activity and by the sloshing motion of the cluster' gas. In addition, it has a filamentary structure similar to that seen in radio relics found in merging clusters. These results illustrate the high-quality images that can be obtained with the new JVLA at low radio-frequencies.

Keywords. galaxies: clusters: Perseus cluster, cooling flows, radio continuum: galaxies, X-rays: galaxies: clusters

---

1. Introduction

Jets created from accretion onto supermassive black holes (SMBHs) release relativistic particles on large distances. These strongly affect the intracluster medium (ICM) when located in the center of a brightest cluster galaxy (BCG). These jets inflate bubbles, displacing the ICM and creating regions of depleted X-ray emission. The bubbles, filled with relativistic plasma (i.e., radio lobes), are often discernible at ~GHz frequencies. On the other hand, the hierarchical merging of subclusters and groups, from which cluster originate, also generates perturbations into the ICM through shocks and turbulence, constituting a potential source of reacceleration for these particles.

In the Perseus cluster, we observe several types of disturbances in the ICM that have been probed in detail by extensive X-ray observations. At least two pairs of X-ray cavities have been identified: the first pair, located at $5 < r < 20$ kpc from the AGN, is filled with radio emission (Böhringer *et al.* 1993) and the second, located further out at $25 < r < 45$ kpc from the AGN, is devoid of radio emission above ~ 1 GHz (Branduardi-Raymont *et al.* 1981; Fabian *et al.* 1981; Churazov *et al.* 2000). Inflated by generations of outbursts from the central AGN, these constitute the imprint of radio mode feedback, injecting energy into the ICM and compensating its radiative losses (e.g. Bîrzan *et al.* 2004; Dunn & Fabian 2006; Rafferty *et al.* 2006). Beyond cavities are quasi-spherical ripples interpreted as sound waves (Fabian *et al.* 2003), as well as a semicircular cold front, two new elliptical cavities interpreted as potential ghost bubbles, two large regions of weak X-ray luminosity (the northern trough and the southern bay) and a loop-like structure above a long Hα filament (Fabian *et al.* 2011).

The Perseus cluster harbours a mini-halo (Soboleva *et al.* 1983; Pedlar *et al.* 1990; Burns *et al.* 1992; Sijbring 1993), a faint diffuse source of radio emission detected so far in about thirty cool core clusters (Giacintucci *et al.* 2017, see Feretti *et al.* 2012 for a review). The X-rays and mini-halo structure of Perseus have been compared in Fabian *et al.* (2011). This emission differs from that filling the X-ray cavities, being $\gtrsim 3$ times radio fainter and having a steeper spectral index ($\alpha < -1$ for $S(\nu) \propto \nu^{\alpha}$, where S is the flux density and ν is the frequency, Giacintucci *et al.* 2014a). Since the radiative timescale of the electrons is much shorter than the time required for them to reach the extent of these sources, the origin of mini-halos remains unclear. Two possible mechanisms have been proposed in the literature to explain the mini-halo emission: it might originate from the reacceleration of pre-existing electrons by turbulence (Gitti *et al.* 2002, 2004) or from the generation of new particles from inelastic collisions between relativistic cosmic-ray protons and thermal protons (e.g. Pfrommer & Enßlin 2004). Simulations seem to suggest that turbulence created by sloshing motions of the cold gas in the core region is sufficient to re-accelerate electrons (e.g. ZuHone *et al.* 2013). A key prediction of these simulations is that mini-halos should be bounded by cold fronts.

Here we summarize results presented in Gendron-Marsolais *et al.* (2017) on new, deep Karl G. Jansky Very Large Array (JVLA) observations of the Perseus cluster in the P-band (230-470 MHz). The resolution and sensitivity of these data provide a detailed and extended view of the mini-halo structure, on which we focus our analysis. The recent update of the facilities with the EVLA project offer new abilities to study this structure.

We assume a redshift of $z = 0.0183$ for NGC 1275 corresponding to a luminosity distance of 78.4 Mpc, assuming $H_0 = 69.6$ km s^{-1}Mpc^{-1}, $\Omega_{\rm M} = 0.286$ and $\Omega_{\rm vac} = 0.714$. This corresponds to an angular scale of 22.5 kpc arcmin^{-1}.

2. JVLA observations of the Perseus cluster

We obtained a total of 5 hours in the P-band (230-470 MHz) of the Karl G. Jansky Very Large Array in B-configuration on November 24$^{\rm th}$, 2013 (project $13B-026$). With the new broadband low frequency receivers, the P-band bandwidth has been widened from 300-340 MHz to 230-470 MHz, increasing significantly the sensitivity of the telescope. The resolution of these observations probes the faint extended emission of the mini-halo structure in the Perseus cluster. The data reduction was performed with CASA (Common Astronomy Software Applications, version 4.6). A pipeline was specifically developed to account for the strong presence of radio frequency interference (RFI) at low-frequencies and the extremely bright central AGN in Perseus outshining faint structures.

3. Results

Figure 1 shows the central part of the B-configuration final map obtained from the data reduction and imaging process described in Section 2. The resulting image has a rms of 0.35 mJy/beam, peak at 10.63 Jy/beam and a beam size of $22.1'' \times 11.3''$. Beyond the central emission from NGC 1275, the large field of view includes NGC 1265, a wide-angle tail radio galaxies, and IC 310, a radio galaxy with a potential blazar-like behavior (Aleksić *et al.* 2014b; Ahnen *et al.* 2017), both discovered by Ryle & Windram (1968).The

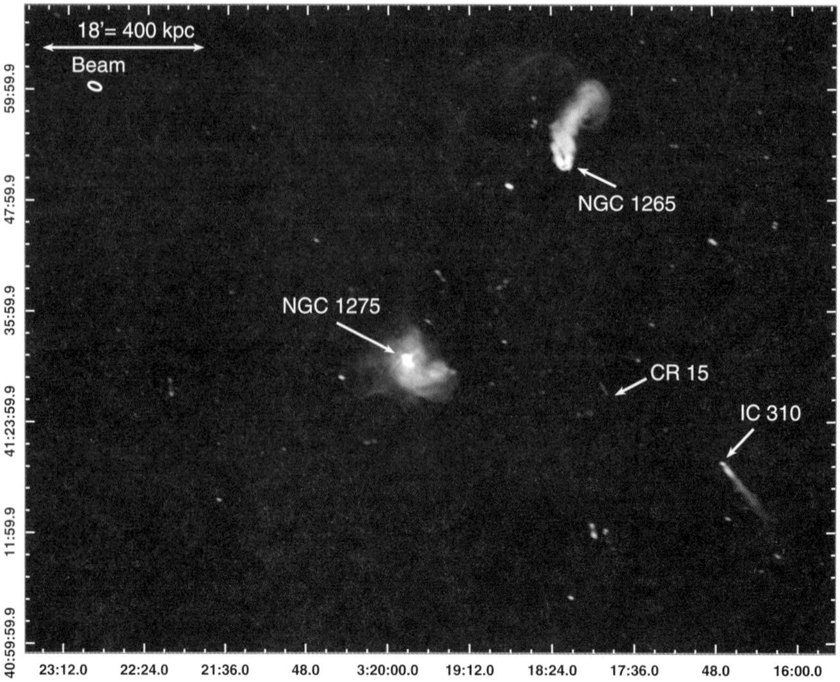

Figure 1. The central 2 ° × 1.5 ° of the total field of view of the JVLA 230-470 MHz radio map obtained in B-configuration. NGC 1275 is the bright source in the middle of the image. Tow wide-angle tail radio galaxies, NGC 1265 (NNW of NGC 1275) and CR 15 (between NGC 1275 and IC 310) as well as IC 310 (WSW of NGC 1275) are clearly visible. The resulting image has a rms noise of 0.35 mJy/beam, a beam size of 22.1″ × 11.3″ and a peak of 10.63 Jy/beam.

analysis of their complex morphologies will be presented in future work (Gendron-Marsolais *et al.* 2017 in prep.). The smaller head-tail source CR 15 is also found between NGC 1275 and IC 310, the tail pointing in a northeast direction (Miley *et al.* 1972).

Figure 2 shows a zoom of the central radio emission surrounding NGC 1275. Features running roughly north-west/south-east through the core are the remaining artifacts due to some problematic antennas all located in the same arm of the JVLA. The diffuse mini-halo structure extends up to ~ 150 kpc from the AGN and shows a complex structure. The general shape of the mini-halo in the Perseus cluster has an irregular morphology, curving counterclockwise. It is also elongated in the direction of the radio bubble system. In addition to this large-scale structure shape, several fine structure details in the emission have been identified: two spurs are seen to the east and southeast of the AGN, an extension to the north, an edge to the south-west, a plume of emission to the south-south-west and a concave edge to the south.

4. Discussion

4.1. *Comparison with previous radio observations*

Figure 3 shows the 74 MHz, 235 MHz, 230-470 MHz and 610 MHz radio contours overlaid on the fractional residual X-ray image. The mini-halo is seen clearly at 230-470 MHz and 610 MHz, while only spurs of radio emission extend outside the X-ray cavities at 74 MHz and almost no emission is seen outside the inner cavities at 235 MHz with the GMRT image at these sensitivity levels. Mini-halos generally have steep spectral indexes but the noise levels of the 74 and 235 MHz images (80 and 10mJy/beam, respectively) are

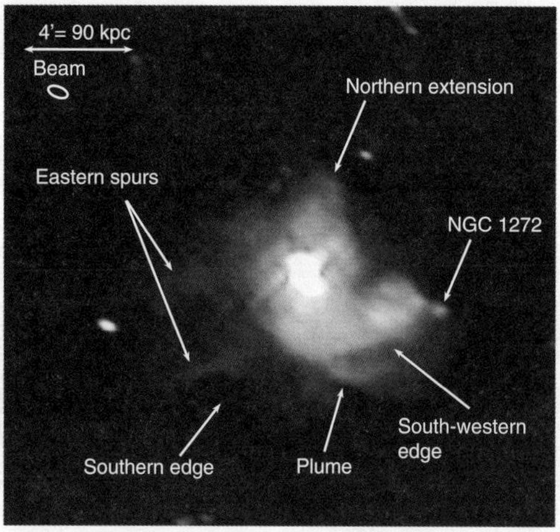

Figure 2. A zoom on the emission surrounding NGC 1275 from the 230-470 MHz radio map seen in figure 1. The main structures of the mini-halo are identified: the northern extension, the two eastern spurs, the concave edge to the south, the south-western edge and a plume of emission to the south-south-west. The small knob at the end of the western tail is the galaxy NGC 1272.

too high for the faint mini-halo to be detected. The new JVLA facilities have produced an order of magnitude deeper image than the previous 330 MHz VLA data ($\sigma_{\rm rms} =$ 7mJy/beam) allowing the detection of the mini-halo emission to much larger radii and in much finer detail. The northern extension of emission as well as hints of the presence of the southern and south-western edges were already visible in 610 MHz WSRT contours from Sijbring (1993). The synthesized beam size of these observations ($29'' \times 44''$) being about five times larger than the beam size of our 230-470 MHz observations ($22.1'' \times 11.3''$), it only probed blurred emission from the eastern spurs, the plume and the southern edge, not fine filaments or sharp edges compared to the new JVLA observations.

4.2. *Large-scale structure of the mini-halo*

The mini-halo - sloshing cold front correspondences found in many clusters support the reacceleration hypothesis according to which cooled relativistic electrons injected by past AGN activity are reaccelerated by turbulence that may be produced by sloshing motion (Gitti *et al.* 2002, 2004; Mazzotta & Giacintucci 2008; ZuHone *et al.* 2013; Giacintucci *et al.* 2014a). The B-configuration radio contours starting at 5σ are overlaid on *Chandra* X-ray image (Fabian *et al.* 2011) on Figure 4- left. One of the most striking features in the deep X-ray images of the Perseus cluster is the spiral pattern of the X-ray emitting gas. A similar trend is also clearly seen in the temperature map (Fabian *et al.* 2006, see figure 4- right) and the western part of it was identified as a cold front (Fabian *et al.* 2011), a sharp contact discontinuity between gas regions with different temperatures and densities. The characteristic spiral pattern of cold fronts are created by the sloshing of gas in a gravitational potential perturbed by a minor merger (see Markevitch & Vikhlinin 2007 for a review). High-resolution simulations of cluster mergers also show that cold fronts are produced by minor mergers and can persist over gigayear time scales (Ascasibar & Markevitch 2006). In the case of the Perseus cluster, a chain of bright galaxies visible to the west of the BCG NGC 1275 have been identified as the possible source of disturbance (Churazov *et al.* 2003). Interestingly, the curving shape of the mini-halo seems to be well aligned with the sloshing pattern. It matches

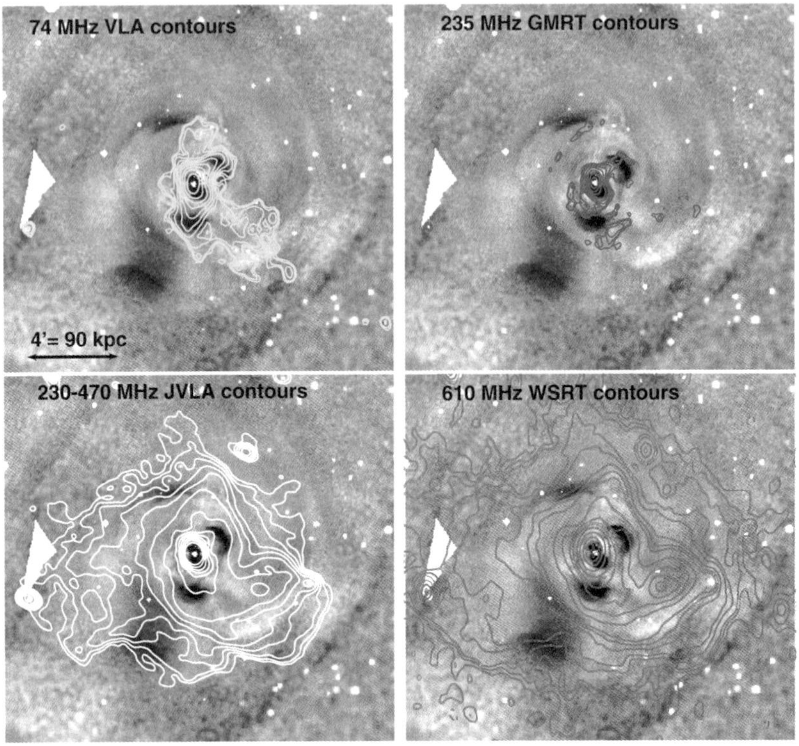

Figure 3. *Chandra* final composite fractional residual image from Fabian *et al.* (2011) in the 0.5-7 keV band (total of 1.4 Ms exposure) with radio contours at different frequencies overlaid. Top-left: 74 MHz A configuration VLA contours (synthesized beamwidth of 24″, $\sigma_{\rm rms} = 80$ mJy/beam). A total of 11 contours are drawn, increasing logarithmically from 0.3 Jy/beam to 36.2 Jy/beam (Blundell *et al.* 2002). Top-right: 235 MHz GMRT contours (synthesized beamwidth of 13″). 10 contours are drawn, increasing logarithmically from $5\sigma_{\rm rms} = 50$ mJy/beam to 9 Jy/beam. Bottom-left: 230-470 MHz contours from the new JVLA B-configuration (beamwidth of 22.1″ × 11.3″). A total of 13 contours are drawn, also increasing logarithmically from $3\sigma = 1.05$ mJy/beam to 1 Jy. Bottom-right: 610 MHz WSRT contours from Sijbring (1993), synthesized beamwidth of 29″ × 44″ and $\sigma_{\rm rms} = 0.4$ mJy/beam). The contours levels are -0.8 (dashed), 0.8, 1.6, 2.4, 5, 7.5, 15, 22.5, 30, 60, 90, 120, 150, 300, 850, 2500, 5000 and 10000 mJy/beam.

both the size and the direction of curvature (counterclockwise). However, the mini-halo is also elongated in the direction of the cavity system. This spatial correlation is consistent with the scenario that AGN feedback could contribute to the injection of turbulence in the ICM and reaccelerate the relativistic particles responsible for the mini-halo emission (Cassano *et al.* 2008).

Figure 4- middle shows the X-ray GGM filtered image with the same radio contours. The position of the western cold front is indicated by the arrows. This image shows very clearly how the mini-halo emission is mostly contained behind the cold front: there is a sharp edge in the radio image associated with the mini-halo, but the particles appear as well to leak out as there is an even fainter (2 to 3 times fainter) part of the mini-halo that extends beyond the cold front in the south-western direction.

Another interesting large scale structure present in the mini-halo of the Perseus cluster is the southern edge identified in figure 2. Compared with the X-ray observations (see figure 4- left), the radio emission seems to avoid the southern bay, an intriguing feature located about 100 kpc south of the nucleus, first reported in Fabian *et al.* (2006).

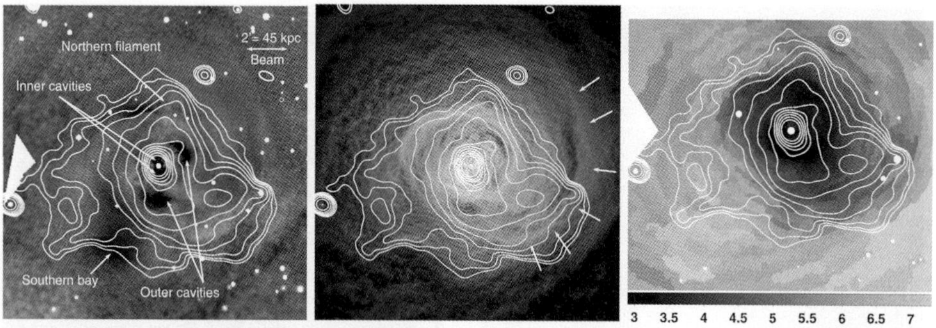

Figure 4. Left - *Chandra* final composite fractional residual image from Fabian *et al.* (2011) in the 0.5-7 keV band (total of 1.4 Ms exposure) with 230-470 MHz contours from $5\sigma = 1.75$ mJy/beam to 1 Jy overlaid from JVLA B-configuration. Middle - GGM filtered image of the merged X-ray observations with Gaussian width $\sigma = 4$ pixels (Sanders *et al.* 2016) with the same 230-470 MHz JVLA contours. The position of the western cold front is indicated with white arrows. Right - Central part of the temperature map of the Perseus cluster from Fabian *et al.* (2011) with signal-to-noise ratio of 150 with the same 230-470 MHz JVLA contours. Units are keV.

This 'bay' structure, founded in three nearby relaxed clusters (Perseus, Centaurus and Abell 1795, Walker *et al.* 2017), behaves like cold fronts but has the opposite curvature toward the interior of the cluster. According to simulations of gas sloshing, they might be resulting from Kelvin-Helmholtz instabilities (Walker *et al.* 2017).

4.3. *Filamentary structure*

Very few filamentary structures like the ones present in the P-band JVLA observations of the Perseus cluster have been observed before in mini-halos. In Abell 2626, two elongated, ~ 5 kpc thick, arc-like radio features with longitudinal extensions of ~ 70 kpc are detected in its mini-halo (Gitti *et al.* 2004; Gitti 2013). In the case of Perseus, the two eastern spurs identified in figure 2 are ≥ 10 kpc thick and extend over ~ 150 kpc in scale. Interestingly, similar filaments are found in radio relics (e.g. the relic in Abell 2256, Owen *et al.* 2014), large elongated diffuse polarized radio sources located at cluster peripheries. Relics could result from synchrotron emission of electrons reaccelerated by mergers or accretion shocks (Ensslin *et al.* 1998; Brunetti & Jones 2014; van Weeren *et al.* 2017). In Perseus though, no shocks corresponding with the position of the filaments are known. We can also speculate that these filaments trace regions of enhanced magnetic fields or locally enhanced turbulence. Alternatively, they could reflect the original distribution of fossil plasma, for example from an old AGN outburst (up to a Gyr ago) that are re-accelerated by turbulence or weak shocks.

As shown in figure 5, the northern extension of the mini-halo also matches the position of the northern filament seen in the Hα map from Conselice *et al.* (2001). The loop-like X-ray structure extending at the end of the northern filament has been interpreted as fallback gas dragged out to the north by previously formed bubbles (Fabian *et al.* 2011). As for the elongated shape of the mini-halo aligned with the cavity system, the northern extension shows another correlation of the mini-halo with the relativistic jets. Therefore, the shape of mini-halos seems to originate both from sloshing and from past AGN activity.

4.4. *Implications for our understanding of mini-halos*

The deep JVLA observations of the Perseus cluster combined with the cluster's properties (proximity, brightness and sloshing plane inclination) have revealed lots of structures,

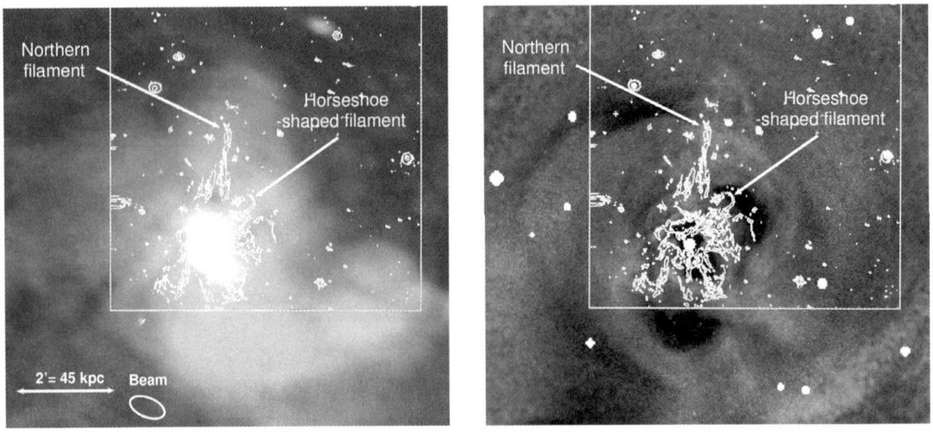

Figure 5. Left - JVLA 230-470 MHz B-configuration image with Hα contours (in green) of the continuum-subtracted Hα map from Conselice *et al.* (2001). The green region indicates the edge of the Hα image. The horseshoe-shaped filament and the northern filament are identified. Right - The fractional residual X-ray image with the same Hα contours.

unlike the present observations of most mini-halos which appear to be of fuzzy and uniform emission. This could be due to the resolution and sensitivity of the radio observations of mini-halos. Few mini-halos also present structures, e.g. the arc-like radio features in Abell 2626 (Gitti *et al.* 2004; Gitti 2013) or the spiral-shaped tail of emission in RX J1720.1+2638 with a length of $\sim$230 kpc (Giacintucci *et al.* 2014b). Perseus's mini-halo has similar size but a much higher flux density, 3 to 4 orders of magnitude higher than other mini-halos (Giacintucci *et al.* 2014a). Even with this level of detail, the emission is still mostly constrained behind the sloshing cold front, delimited by a sharp radio edge, providing a qualitative test of the reacceleration hypothesis. However, faint emission is also seen beyond this edge as if particles appear to leak out (see figure 4- middle). Again, as larger structures are resolved out in the B-array observations, we could still miss a large-scale diffuse mini-halo that would extend well beyond the cold front edge indicating that there must be a source of acceleration that goes beyond the cold front.

5. Conclusions

We present a detailed radio map of the Perseus cluster obtained from 5 h of observations with the JVLA at 230-470 MHz in the B-configuration. In summary, we conclude the following.

(*a*) Several structures have been identified in the mini-halo emission: the northern extension, two filamentary spurs to the east and a clear edge avoiding the X-ray southern bay. The general shape of the mini-halo is curving counterclockwise and is elongated in the direction of the cavity system. At 230-470 MHz, Perseus's mini-halo extends up to 135 kpc from the nucleus and has a total flux density of 12.64 Jy.

(*b*) The comparison of the 230-470 MHz map with deep *Chandra* observations has shown that the mini-halo is enclosed mostly behind the western sloshing cold front, qualitatively supporting the reacceleration hypothesis. However, fainter emission is also seen beyond, as if particles leaking out.

(*c*) The large-scale and fine structure show a correlation of the mini-halo emission with both the sloshing motion and the relativistic jets of the AGN.

(*d*) Mysterious filamentary spurs of emission are found to the east, similar to radio relics, but no shocks corresponding with the position of the filaments are known.

(*e*) The shape of the mini-halo resembles the central simulated synchrotron radiation in magnetohydrodynamic simulations of gas sloshing in galaxy clusters for example from ZuHone *et al.* (2013).

These results illustrate the high-quality images that can be obtained with JVLA at low radio-frequencies, as well as the necessity to obtain deeper, higher-fidelity radio images of extended emission in clusters to further understand their origin.

References

Ahnen, M. L. *et al.* 2017, *A&A*
Aleksić, J. *et al.* 2014b, *A&A*, 563, A91
Ascasibar, Y., & Markevitch, M. 2006, *ApJ*, 650, 102
Bîrzan, L., Rafferty, D. A., McNamara, B. R., Wise, M. W., & Nulsen, P. E. J. 2004, *ApJ*, 607, 800
Blundell, K. M., Kassim, N. E., & Perley, R. A. 2002, in Proc. IAU Coll. 199. p. 189
Böhringer, H., Voges, W., Fabian, A. C., Edge, A. C., & Neumann, D. M. 1993, 264, L25
Branduardi-Raymont, G., Fabricant, D., Feigelson, E., Gorenstein, P., Grindlay, J., Soltan, A., & Zamorani, G. 1981, 248, 55
Brunetti, G., & Jones, T. W. 2014, 23, 1430007
Burns, J. O., Sulkanen, M. E., Gisler G. R., & Perley, R. A. 1992, 388, L49
Cassano, R., Gitti, M., & Brunetti, G. 2008, 486, L31
Churazov, E., Forman, W., Jones, C., & Böhringer, H. 2000, *A&A*, 356, 788
Churazov, E., Forman, W., Jones, C., & Böhringer, H. 2003, *ApJ*, 590, 225
Conselice, C. J., Gallagher, III J. S., & Wyse, R. F. G. 2001, *ApJ*, 122, 2281
Dunn, R. J. H., & Fabian, A. C. 2006, *MNRAS*, 373, 959
Ensslin, T. A., Biermann, P. L., Klein, U., & Kohle, S. 1998, *A&A*, 332, 395
Fabian, A. C., Hu, E. M., Cowie, L. L., & Grindlay, J. 1981, *ApJ*, 248, 47
Fabian, A. C., Sanders, J. S., Allen, S. W., Crawford, C. S., Iwasawa, K., Johnstone, R. M., Schmidt, R. W., & Taylor, G. B. 2003, *MNRAS*, 344, L43
Fabian, A. C., Sanders, J. S., Taylor, G. B., Allen, S. W., Crawford, C. S., Johnstone, R. M., & Iwasawa, K. 2006, *MNRAS*, 366, 417
Fabian, A. C., *et al.* 2011, *MNRAS*, 418, 2154
Feretti, L., Giovannini, G., Govoni, F., & Murgia, M. 2012, *A&ARv*, 20, 54
Gendron-Marsolais, M., *et al.* 2017, *MNRAS*, 469, 3872
Giacintucci, S., Markevitch, M., Venturi, T., Clarke, T. E., & Cassano, R. Pasquale Mazzotta 2014a, *ApJ*, 781, 9
Giacintucci, S., Markevitch, M., Brunetti, G., ZuHone, J. A., Venturi, T., Mazzotta, P., & Bourdin, H. 2014b, *ApJ*, 795, 73
Giacintucci, S., Markevitch, M., Cassano, R., Venturi, T., Clarke, T. E., & Brunetti, G. 2017, *ApJ*
Gitti, M. 2013, *MNRAS*, 436, L84
Gitti, M., Brunetti, G., & Setti, G. 2002, *A&A*, 386, 456
Gitti, M., Brunetti, G., Feretti, L., & Setti, G. 2004, *A&A*, 417, 1
Markevitch, M., & Vikhlinin, A. 2007, *ApJ*, 443, 1
Mazzotta, P., & Giacintucci, S. 2008, *ApJ*, 675, L9
Miley, G. K., Perola, G. C., van der Kruit, P. C., & van der Laan, H. 1972, Nature, 237, 269
Owen, F. N., Rudnick, L., Eilek, J., Rau, U., Bhatnagar, S., & Kogan, L. 2014, *ApJ*, 794, 24
Pedlar, A., Ghataure, H. S., Davies, R. D., Harrison, B. A., Perley, R., Crane, P. C., & Unger, S. W. 1990, *MNRAS*, 246, 477
Pfrommer, C., & Enßlin, T. A. 2004, *A&A*, 413, 17
Rafferty, D. A., McNamara, B. R., Nulsen, P. E. J., & Wise, M. W. 2006, *ApJ*, 652, 216
Ryle, M., & Windram, M. D. 1968, *MNRAS*, 138, 1
Sanders, J. S., Fabian, A. C., Russell, H. R., Walker, S. A., & Blundell, K. M. 2016, *MNRAS*, 460, 1898
Sijbring, L. G. 1993, PhD thesis, Groningen Univ.

Soboleva, N. S., Temirova, A. V., Timofeeva, G. M., & Aliakberov, K. D. 1983, Soviet Astronomy Letters, 9, 305
Walker S. A., Hlavacek-Larrondo J., Gendron-Marsolais M., Fabian A. C., Intema H., Sanders, J. S., Bamford, J. T., & van Weeren, R. 2017, *MNRAS*, 468, 2506
ZuHone, J., Markevitch, M., Brunetti, G., & Giacintucci, S. 2013, *ApJ*, 762, 78
van Weeren, R. J. *et al.* 2017, Nature Astronomy, 1, 0005

Perseus in Sicily: From Black Hole to Cluster Outskirts
Proceedings IAU Symposium No. 342, 2018
K. Asada, E. de Gouveia Dal Pino, M. Giroletti,
H. Nagai & R. Nemmen, eds.

doi:10.1017/S1743921318008463

High-dynamic-range 21 cm JVLA observations of the Perseus Cluster

Charles L. H. Hull[1,2,3]

[1]National Astronomical Observatory of Japan, NAOJ Chile Observatory, Alonso de Córdova 3788, Office 61B, 7630422, Vitacura, Santiago, Chile

[2]Joint ALMA Observatory, Alonso de Córdova 3107, Vitacura, Santiago, Chile

[3]NAOJ Fellow

email: chat.hull@nao.ac.jp

Abstract. We present 21 cm (L-band) JVLA observations of the Perseus Cluster, comprising data from all four array configurations. The resulting images are nearly an order of magnitude deeper than any other image made of NGC 1275 and its environs at this frequency.

Keywords. galaxies: clusters: individual (Perseus), (galaxies:) cooling flows, (galaxies:) intergalactic medium, galaxies: ISM

1. Summary

The center of the Perseus galaxy cluster displays a spectacular array of intricate, filamentary nebulae. These filaments typically lie between 10 and 50 kpc from the center of the cluster, with lengths $\lesssim 10$ kpc. First discovered in H_α by Minkowski (1957), they have now been studied extensively in the radio (Salomé *et al.* 2008a, 2008b), infrared (Werner *et al.* 2013), optical (McDonald *et al.* 2010), UV (Canning *et al.* 2010), and soft x-ray (Sparks *et al.* 2009), as well as in hydrodynamic and magnetohydrodynamic simulations (e.g., Revaz *et al.* 2008; Sharma *et al.* 2010).

Despite more than five decades of study, however, several features of the filaments are extremely difficult to explain. Though immersed in the $\sim 10^8$ K intracluster medium (ICM), these filaments emit strongly in H_α, indicating an internal temperature $\lesssim 10^4$ K. In some filaments, the detection of molecular CO and HCN suggests even lower internal temperatures $\lesssim 2000$ K. Thus, these nebulae have temperatures orders of magnitude lower than the surrounding ICM, yet have radii smaller than the mean free path of ICM electrons of ~ 100 pc. If thermal conduction is not suppressed by the local magnetic field, the timescale for such thin filaments to evaporate could be as short as 100 years!

HST observations of the filaments show that some, but not all, of the filaments, are actively forming stars, representing a unique case of star formation taking place outside of a galaxy. A by-product of star formation in star forming galaxies is synchrotron emission, produced through the ejection of cosmic ray electrons into the ISM by supernovae, which then interact with the intergalactic magnetic field.

Measuring the synchrotron emission from the filaments would yield information about the cosmic ray population and would confirm the presence of a strong magnetic field. Comparing the synchrotron flux to existing far-infrared (FIR) observations would also probe the universality of star formation.

This study is still in progress, and involves making the highest-ever-dynamic-range L-band (1.4 GHz) image of the cD galaxy of the Perseus cluster, NGC 1275. The data involved are from all four (A, B, C, D) configurations of the Karl G. Jansky Very Large

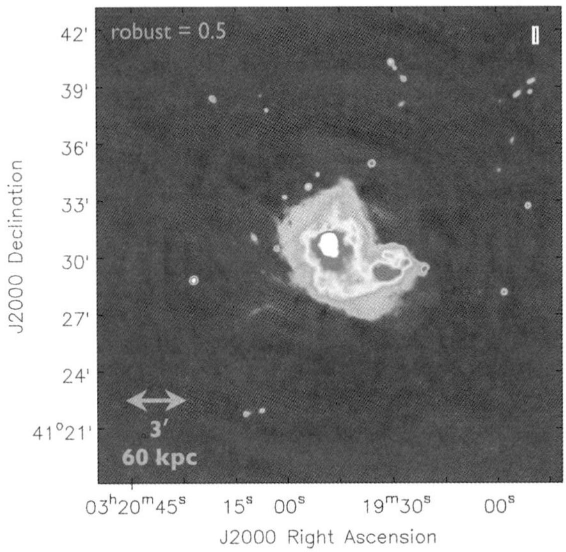

Figure 1. L-band (1.4 GHz) image of NGC 1275 made using a combination of C and D-configuration data from the VLA. The central extended source is the famous "mini-halo" surrounding NGC 1275. A wealth of background sources and intricate structure can be seen in this high-dynamic-range image; these structures will become clearer upon combination with the higher-resolution A and B-configuration data.

Array (VLA), and are revealing intricate structures in the galaxy itself, as well as a wealth of intriguing background sources. A preliminary C+D-configuration image can be seen in Figure 1. These data will be combined with A and B-configuration data to make the best possible image of the region.

Acknowledgements

C.L.H.H. thanks the IAU and IAUS 342 committee members for their generous support. C.L.H.H. also acknowledges the support of the JSPS KAKENHI grant 18K13586. ALMA is a partnership of ESO (representing its member states), NSF (USA) and NINS (Japan), together with NRC (Canada), MOST and ASIAA (Taiwan), and KASI (Republic of Korea), in cooperation with the Republic of Chile. The Joint ALMA Observatory is operated by ESO, AUI/NRAO and NAOJ. The National Radio Astronomy Observatory is a facility of the National Science Foundation operated under cooperative agreement by Associated Universities, Inc.

References

Canning, R. E. A., Fabian, A. C., Johnstone, R. M., *et al.* 2010, *MNRAS*, 405, 115
McDonald, M., Veilleux, S., Rupke, D. S. N., & Mushotzky, R. 2010, *ApJ*, 721, 1262
Minkowski, R. 1957, in IAU Symposium, Vol. 4, Radio astronomy, ed. H. C. van de Hulst, 107
Revaz, Y., Combes, F., & Salomé, P. 2008, *A&A*, 477, L33
Salomé, P., Combes, F., Revaz, Y., *et al.* 2008a, *A&A*, 484, 317
Salomé, P., Revaz, Y., Combes, F., *et al.* 2008b, *A&A*, 483, 793
Sharma, P., Parrish, I. J., & Quataert, E. 2010, *ApJ*, 720, 652
Sparks, W. B., Pringle, J. E., Donahue, M., *et al.* 2009, *ApJ*, 704, L20
Werner, N., Oonk, J. B. R., Canning, R. E. A., *et al.* 2013, *ApJ*, 767, 153

Perseus in Sicily: From Black Hole to Cluster Outskirts
Proceedings IAU Symposium No. 342, 2018
K. Asada, E. de Gouveia Dal Pino, M. Giroletti,
H. Nagai & R. Nemmen, eds.

doi:10.1017/S1743921318007226

A wide and collimated radio jet in 3C 84

Gabriele Giovannini

Physics and Astronomy Dpt, Bologna University
via Gobetti 93/2
40129 Bologna, Italy and
Radio Astronomy Institute-INAF
via Gobetti 101
40129 Bologna, Italy
email: ggiovann@ira.inaf.it

Abstract. I report observations of 3C84 (NGC 1275), the central galaxy of the Perseus cluster, made with an interferometric array including the orbiting radio telescope of the RadioAstron mission. The data transversely resolve the edge-brightened jet in 3C84 only 0.03 mas from the core, which is ten times closer to the central engine than was possible in previous ground-based observations. To better understand physical properties of this peculiar source, I will discuss its present properties in comparison to the past and more recent evolution in the radio band of 3C 84.

Keywords. 3C 84, VLBI

---

1. Introduction and historical activity

The radio source 3C 84 is identified with NGC 1275, the central galaxy of the cooling flow Perseus Cluster (Abell 426) at z = 0.0176. This FR-I type radio source has been studied extensively since the early days of radio astronomy. At low resolution the radio emission is extended and diffuse with a halo-core structure. The radio flux has been monitored since 1960, and episodes of violent flux increase have been reported (see Fig. 1).

In the mid-1980s, the radio flux became exceptionally bright, more than 60 Jy at centimetre wavelengths, and then it subsequently decreased to about 10 Jy by the early 2000s. Around 2005, it was reported that the radio flux had started to increase again (see Fig. 2).

Thanks to VLBI images it is possible to see that inside the central 10 parsecs, there is a bright core region and two symmetric lobes, with evidence for free-free absorption in the northern one (Walker *et al.* 2000). These mini-lobes were probably formed by the jet activity originating in the 1959-1960 radio outburst and are expanding at sub-luminal velocity. At higher angular resolution, the bright core region is resolved into a compact core and an one-sided complex jet emission. VLBI observations obtained at the epoch of the highest flux density (around 1985) show the presence of a compact bright component identified with the core (C or C1) and an one-sided short and bright emission named C2 (see e.g. Venturi *et al.* 1993). In the following years the structure expanded and decreased in flux density, as shown by the 1995.1 image (Romney *et al.* 1995). In early 2000s the total flux density decreased to $\sim$ 10 Jy (see e.g. Nagai *et al.* 2012). Around 2005 it was reported that the radio flux had started to increase again (Abdo *et al.* 2009). VLBI observations revealed that this flux density increase originated in the central parsec region and it was accompanied by the ejection of a new component named C3 (see Suzuki *et al.* 2012 and Nagai *et al.* 2014 and references therein).

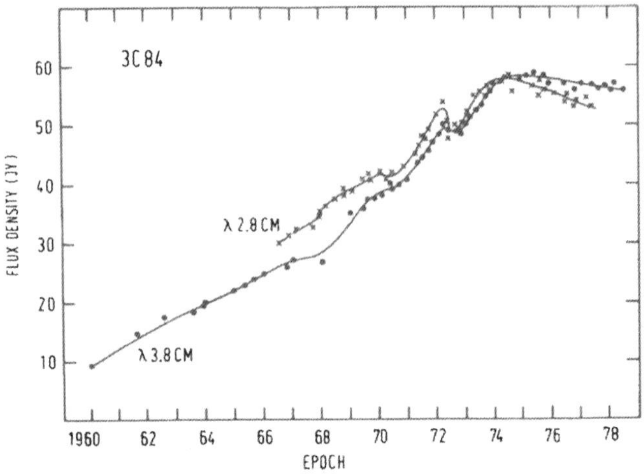

Figure 1. The total flux density of 3C 84 at 2.8 and 3.8 cm wavelength as a function of time. The Figure is taken from Preuss *et al.* (1979)

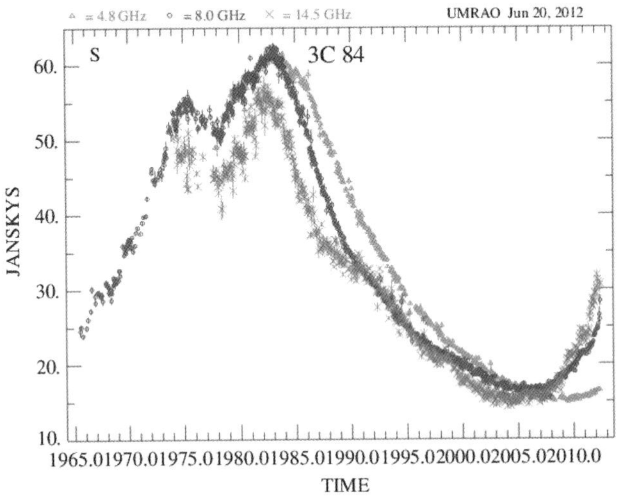

Figure 2. The total flux density of 3C 84 from the University of Michigan Radio Astronomy Observatory (UMRAO) data base at 4.8, 8.0 and 14.5 GHz

Component C3 (see Fig. 3) was first detected on 2003.11 and it is moving from the core with an apparent velocity of about 0.3 - 0.5 c and increasing in flux density. As shown in Fig. 3 it is unrelated to the old component C2 and its motion is at a different Position Angle (PA).

In high resolution VLBI images at 43 and 86 GHz obtained before 2003, the inner structure is dominated by the core C1 and the component C2 at about 1.2 mas. In these images the jet morphology was centrally peaked. In the images obtained a few years later (2005-2007) the new component C3 is well visible and became the dominant component of the inner 3C 84 nuclear structure. Component C3 is advancing with an increasing apparent velocity (0.47c in November 2008). The morphology of the jet connecting C1 to C3 is limb-brightened and the jet PA is $\sim 170\,^{\circ}$ quite different from the C2 structure.

3C 84 was detected in γ-rays by the Fermi Large Area Telescope (LAT) in Aug. 2008 (Abdo *et al.* 2009). This detection is particularly relevant because 3C 84 was not detected by EGRET (Energetic Gamma Ray Experiment Telescope) of the Compton Gamma Ray

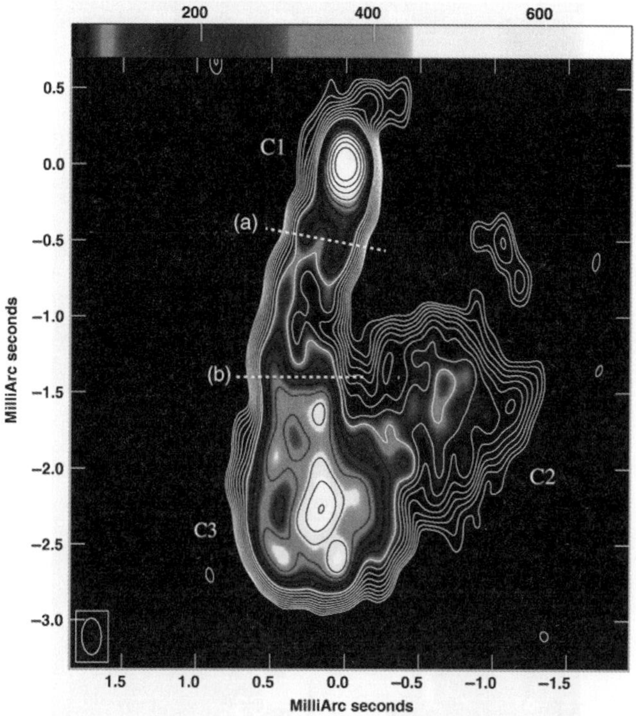

Figure 3. 43 GHz total intensity map of 3C 84. The contours are plotted at the level of 5.43×(1.41, 1, 1.41, 2.83, 4, 5.66, 8, 11.3, 16, 22.6, 32, 45.3, 64, 90.5, 128, 181, 256) mJy/beam. The ellipse shown at the bottom left corner of the image indicates the FWHM of the convolved beam. The FWHM of the convolved beam is 0.24 × 0.13 mas at the position angle 0.69°. The white dashed lines (a) and (b) indicated in the figure are the slice locations, (from Nagai *et al.* 2014).

Observatory in the 1990s (Reymer *et al.* 2003). The γ-ray flux density measured by Fermi is about seven times higher than the upper limit from EGRET observations.

This suggest the presence of two different activities in the nuclear region of 3C84: a first phase with a flux density decreasing, evolution of C2 component, centrally peaked jet structure, and no γ-ray emission and a second phase with an increasing of the flux density, the evolution of new component C3, a limb-brightened jet structure and γ-ray emission.

2. Space VLBI

3C84 was observed by the RadioAstron space radio telescope and an array of 29 ground radio telescopes from September 21, 2013 at 15:00 (UT) to September 22, 13:00 (UT). Observations were obtained in the C- and K-band. Observations were obtained during a perigee passage with projected space baselines from 0.2 to 10.4 Earth diameters. Here I am presenting only K band data. A positive signal was detected up to 8 Earth Diameters corresponding to 0.027 mas fringe spacing (for more details see Giovannini *et al.* 2018.

Figure 4 presents our 22 GHz space-VLBI image of the innermost parsec of 3C 84 at the angular resolution of 0.10 × 0.05 mas in PA 0°.

The bright and compact emission at about 2.2 mas north of the image reference center is identified with the radio core, from where a faint and short counter-jet and a 3-mas-long jet depart, in N-S direction. The main jet ends in a bright spot with a surrounding

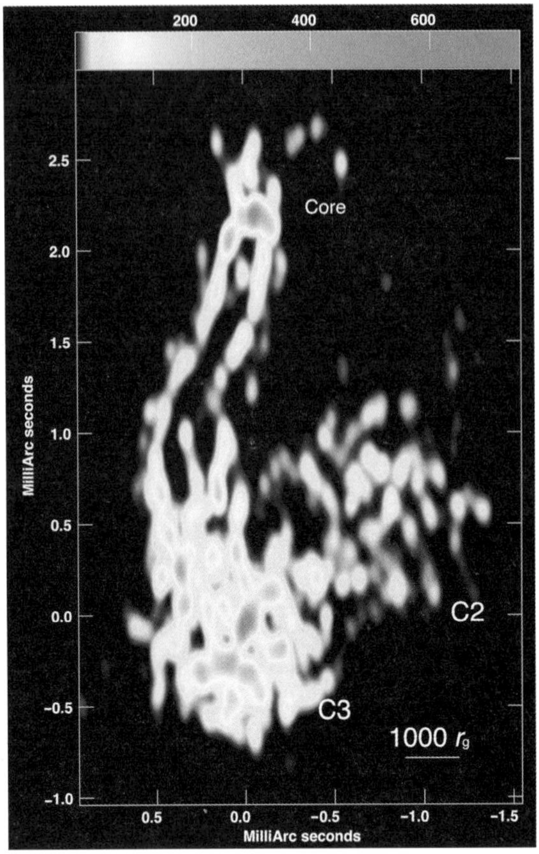

Figure 4. Radio image of the central parsec in 3C84 obtained with the space VLBI array. The x and y axes show the distance (in mas) from the image reference centre. The HPBW is 0.10×0.05 mas at PA $= 0°$. The noise level is 1.4 mJy/beam and the peak intensity is 0.75 Jy/beam.

diffuse emission identified with the emission feature called "C3" (Nagai *et al.* 2014). The diffuse emission feature, called "C2", is visible on the western side of the jet.

Main results from space VLBI images, are discussed below.

2.1. *The jet structure*

The jet is edge-brightened already at 0.03 mas from the core corresponding to 350 gravitational radii (r_g) de-projected. The large intensity ratio between the bright sheath and the central jet spine can be due to a specific velocity structure combination (Komissarov 1990), or by an intrinsic brightness difference, or both (see also Giovannini *et al.* 2018).

2.2. *The jet opening angle*

The limb-brightened jet connecting the core and C3 shows a large initial opening angle (the largest opening angle observed in any astrophysical jet). This large initial opening angle is followed by a rapid collimation to a quasi-cylindrical shape. This quasi-cylindrical structure is already present at a few hundred r_g from the core. This implies a strong jet collimation inside a few hundred r_g from the central engine. If the bright outer jet layer is launched by the BH ergosphere, it has to rapidly expand laterally followed by an almost cylindrical collimation. If this is not the case the jet sheath is likely launched from the accretion disk.

2.3. *Jet evolution and the C3 structure*

The component C3 is advancing with a position angle of $\sim 170°$. This direction is clearly unrelated to that of comonent C2. The limb-brightened jet appears to terminate in the brightest region of C3 component, slightly resolved in our images. The surrounding region rembers the shape of radio lobes in radio galaxies suggesting that is is due to the interaction of the jet with the surrounding medium. This is like a young source in the expanding phase.

Recently, VLBA images at 43 GHz from the Boston University Blazar Program have pointed out that the C3 hotspot is not stable but its position is not only moving out from the core with an apparent sub-luminal velocity but is also 'oscillating' from East to West. Polarization, flux density and morphology are also not constant and possibly these changes are correlated each other (see Nagai *et al.* 2017 and Kino *et al.* 2018). These behaviors suggest a strong interaction between the jet and ambient medium. More detailed studies will allow to derive the physical properties in this region.

3. Conclusions

High resolution space VLBI data have allowed us to study the inner jet region of 3C 84. The data transversely resolve the edge-brightened jet at only 0.03 mas from the core. The jet collimation profile is known from $\sim 10^2$ to 10^4 gravitational radii. A comparison between M87 and 3C84 jets suggests a strong relation with the restarted nature and young age of the 3C84 jet. A comparison with previous activity of 3C 84 shows a change in the jet properties and activity with time. The possibility to follow a young jet activity from its beginning can give a unique record of the early jet evolution in a galaxy.

Acknowledgements

I would like to thank all authors of the letter published on Nature Astronomy (Giovannini *et al.* 2018) for their contribution.

The RadioAstron project is led by the Astro Space Center of the Lebedev Physical Institute of the Russian Academy of Sciences and the Lavochkin Scientific and Production Association under a contract with the State Space Corporation ROSCOSMOS, in collaboration with partner organizations in Russia and other countries. The National Radio Astronomy Observatory is a facility of the National Science Foundation operated under cooperative agreement by Associated Universities. The European VLBI Network is a joint facility of independent European, African, Asian and North American radio astronomy institutes. The KVN is a facility operated by the Korea Astronomy and Space Science Institute. The KVN operations are supported by the Korea Research Environment Open NETwork, which is managed and operated by the Korea Institute of Science and Technology Information. This research is based on observations correlated at the Bonn Correlator, jointly operated by the Max-Planck-Institut fr Radioastronomie and the Federal Agency for Cartography and Geodesy.

This study makes use of 43 GHz VLBA data from the VLBA-BU Blazar Monitoring Program (VLBA-BU-BLAZAR; http://www.bu.edu/blazars/VLBAproject.html), funded by NASA through the Fermi Guest Investigator Program.

References

Abdo, A. A., *et al.* 2009, *ApJ* 699, 31
Giovannini, G.,, *et al.* 2018, *Nature Ast.*, 2, 472
Kino, M., *et al.* (2018) 2018, arXiv:180808855
Ko missarov (1990) 1990, *Sov. Ast. Lett.* 16, 284
Nagai, H., *et al.* 2012, *MNRAS*, 423, L122

Nagai, H., *et al.* 2014, *ApJ*, 785, 53
Nagai, H., *et al.* 2017, *ApJ*, 849, 52
Preuss, E., Kellerman, K. I., Pauliny-Toth, I. I. K., Witzel, A., & Shaffer, D. B. 1979, *A&A*, 79, 268
Reimer, O., Pohl, M., Sreekumar, P., & Mattox, J. R. 2003, *ApJ*, 588, 155
Romney, J. D., Benson, J. M., Dhawan, V., Kellerman, K. I., Vermeulen, R. C., & Walker, R. C. 1995, *Proc. Natl. Acad. Sci. USA*, 92, 11360
Suzuki, K., *et al.* 2012, *ApJ*, 746, 140
Venturi, T., Readhead, A. C. S., Marr, J. M., & Backer, D. C. 1993, *ApJ*, 411, 552
Walker, R. C., Dhawan, V., Romney, J. D., Kellerman, K. I., & Vermeulen, R. C. 2000, *ApJ*, 530, 233

Perseus in Sicily: From Black Hole to Cluster Outskirts
Proceedings IAU Symposium No. 342, 2018
K. Asada, E. de Gouveia Dal Pino, M. Giroletti,
H. Nagai & R. Nemmen, eds.

doi:10.1017/S1743921318008153

Extragalactic relativistic jets

Monica Orienti

INAF-IRA, Via Gobetti 101, 40129 Bologna, Italy
email: orienti@ira.inaf.it

Abstract. Relativistic jets are one of the most powerful manifestations of the release of energy produced around supermassive black holes at the centre of active galactic nuclei (AGN). Their emission is observed across the entire electromagnetic spectrum, from the radio band to gamma rays. Despite decades of efforts, many aspects of the physics of relativistic jets remain elusive. In particular, the location and the mechanisms responsible for the high-energy emission and the connection of the variability at different wavelengths are among the greatest challenges in the study of AGN. Recent high resolution radio observations of flaring objects locate the high energy emitting region downstream the jet at parsec scale distance from the central engine. Furthermore, monitoring campaigns of the most active blazars indicate that not all the high energy flares have the same characteristics in the various energy bands, even from the same source, making the interpretation of the mechanism responsible for the high-energy emission not trivial. Here I will discuss gamma-ray properties of blazars obtained by *Fermi* Large Area Telescope observations and the connection between radio and high-energy emission in relativistic jets, and I will focus on the importance of high angular resolution observations.

Keywords. radiation mechanisms: nonthermal, galaxies: active, galaxies: jets, gamma rays: observations

1. Introduction

Relativistic jets are one of the most powerful manifestation of the energy released by a super massive black hole at the centre of an active galactic nucleus (AGN). Only 10 per cent of AGN have powerful emission in the radio band connected with the presence of relativistic jets. These objects, termed jetted-AGN by Padovani (2016), are characterized by a bipolar outflow of relativistic particles that are produced in the central region of the AGN and may extend up to a few Mpc, well beyond the host galaxy.

The emission of relativistic jets is produced by non-thermal radiation processes and is observable throughout the electromagnetic spectrum, from radio wavelengths to gamma-ray energies. Their broad-band spectral energy distribution (SED) is characterized by two peaks: one peaking between infrared (IR) and soft X-rays, and the other at high energies (Fossati *et al.* 1998; Donato *et al.* 2001). The first hump is produced by synchrotron radiation from relativistic electrons in the jet magnetic field, while the second hump is likely due to inverse Compton (IC) scattering off photons of the radiation field by the same relativistic electron population responsible for synchrotron emission, indicating a possible correlation between the low-energy and high-energy emission. The origin of the seed photons for the IC scattering is still an open issue. A possibility is that seed photons are those produced by synchrotron by the same population of relativistic electrons. In this case the radiation mechanism is termed Synchrotron-self Compton (SSC). On the other hand, the process is termed external Compton if the seed photons are from an external radiation field, like the ultraviolet radiation from the accretion disk and/or broad line region, the IR emission from the dusty torus, cosmic microwave background (CMB) radiation, or synchrotron photons produced by a different population of relativistic electrons (see e.g. Joshi *et al.* 2014).

When the jet is observed at small angles to the line of sight its luminosity is boosted by projection and beaming effects, giving rise to the 'blazar phenomenon'. As a consequence of their small viewing angle, blazars look brighter than what they are, facilitating their detection. Depending on the optical spectrum, blazars are divided into two classes: flat spectrum radio quasars (FSRQ) and BL Lac objects. FSRQ have broad emission lines in their optical spectra, while the continuum spectra in BL Lacs is almost featureless. This is supposed to be directly connected to the efficiency of the accretion onto the super massive black hole, with the accretion being highly efficient in FSRQ which are characterized by high luminosity, while the accretion should be inefficient in BL Lacs, which also have low luminosity (Ghisellini *et al.* 2011). In the unified model, radio galaxies are the parent population of blazars at large viewing angles, being Fanaroff-Riley Type-I and Type-II radio sources the debeamed counterparts of BL Lacs and FSRQ, respectively (Urry & Padovani (1995); see also Torresi 2018, these Proceedings).

Since the EGRET era, jetted-AGN have dominated the extragalactic population of objects detected in gamma rays (Hartman *et al.* 1999). The advent of the Large Area Telescope (LAT) on board the *Fermi* satellite has dramatically increased the number of gamma-ray sources. *Fermi*-LAT is a pair-conversion telescope with a large field of view of $\sim$2.4 sr sensitive to energies from 20 MeV to energies above 300 GeV (Atwood *et al.* 2009). It operates in survey mode and completes an all-sky map every three hours. Giving its huge improvement in sensitivity and angular resolution, *Fermi*-LAT detected about 200 gamma-ray sources in only three months (Abdo *et al.* 2009a) and more than half of them are unambiguously associated with jetted-AGN (Abdo *et al.* 2009b). These numbers increased in the next release of *Fermi*-LAT catalogues which considered 11-month (First LAT AGN catalogue (1LAC), Abdo *et al.* 2010a) and 2-yr of *Fermi* observations (2LAC, Ackermann *et al.* 2011a), up to reaching 1591 extragalactic sources reported in the 3LAC (Ackermann *et al.* 2015) out of 3033 gamma ray sources detected in 4 years of sky surveying (Acero *et al.* 2015). It is worth noticing that during the EGRET era and the first period of *Fermi*-LAT observations, FSRQ was the dominant population, while BL Lacs started to dominate in the 2LAC, and this is likely a consequence of the improving sensitivity with observing time. Remarkably, the number of misaligned jetted-AGN (i.e. those radio sources whose jet forms a large angle to our line of sight, and therefore not subject to significant boosting effects) has increased, going from the detection of only Centaurus A by EGRET, up to about 30 objects listed in the 3LAC.

2. Radio and gamma-ray connection

As mentioned earlier, the same population of relativistic particles is responsible for both synchrotron emission at low energies and inverse Compton scattering at high energy. Therefore, radio and gamma-ray emission is expected to be connected. Several works have been undertaken with the aim of establishing a radio/gamma-rays correlation. Making use of the sources detected by *Fermi*-LAT during the first three months of operation, Abdo *et al.* (2009b) found a tentative correlation. Significant correlation was found between gamma-ray flux and radio flux density for the blazars in the 1LAC with a radio counterpart (e.g. Ghirlanda *et al.* 2010, Ackermann *et al.* 2011b). A similar correlation is also found for blazars in the 3LAC (Ackermann *et al.* 2015). Interestingly, the correlation holds also if FSRQ and BL Lacs are considered separately at least when gamma-ray flux between 0.3 and 300 GeV are considered. In fact, a dependence of the correlation strength on the considered gamma-ray energy band is also present and only BL Lacs with a synchrotron peak occurring in the soft X-rays seem to preserve a radio and gamma-ray correlation when energies above 10 GeV is considered (see e.g. Lico *et al.* 2017).

Mahony *et al.* (2010) pointed out that radio/gamma-ray correlation gets slightly stronger when non-variable sources are considered.

2.1. *Variability*

Variability is a key feature of blazars. Among the flaring gamma-ray sources detected by *Fermi*-LAT in 7.4 years of observations with a clear identification, the majority are blazars (Abdollahi *et al.* 2017). In particular, 69% of FSRQ is variable, while the percentage for BL Lacs is much lower on average (23%) and decreases with the increase of synchrotron peak frequency (Ackermann *et al.* 2015).

Changes in the luminosity are observed at all wavelengths and related to periods of different activity: quiescent states are interleaved with periods of enhanced activity. Sometimes abrupt outbursts occur and may be observed across the electromagnetic spectrum, or in a single band ('orphan' flares).

Variability timescale and the time-delay observed at different wavelength provide clues on the emitting region (size, location, seed photons) and the physics involved. A cross-correlation study between gamma-ray and radio light curves obtained at different frequencies of a sample of *Fermi*-LAT 54 blazars points out a time lag between the variability episodes in there two bands, with the gamma rays leading the radio variability. Furthermore, the time lag decreases with wavelengths, from 76±23 days at 11 cm to a few days or less at 3 mm (Fuhrmann *et al.* 2014). Similar results are found by dedicated multi-wavelength studies of individual blazars (e.g. Hughes *et al.* 2011, Orienti *et al.* 2013). These time lags are reflected into a spatial separation between the emitting regions, being the gamma-ray and the 3-mm emission almost co-spatial, whereas the emission at 11 cm arises a few parsecs downstream along the jet. The increase of the time lag with wavelength is consistent with the shock-in-jet model, in which compact regions at the base of the jet are self-absorbed in radio band, and their emission becomes optically thin at longer and longer wavelengths as the regions expand and move downstream the jet (e.g. Marscher & Gear 1985).

A different result was reported by León-Tavares *et al.* (2011) who found that for FSRQ the onset of the flare at 1 cm leads the peak of the gamma-ray flare, suggesting that the gamma-ray emitting region is located at several pc downstream the jet. On the other hand, no obvious radio/gamma-ray correlation is found for BL Lacs.

2.2. *VLBI observations of relativistic jets*

The studies mentioned earlier are based on single-dish observations where it is not possible to disentangle the contribution from different regions of the radio source, and changes in the pc-scale structure may be washed out by the contribution of stationary components. To circumvent this issue, high angular resolution observations are crucial. With (sub-)milliarcsecond resolution Very Long Baseline Interferometry (VLBI) observations provide insights into the innermost regions of relativistic jets and are fundamental for locating the site in which flares are produced.

VLBI monitoring campaigns of bright blazars during the EGRET era pointed out a connection between gamma-ray flares and the ejection of superluminal jet components (Jorstad *et al.* 2001). In the context of the shock-in-jet model, the superluminal components are the observable manifestation of shocks that are moving downstream along the jet. Given its all-sky scanning mode, *Fermi*-LAT has proved to be optimal in detecting flares from blazars. As a consequence, VLBI monitoring campaigns of selected objects, like the MOJAVE programme (Lister *et al.* 2018) and the VLBA-BU-BLAZAR Program (Jorstad *et al.* 2017), as well as triggered follow-up VLBI campaigns of individual objects caught by *Fermi*-LAT during a flare (e.g. 3C 454.3, Jorstad *et al.* 2013; PKS 1510-089,

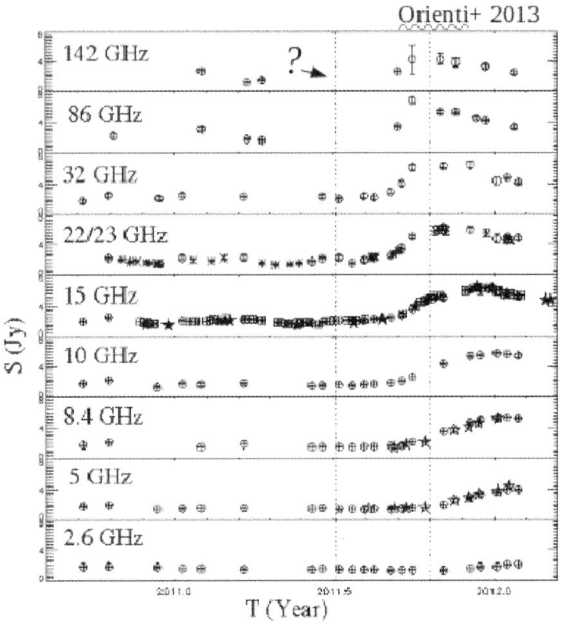

Figure 1. Multi-frequency light curves of PKS 1510−089. Dashed lines represent the epoch of gamma-ray flares. Adapted from Orienti *et al.* (2013).

Orienti *et al.* 2013, TXS 0536+145, Orienti *et al.* 2014; JVAS B0218+357, Spingola *et al.* 2016; BL Lacertae, Wehrle *et al.* 2016; 3C 273, Lisakov *et al.* 2017) have been undertaken. Remarkably, gamma-ray flares do show different multi-band properties even if they are from the same source (Fig. 1). For example, not all the gamma-ray flares are associated with a radio outburst and the ejection of a superluminal component, like in the case of BL Lacertae (Marscher *et al.* 2008), 3C 454.3 (Jorstad *et al.* 2013) and PKS 1510+089 (Orienti *et al.* 2013). In the case of the Narrow Line Seyfert 1 SBS 0846+513 there seems to be an anti-correlation between the radio and gamma-ray outbursts (D'Ammando *et al.* 2012). All these pieces of evidence suggests that the location of the gamma-ray emitting region may vary even within the same source. An outstanding example is the jet of the radio galaxy M87 where gamma-ray photons up to TeV energy may originate in the core (Hada *et al.* 2014), as well as in a jet region, HST-1, about 120 pc from the central engine (Cheung *et al.* 2007).

3. The gamma-ray emitting region

Locating the region where gamma-ray flares are produced is not trivial. As mentioned earlier, the time lag observed between radio and gamma rays can help us locate the site of the gamma-ray emitting region and determine the seed photons involved in the scatter process.

In the shock-in-jet model a perturbation is likely produced in the jet triggering an outburst. If the shock happens close to the central engine, well within the broad line region, the emission is still opaque at radio wavelengths. Therefore we expect to observe either a significant time lag between gamma rays and radio even at mm wavelengths, or no radio outburst at all (e.g. BL Lacertae Marscher *et al.* 2008; 3C 279 Abdo *et al.* 2010). On the contrary a close connection between gamma-rays and optical emission is observed, with the optical polarization angle experiencing significant rotation close in time with the gamma-ray outburst, that may be explained in terms of a shock that is moving in a helical jet (e.g. Marscher *et al.* 2010). In this case the seed photons should be the UV from BLR. However, it is worth mentioning that in a recent work Costamante *et al.* (2018) questioned the BLR as the site of gamma-ray emission because no sign of γ-γ absorption is found in the gamma-ray spectra of a sample of 100 objects from the 3LAC. However, in their work Costamante *et al.* (2018) consider an average activity state of the sources, and a different location of gamma-ray emission during outbursts cannot be excluded.

If no significant delay is observed between the outburst in gamma rays and in radio, at least in the mm regime, the emitting site should be at a distance in which the emission is already optically thin to short radio wavelengths. In this case the gamma-ray emitting region is at pc distance from the central engine and seed photons may be IR from the dusty torus (e.g. Sikora *et al.* 2008), or synchrotron photons produced by either the same population of relativistic electrons (SSC) or a different population of relativistic electrons (e.g. Marscher *et al.* 2010). A location about a few pc downstream the jet may be difficult to reconcile with the short (sub-daily) variability timescales observed in flaring blazars. However, the size of emitting region may be smaller than the jet cross section (e.g. Lister *et al.* 2013), and the outburst may be caused by turbulent relativistic flow crossing a standing conical shock (Marscher 2014), or alternatively by turbulent fast magnetic reconnection process (de Gouveia Dal Pino *et al.* 2018 these Procs. and references therein).

The shock-in-jet model is one of the leptonic models that have been proposed to explain high energy emission from relativistic jets. Other models assume that the jet has a velocity structure either transverse (e.g. Ghisellini *et al.* 2005) or radial (e.g. Georganopoulos & Kazanas 2003), and the synchrotron photons of the fast moving component are up-scattered by the electrons in the slower one.

In the spine-sheath scenario a fast spine is surrounded by a slower layer. The gamma-ray emission is produced by SSC and EC and no trivial correlation between gamma rays and radio variability is expected (Tavecchio & Ghisellini 2008). Jets with a limb-brightened structure, like that expected in spine-sheath model, have been observed in several BL Lacs and radio galaxies, supporting the structured jet scenario (e.g. Mrk 501, Giroletti *et al.* 2008; Mrk 421, Lico *et al.* 2014; M87, Hada *et al.* 2013). An outstanding example of limb-brightened jet structure is 3C 84.

4. The case of 3C 84

The radio source 3C 84 is an FR-I radio galaxy and is hosted by the early-type cD galaxy NGC 1275 in the Perseus cluster. Since 1965 the source had been monitored by the University of Michigan Blazar monitoring group and the long-term light curves point out different periods of activity with a major outburst at the beginning of the '80s, followed by a decrease of activity until about 2006 when the radio flux started to increase again (Aller *et al.* 2017). The recurrent radio activity pointed out by the long-term light curves seem connected with changes in the radio structure with the emergence of new components close in time with the beginning of high activity periods (e.g. Asada *et al.* 2006, and Nagai *et al.* 2010).

There seems to be a connection between high activity states observed in radio with those in X-rays and gamma rays (Dutson *et al.* 2014, Fabian *et al.* 2015). Remarkably, in the '90s the source was not detected by EGRET. However, *Fermi*-LAT could detected the source in 4 months of observations, with a flux that was significantly higher than the upper limit set by EGRET (Abdo *et al.* 2009c). VLBI observations of the central pc-scale jet indicate that during the EGRET era, before the emission of the new jet component, the jet had a ridge-brightened structure (Fig. 2), whereas in the *Fermi*-LAT era it shows a clear limb-brightened structure with a wide opening angle (Nagai *et al.* 2014, Giovannini *et al.* 2018).

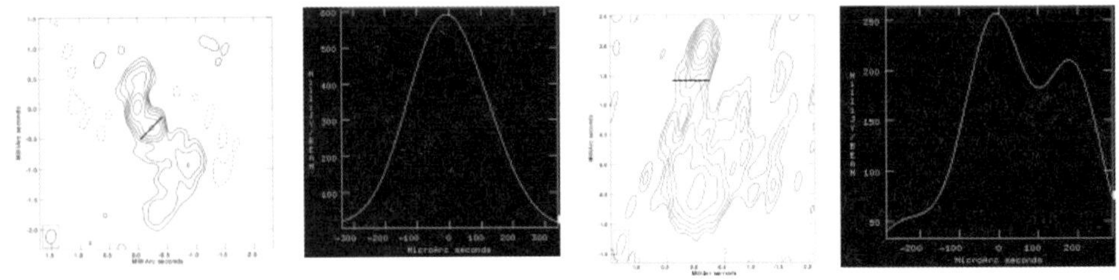

Figure 2. VLBI images at 43 GHz and transverse jet profile of 3C 84 obtained in 1995 (*left*) and in 2013 *right*. The solid line indicates the slice used to derive the jet profile.

Several works on radio, X-rays and gamma-ray variability suggests that short-term and long-term variability may be produced in different regions of the source. In particular, short-term variability seems related to injection of fresh particles accelerated in a shock in the core region, whereas long-term variability is more likely connected with the jet structure (e.g. Fukazawa *et al.* 2018, Hodgson *et al.* 2018). This scenario is supported by the change of gamma-ray hardness ratio around the beginning of 2011. Before 2011, the variability of hardness ratio is consistent with injection of fresh particles, while after 2011 the variability is better explained by a change in the Doppler factor and/or in the angle to our the line of sight (Tanada *et al.* 2018). Results of spectral energy distribution (SED) modelling using a SSC and considering data before and after 2011 support this scenario. It is also worth mentioning that the SED modelling is not unique, and the SED can be reproduced with a spine-sheath model, in agreement with the limb-brightened jet structure observed in VLBI images, in which the angle to our line of sight is about 18 degrees (Tavecchio & Ghisellini 2014).

5. Summary

Since the EGRET era, the extragalactic gamma-ray sky has been dominated by the population of jetted-AGN. The sub-population of blazars represents the largest fraction. Giving their characteristic multi-band variability, blazars have been studied to search for connection between the emission at low and high energy. Statistical studies of gamma-ray emitting blazars indicate significant correlation between radio and gamma-ray flux, indicating a common origin. When multi-wavelength light curves are compared, a time lag is usually observed with the gamma-ray leading the radio variability. Furthermore, the time lag increases with increasing the wavelength, as expected in case of opacity effects. The determination of the time lag between radio and gamma-ray is a useful tool for setting constraints on the location of the region responsible for the gamma ray emission: significant time lags or absence of a radio counterpart would locate the gamma ray emitting region within the broad line region, where the emission is opaque even at the short radio wavelengths. On the other hand, if no time lag is observed, the gamma-ray emission would take place at parsec distance from the central engine, in a region where the jet is already optically thin to radio emission.

Gamma-ray emission may be produced by perturbations that propagates downstream along the jet. In this scenario superluminal components, which are the observable manifestation of a propagating shock, should be ejected close in time with a gamma-ray flare. On the other hand, gamma-ray emission may be also produced in a structured jet. Observations of jets with a limb-brightened structured strongly support the existence of structured jet.

A peculiar case is represented by 3C 84. Since the '60s, this radio sources has proved to be highly variable. Gamma-ray emission shows both short-term and long-term variability which are likely to be produced in different regions: the former in the core region, whereas the latter in the jet.

All these pieces of evidence clearly point out that a unique scenario capable to reproduce the gamma-ray emission and its connection with the other energy band in blazars in general, and in 3C 84 in particular, is still far to be achieved, and multi-band monitoring campaigns have proved to be the first and necessary step for shedding a light on the 'blazar' phenomenon.

Acknowledgments

The VLBA is operated by the US National Radio Astronomy Observatory which is a facility of the National Science Foundation operated under a cooperative agreement by Associated Universities, Inc. The Fermi LAT Collaboration acknowledges support from a number of agencies and institutes for both the development and the operation of the LAT as well as scientific data analysis. These include NASA and DOE in the United States, CEA/Irfu and IN2P3/CNRS in France, ASI and INFN in Italy, MEXT, KEK, and JAXA in Japan, and the K. A. Wallenberg Foundation, the Swedish Research Council and the National Space Board in Sweden. Additional support from INAF in Italy and CNES in France for science analysis during the operations phase is also gratefully acknowledged. This work was supported by the Korea's National Research Council of Science & Technology (NST) granted by the International joint research project (EU-16-001).

References

Abdo, A. A., *et al.* 2009a, *ApJS*, 183, 46
Abdo, A. A., *et al.* 2009b, *ApJ*, 700, 597
Abdo, A. A., *et al.* 2009c, *ApJ*, 699, 31
Abdo, A. A., *et al.* 2010a, *ApJ*, 715, 429
Abdo, A. A., *et al.* 2010b, *Nature*, 463, 919
Abdollahi, S., *et al.*, 2017, *ApJ*, 846, 34
Acero, F., *et al.* 2015, *ApJS*, 218 23
Ackermann, M., *et al.* 2011a, *ApJ*, 743, 171
Ackermann, M., *et al.* 2011b, *ApJ*, 741, 30
Ackermann, M., *et al.* 2015, *ApJ*, 810, 14
Aller, M., Aller, H., & Hughes, P. 2017, *Galaxies*, 5, 75
Asada, K., *et al.* 2006, *PASJ*, 58, 261
Atwood, W. B., *et al.* 2009, *ApJ*, 697, 1071
Cheung, C. C.; Harris, D. E.; & Stawarz, L. 2007, *ApJ* (Letter), 663, 65
Costamante, L., Cutini, S., Tosti, G., Antolini, E., & Tramacere, A. 2018, *MNRAS*, 477, 4749
D'Ammando, F., *et al.* 2012, *MNRAS*, 426, 317
Donato, D., Ghisellini, G., Tagliaferri, G., & Fossati, G. 2001, *A&A*, 375, 739
Dutson, K. L., *et al.* 2014, *MNRAS*, 442, 2048
Fabian, A. C., Walker, S. A., Pinto, C., Russell, H. R., & Edge, A. C. 2015, *MNRAS*, 451, 3061
Fossati, G., Maraschi, L., Celotti, A., Comastri, A., & Ghisellini, G. 1998, *MNRAS*, 299, 433
Fukazawa, Y., *et al.* 2018, *ApJ*, 855, 93
Fuhrmann, L., *et al.* 2014, *MNRAS*, 441, 1899
Georganopoulos, M., & Kazanas, D. 2003, *ApJ* (Letter), 594, 27
Ghirlanda, G., Ghisellini, G., Tavecchio, F., Foschini, L., & Bonnoli, G. 2010, *MNRAS*, 413, 852
Ghisellini, G., Tavecchio, F., & Chiaberge, M. 2005, *A&A*, 432, 401
Ghisellini, G., Tavecchio, F., Foschini, L., & Ghirlanda, G. 2011, *MNRAS*, 414, 2674
Giovannini, G., *et al.* 2018, *Nature Astronomy*, 2, 472
Giroletti, M., *et al.* 2008, *A&A*, 488, 905
Hada, K., *et al.* 2013, *ApJ*, 775, 70
Hada, K., *et al.* 2014, *ApJ*, 788, 165

Hartman, R. C.; Bertsch, D. L.; Bloom, S. D.; Chen, A. W.; Deines-Jones, P., *et al.* 1999, *ApJS*, 123, 79
Hodgson, J. A., *et al.* 2018, *MNRAS*, 475, 368
Hughes, P. A., Aller, M. F., & Aller, H. D. 2011, *ApJ*, 735, 81
Jorstad, S. G., *et al.* 2001, *ApJ*, 556, 738
Jorstad, S. G., *et al.* 2013, *ApJ*, 773, 147
Jorstad, S. G., *et al.* 2017, *ApJ*, 846, 98
Joshi, M., Marscher, A. P., & Böttcher, M. 2014, *ApJ*, 785, 132
León-Tavares, J., Valtaoja, E., Tornikoski, M., Läthteenmäki, A., & Nieppola, E. 2011, *A&A*, 532, 146
Lico, R., *et al.* 2014, *A&A*, 571, 74
Lico, R., *et al.* 2017, *A&A*, 606, 138
Lisakov, M. M., Kovalev, Y. Y., Savolainen, T., Hovatta, T., & Kutkin, A. M. 2017, *MNRAS*, 468, 4478
Lister, M. L., *et al.* 2013, *AJ*, 146, 120
Lister, M. L., *et al.*, 2018, *ApJS*, 234, 12
Mahony, E. K., *et al.* 2010, *ApJ*, 718, 587
Marscher, A. P. & Gear, W. K. 1985, *ApJ*, 298, 114
Marscher, A. P., *et al.* 2008, *Nature*, 452, 966
Marscher, A. P., *et al.* 2010, *ApJ* (Letter), 710, 126
Marscher, A. P 2014, *ApJ*, 780, 87
Nagai, H., *et al.* 2010, *PASJ* (Letter), 62, 11
Nagai, H., *et al.* 2014, *ApJ*, 785, 53
Orienti, M., *et al.* 2013, *MNRAS*, 428, 2418
Orienti, M., *et al.* 2014, *MNRAS*, 444, 3040
Padovani P. 2016, *A&ARv*, 24, 13
Sikora, M., Moderski, R., & Madejski, G. M. 2008, *ApJ*, 675, 71
Spingola, C., *et al.* 2016, *MNRAS*, 457, 2263
Tanada, K., *et al.* 2018, *ApJ*, 860, 74
Tavecchio, F. & Ghisellini, G. 2008, *MNRAS* (Letter), 385, 98
Tavecchio, F. & Ghisellini, G. 2014, *MNRAS*, 443, 1224
Urry, C. M., & Padovani, P. 1995, *PASP*, 107, 803
Wehrle, A. E., *et al.* 2016, *ApJ*, 816, 53

Perseus in Sicily: From Black Hole to Cluster Outskirts
Proceedings IAU Symposium No. 342, 2018
K. Asada, E. de Gouveia Dal Pino, M. Giroletti, H. Nagai & R. Nemmen, eds.

doi:10.1017/S1743921318005744

Inflow and Outflow (Jets) in NGC 1275

Hiroshi Nagai[1,2]

[1]National Astronomical Observatory of Japan, Osawa 2-21-1, Mitaka, Tokyo 1818-8588, Japan
email: hiroshi.nagai@nao.ac.jp

[2]The Graduate University for Advanced Studies (SOKENDA), Osawa 2-21-1, Mitaka, Tokyo 1818-8588, Japan

Abstract. NGC 1275 is one of the best targets to study the high energy emission mechanism in radio galaxies and the accretion flow properties using a young re-started jet 3C 84 as a prober of subpc-to-pc scale environment. In this proceeding, we review the observation results from a series of our VLBI observations and discuss on the origin of gamma-ray emission and accretion flow properties. We also briefly present the preliminary results from our recent ALMA observations.

Keywords. galaxies: active, galaxies: individual (NGC 1275), galaxies: jets

1. Introduction

The Perseus cluster is a nearby (z=0.017) cluster of galaxies and has been the subject of extensive research over years at all wavelengths. The cluster harbors the giant elliptical and radio galaxy NGC 1275 at its center, and this galaxy shows an intermittent jet activity in which the most recent activity started in ~2005 (Abdo *et al.* (2009); Nagai *et al.* (2010); Suzuki *et al.* (2012)). This system is the brightest X-ray cluster of galaxies and best example of a prototypical "cooling core" cluster with a signature of the interaction between the jet and the intra-cluster medium (ICM) (e.g., Fabian *et al.* (2002)).

Strong time variability of the nucleus of NGC 1275 is observed at multiwavelengths from radio to γ-ray as well as the jet intermittency. In particular, this source shows a remarkable flux variation in γ-ray band (Dutson *et al.* (2014)). Since *Fermi* started to monitor this source, a monotonic increase of the γ-ray flux has been observed in addition to rapid time variation with a timescale of days-weeks. This long-term flux increase is not commonly observed in blazars, and thus it might give a hint to understand the γ-ray emission mechanism in non-blazar sources, which is still disputed.

The observed time variability indicates that the black hole accretion is time variable, and perhaps the accretion flow can be strongly inhomogeneous. It is suggested by numerical simulations of giant elliptical galaxies that the growth of thermal instabilities in the hot halo/cluster gas leads to the condensation of cold clouds and filaments (Gaspari *et al.* (2013)). As a result, the accretion flow from cluster scale to the central black hole is dominated by cold and chaotic accretion. A large amount of cold molecular gas is in fact observed in NGC 1275 in the form of filaments of CO gas on kpc scales (Lim *et al.* (2008)). However, it is not clear how the large-scale molecular gas is settling into the circumnuclear region because of the lack of spatial resolution in previous observations.

In this paper, we review our previous VLBI observations and present new ALMA observations to discuss on three open questions: i) What is the γ-ray emission mechanism?, ii) What is the accretion flow properties of NGC 1275?, iii) Is the cold accretion dominant within 10-100 pc scale around the nucleus?.

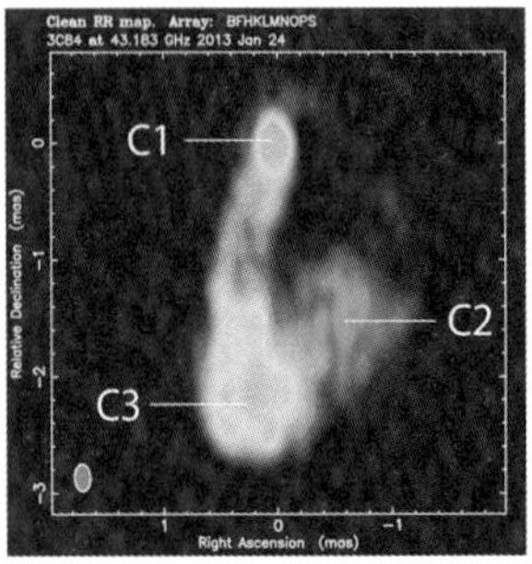

Figure 1. VLBI image of NGC 1275 (3C 84) on 2013 January 24.

2. Connection between Radio and γ-rays: A hint for the γ-ray emission mechanism

As we can see from the light curve available at the Submillimeter Array (SMA) and *Fermi*-LAT homepages (SMA: http://sma1.sma.hawaii.edu/callist/callist.html?plot=0319%2B415, *Fermi*: https://fermi.gsfc.nasa.gov/ssc/data/access/lat/msl_lc/source/NGC_1275), both radio and γ-ray light curves have shown flux increases since the beginning of *Fermi* operation on the timescale of years. It was shown that the radio brightening was caused by the re-started jet activity in ~2005 (Abdo *et al.* (2009); Nagai *et al.* (2010); Suzuki *et al.* (2012)). This likely indicates that the recent high flux level of γ-ray originates in the re-started jet activity. However, the correlation in radio and γ-ray light curves is not evident on shorter timescales. We can see several small flares in radio and γ-ray on the timescale of weeks-months, but no clear one-to-one correspondences has been observed for each flare (e.g., Nagai *et al.* (2012)).

The absence of radio counterpart of the γ-ray flare has been conventionally understood as the strong self-absorption of the flaring region in radio bands. Several authors attempted to do modeling the observed spectral energy distribution (SED) from radio to γ-ray using one-zone synchrotron self-Compton (SSC) model, and radio data were discarded because of the same reason. However, the large separation between two bumps of SED in ν-νF_ν plane cannot be well reproduced by one-zone SSC (see discussion by Tavecchio & Ghisellini (2008)) even if we discard the radio data. Following the discovery of limb-brightened structure in the subpc-scale jet (Nagai *et al.* (2014) and see Fig. 1), the SED modeling in the framework of the spine-layer scenario is adopted to NGC 1275 (Tavecchio & Ghisellini (2014)), which seems to rather reproduce the observed SED under the condition of the jet viewing angle of 18°. On the other hand, Fujita & Nagai (2017) discovered the counter jet component and derived the jet viewing angle of $65° \pm 16°$ from the apparent length ratio between the jet and counter jet. It seems that the observations do not favor the misaligned blazar model. One more complication is that the total flux density of the hotspot/lobe (C3) is comparable to that of the core (C1), and thus the SSC emission from C3 may not be negligible. This could be a reason that the SED cannot be explained by one-zone misaligned blazar model (see also discussion by Hodgson *et al.* (2018)). We should also keep in mind that the entire structure of C1+C2+C3 is very small (~ 1 pc) and thus resides in the dense external photon fields from accretion disk and circumnuclear disk/torus. Inverse Compton upscattering of various surrounding photons can also contribute to the high energy emission of NGC 1275 (e.g., Stawarz *et al.* (2008)).

3. Accretion Flow Properties

The entire radio structure associated with the most recent restarted jet can be used as a prober of the environment in subpc to pc scales. Fujita & Nagai (2017) recently

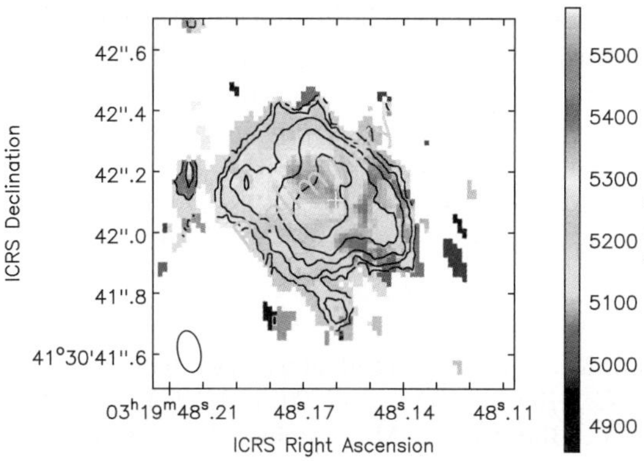

Figure 2. The contour of velocity integrated intensity overlaid on the velocity distribution of HCN(3-2) emission. The plus symbol in the image indicate the AGN position.

discovered the counter jet component at ~ 0.8 pc north of the core, which is presumably associated with the re-started jet activity in ~2005. The counter jet component shows a strong inverted spectrum between 15 GHz and 43 GHz, which is similar spectral behavior observed in larger scale counter jet (Walker *et al.* (2000)). The most likely explanation for the inverted spectrum is free-free absorption. Walker *et al.* (2000) discussed that there is an accretion disk extending to the pc scales and that the ionized part of the accretion disk is responsible for the absorption. The observed absorption by Fujita & Nagai (2017) is likely caused by the further inner part of the accretion disk.

Fujita & Nagai (2017) found that the frequency dependence of the free-free absorption is $\tau_{\rm ff} \propto \nu^{-0.6}$, which is different from the theoretical expectation $\tau_{\rm ff} \propto \nu^{-2}$. They argued that the absorbing medium is highly inhomogeneous and that it consists of regions of $\tau_{\rm ff} \gg 1$ and $\tau_{\rm ff} \ll 1$.

Nagai *et al.* (2017) reported the abrupt change in the hotspot position as well as the enhancement of the polarized emission by VLBI observations. The change in the hotspot position can be caused by the interaction between the jet and inhomogeneous/clumpy ambient medium. They also reported the detection of Faraday rotation. The derived rotation measure (RM) is at most $(6.3 \pm 1.9) \times 10^5$ rad m^{-2}, which shows an agreement with the RM reported by the CARMA and SMA observations within a factor of 2 (Plambeck *et al.* (2014)). One possible origin of the Faraday rotation is the accretion flow. However, Nagai *et al.* (2017) discussed that similar RM values between the VLBI observations and the CARMA/SMA observations cannot be consistently explained by the spherical accretion flow with a power-law profile, such as radiatively inefficient accretion flow (RIAF), which is the favored model for the low luminosity AGNs such as NGC 1275. They claimed that the dense clump could be responsible for the observed Faraday rotation by VLBI observations.

Overall results indicate that the subpc-pc scale environment can be quite inhomogeneous and is not consistent with the RIAF-like quasi-spherical accretion flow.

4. Cold Accretion

We recently observed NGC 1275 with ALMA to study the cold gas properties of circumnuclear region in 10-100 pc scale. We detected CO(2-1), HCN(3-2), and HCO$^+$(3-2) emissions in this region. Figure 2 shows the velocity integrated intensity overlaid on the velocity distribution of HCN(3-2) emission. The emission clearly shows a central concentration at the AGN position, which suggests that the cold gas is still abundant within the central ~ 10 pc. The velocity gradient is observed in the position angle of ~ 70 deg. Both the morphology and kinematic structure indicate the rotating disk around the AGN position, yet the disk structure is not smooth nor uniform. The velocity gradient is roughly perpendicular to the subpc-scale jet axis (Giovannini *et al.* (2018), see also Fig. 1). This may suggest that the molecular gas disk is physically connected to the accretion disk at further inner region, which is responsible for the jet launching, possibly changing its form from the cold molecular gas to ionized gas. The cold accretion might be a dominant mechanism to characterize the AGN activity of NGC 1275.

References

Abdo, A. A., Ackermann, M., Ajello, M., *et al.* 2009, *ApJ*, 699, 31
Dutson, K. L., Edge, A. C., Hinton, J. A., *et al.* 2014, *MNRAS*, 442, 2048
Fabian, A. C., Celotti, A., Blundell, K. M., Kassim, N. E., & Perley, R. A. 2002, *MNRAS*, 331, 369
Fujita, Y., & Nagai, H. 2017, *MNRAS*, 465, L94
Gaspari, M., Ruszkowski, M., & Oh, S. P. 2013, *MNRAS*, 432, 3401
Giovannini, G., Savolainen, T., Orienti, M., *et al.* 2018, *Nature Astronomy*, 2, 472
Hodgson, J. A., Rani, B., Lee, S.-S., *et al.* 2018, *MNRAS*, 475, 368
Kataoka, J., Stawarz, Ł., Cheung, C. C., *et al.* 2010, *ApJ*, 715, 554
Lim, J., Ao, Y., & Dinh-V-Trung 2008, *ApJ*, 672, 252
Nagai, H., Suzuki, K., Asada, K., *et al.* 2010, *PASJ*, 62, L11
Nagai, H., Orienti, M., Kino, M., *et al.* 2012, *MNRAS*, 423, L122
Nagai, H., Haga, T., Giovannini, G., *et al.* 2014, *ApJ*, 785, 53
Nagai, H., Fujita, Y., Nakamura, M., *et al.* 2017, *ApJ*, 849, 52
Plambeck, R. L., Bower, G. C., Rao, R., *et al.* 2014, *ApJ*, 797, 66
Stawarz, Ł., Ostorero, L., Begelman, M. C., *et al.* 2008, *ApJ*, 680, 911
Suzuki, K., Nagai, H., Kino, M., *et al.* 2012, *ApJ*, 746, 140
Tavecchio, F., & Ghisellini, G. 2008, *MNRAS*, 385, L98
Tavecchio, F., & Ghisellini, G. 2014, *MNRAS*, 443, 1224
Walker, R. C., Dhawan, V., Romney, J. D., Kellermann, K. I., & Vermeulen, R. C. 2000, *ApJ*, 530, 233

Perseus in Sicily: From Black Hole to Cluster Outskirts
Proceedings IAU Symposium No. 342, 2018
K. Asada, E. de Gouveia Dal Pino, M. Giroletti,
H. Nagai & R. Nemmen, eds.

doi:10.1017/S1743921318005173

Observations of nearby relativistic jets with EAVN and EATING VLBI

Kazuhiro Hada[1,2] **and EAVN/EATING VLBI Collaboration**

[1]Mizusawa VLBI Observatory, National Astronomical Observatory of Japan, 2-21-1 Osawa, Mitaka, Tokyo 181-8588, Japan
email: kazuhiro.hada@nao.ac.jp

[2]Department of Astronomical Science, The Graduate University for Advanced Studies (SOKENDAI), 2-21-1 Osawa, Mitaka, Tokyo 181-8588, Japan

Abstract. The international collaboration of VLBI in East Asia (EA) is rapidly growing. Besides the recent successful operation of the KVN and VERA Array (KaVA), the effort of joint VLBI operation in EA is being expanded into China. This forms the East Asian VLBI Network (EAVN), which offers detailed imaging and monitoring of the innermost regions of relativistic jets at cm wavelengths and thus complementary to the Event Horizon Telescope operated at mm wavelengths. In this contribution we present some of the ongoing KaVA/EAVN studies for the inner jets of the two important nearby radio galaxies 3C84 and M87. The observations have revealed detailed kinematics for the parsec-to-subparsec-scale structures of these jets. Future regular observations with the East-Asia-To-Italy-Nearly-Global (EATING) VLBI, a joint effort between EAVN and the Italian VLBI Network, will enable us to resolve and regularly monitor the innermost regions of nearby AGN jets at tens of micro-arcsecond scales.

Keywords. instrumentation: high angular resolution, galaxies: active, galaxies: jets, radio continuum: galaxies

1. Introduction

It is now widely believed that relativistic jets in active galactic nuclei (AGN) are powered by the accretion onto the central supermassive black holes (SMBH). The AGN jets carry a significant fraction of the accretion energy into the interstellar or/and intergalactic space, playing a key role in AGN feedback. However, detailed processes of jet propagation from the central engine to the large scales are still not well understood. The inner jet regions at parsec to subparsec scales are particularly important since a number of key physical processes are expected to take place, such as the flow acceleration, collimation, interaction with circumnuclear medium, and production of high-energy γ-ray emission. VLBI at cm-wavelengths is particularly suited to probe these scales.

The KVN and VERA Array (KaVA) has recently emerged as the first regularly operated international VLBI network in East Asia, consisting of 7 stations available at 22 and 43 GHz. Since KaVA operates for a quasi-full year and has angular resolution down to $\sim$0.5 mas, the array is suitable for monitoring the detailed structural evolution of nearby AGN jets at parsec to sub-parsec scales (Niinuma *et al.* (2014); Oh *et al.* (2015); An *et al.* (2016); Zhang *et al.* (2017)). Moreover, KVN has a unique capability of quasi-simultaneous multi-frequency full-polarimetric receiving system at 22/43/86/129 GHz, which allows us to study high-frequency properties of relativistic jets (e.g., Lee *et al.* (2014)).

Here we introduce some of recent and ongoing studies on the radio galaxies 3C84 and M87 with the VLBI facilities in East Asia. Since these two sources are nearby and the

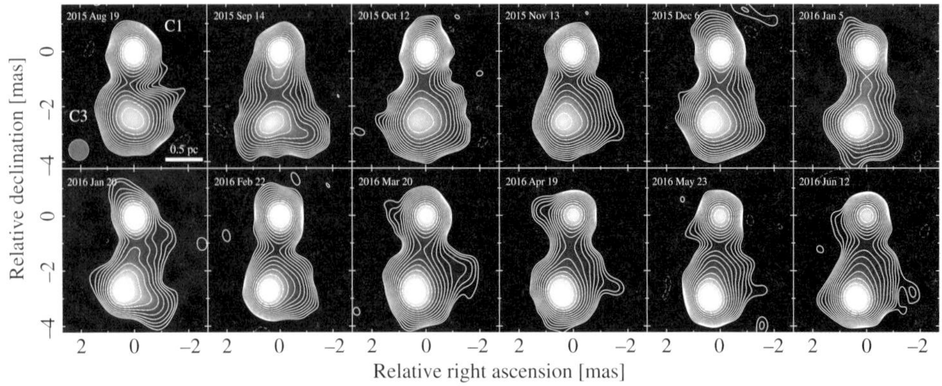

Figure 1. Multi-epoch KaVA 43 GHz images of 3C84 between August 2015 and January 2016 (Kino *et al.* submitted). The bright core is C1 while the moving feature in the southern part of the jet is C3.

jets are quite active over a wide range of energy spectra from radio to γ-ray, we can study in detail the innermost active jet scales near the black boles with high-resolution radio observations (e.g., Giovannini *et al.* (2018)).

2. 3C84

3C84 is a rare object that enables us to study the evolution of the active parsec-scale jet and its interaction with the external medium because of the dense gas-rich nuclear environment. Over the past years the source has intensively been monitored by VERA and KVN (Nagai *et al.* (2010); Suzuki *et al.* (2012); Kim *et al.* (2015); Hodgson *et al.* (2018)). From 2015, we started monthly 43 GHz observations of this source with KaVA (Fig.1). The aim of this program is to monitor the parsec-scale jet evolution including the C3 component, which was ejected from the core after the 2005 outburst (Nagai *et al.* (2010)). Until early 2015, C3 was straightly moving toward the south. However, during mid 2015, we found a sudden change of the C3 position toward the east by 0.4 mas (0.14 parsec in linear scale). After the flip, C3 wobbled there for a few months and then restarted to move to the south. C3 showed a monotonic increase in flux density during this period. This kind of behaviour is hard to be explained solely by the intrinsic motion of the jet, but is in good agreement with the interaction between jet and clumpy ambient that is often seen in some numerical simulations (e.g., Wagner & Bicknell (2011)).

As a complementary program, we are also monitoring the source with KVN at 43 and 86 GHz, where the synchrotron emission is less absorbed. Recently Fujita & Nagai (2017) has discovered a new feature N1 located at 0.8 pc north from the core. Since this feature shows an inverted spectrum between 15 and 43 GHz, it is proposed that N1 is associated with a counter-jet that is affected by the free-free absorption by the nuclear accretion disk. Our KaVA and KVN observations consistently detected this feature (Wajima *et al.* in prep). In particular, this is the first detection of N1 up to 86 GHz. Remarkably, N1 shows a highly inverted spectrum even at 86 GHz, suggesting a high electron density of the absorbing medium. While the presence of a 10-parsec-scale plasma torus was already suggested in the past VLBI studies, the KaVA/KVN results newly suggest a much smaller scale ($\sim$1 parsec) absorbing plasma. These programs are still ongoing, and further monitoring of C3 or N1 will allow us to better understand the geometry and spatial extent of the nuclear environment of 3C84.

3. M87

M87 is a unique jet source that one can directly resolve the jet formation scales within a few tens to hundreds Schwarzschild radii (R_s) by VLBI. Recent high-sensitivity VLBI imaging studies at 86 GHz revealed a limb-brightened, wide-opening-angle jet launching at a scale of $10\,R_s$ (Hada *et al.* (2016); Kim *et al.* (2018)). Moreover, extensive analyses of multi-frequency VLBI images revealed a parabolic collimation over a distance from $z{\sim}100\,R_s$ to $z{\sim}10^5\,R_s$ (Asada & Nakamura (2012); Hada *et al.* (2013)), which is consistent with magnetic collimation processes (e.g., Nakamura & Asada (2013)).

Nevertheless, the kinematics or velocity fields of the inner jet of M87 is still controversial. Previous multi-epoch VLBI studies of this jet report a variety of jet speeds such as quasi-stationary, subluminal and superluminal motions (Kovalev *et al.* (2007); Ly *et al.* (2007); Asada *et al.* (2014); Mertens *et al.* (2016); Walker *et al.* (2018)). Since M87 is one of the very few jets for which the acceleration scales of $10^{2-4}R_s$ are directly resolvable, having a regular monitoring program should be of great value for testing the magnetically driven jet paradigm.

In this context, we started a dense monitoring program of M87 with KaVA from 2014. The initial outcome from the pilot study was reported by Hada *et al.* (2017) The densely sampled multi-epoch KaVA images have traced the detailed structural evolution of this jet. We detected superluminal motions at these scales, together with a trend of gradual acceleration. These results are similar to those obtained by another recent M87 monitoring with VLBA at 43 GHz (Mertens *et al.* (2016); Walker *et al.* (2018)). The independent confirmation at different time, at different frequency and with different instruments suggests that the acceleration near the jet base is not episodic but a fundamental property of this jet. Also, the accelerating signature at these scales imply that the jet collimation and acceleration coexist, which is consistent with magnetic processes (e.g., Komissarov *et al.* (2009)).

From 2016 we have upgraded our KaVA monitoring program of M87 by adding 43 GHz. The dual-frequency strategy allows us to investigate the jet velocity fields over a wider range of distances from the black hole. Also, we can derive spectral index maps and their evolution with time, which will provide additional insights into the underlying particle energetics, density, and magnetic fields. The analysis is actively going on and these will update our knowledge of the M87 jet at these scales.

4. EAVN and EATING VLBI

Beyond KaVA, the international VLBI collaboration in East Asia is now rapidly expanding. In particular, efforts have been made to combine KaVA with some more telescopes in China, Korea and Japan. This ultimately forms the East Asian VLBI Network (EAVN; Wajima *et al.* (2016); An *et al.* (2018)). Thanks to the more number of stations and improved uv-coverage, EAVN will allow us to image and monitor the innermost regions of relativistic jets in much more detail.

From late 2018 EAVN is going to start regular operation at 22/43 GHz by adding the two large telescopes to KaVA – Tianma 65m in Shanghai and NRO 45m in Nobeyama. Thanks the large collecting areas of Tianma and NRO, we can greatly improve the overall array sensitivity (compared to KaVA, a factor of 10 improvement in baseline sensitivity, and a factor of 4 in image sensitivity). Concurrently, we are also commissioning EAVN with Urumqi and JVN at 22 GHz. EAVN test observations at 6.7 GHz are also actively ongoing. We plan to start regular operation of these capabilities in 2019.

Beyond EAVN, we are also expanding the array by collaborating with the Italian VLBI Network (IVN). By combing EAVN and IVN, we can obtain 10000 km baselines, which realizes a nearly global VLBI (East-Asia-To-Italy-Nearly-Global VLBI, or

EATING VLBI). 3C84 and M87 are among the key targets of EATING VLBI. For M87, along with the Event Horizon Telescope (EHT) observations in spring 2017 and 2018, we succeeded in coordinating biweekly EATING observations at 22 GHz. Such a frequent monitoring is generally very difficult to realize with global VLBI. The analysis of these data is still in progress, but our preliminary image resolves the jet base of M87 at scales of $\sim 18\,R_s$, which is quite comparable to the scales of the EHT. This demonstrates that EATING VLBI would be useful as a guide to properly interpret the contemporaneous EHT images. For 3C84, this source is actually a very good target for EATING VLBI because it is a high declination source and there is sufficient common sky between East Asia and Italy. Similarly, other famous high declination sources such as BL Lac, Mrk421, Mrk501, OJ287 and Cygnus A are suitable targets of EATING. Future regular EATING monitoring of these sources at 22/43 GHz will uniquely determine the structural evolution of these jets at scales of $\sim 70\,\mu$as.

References

An, T., *et al.* 2016, *PASJ*, 68, 77
An, T., Sohn, B.-W., & Imai, H. *et al.* 2018, *Nature Astronomy*, 2, 118
Asada, K. & Nakamura, M. 2012, *ApJ* (Letters), 745, L28
Asada, K., Nakamura, M., Doi, A. *et al.* 2014, *ApJ* (Letters), 781, L2
Fujita, Y., & Nagai, H. 2017, *MNRAS*, 62, L11
Giovannini, G., Savolainen, T., Orienti, M., *et al.* 2018, *Nature Astronomy*, 2, 472
Hada, K., Kino, M., Doi, A. *et al.* 2013, *ApJ*, 775, 70
Hada, K., Kino, M., Doi, A. *et al.* 2016, *ApJ*, 817, 131
Hada, K., Park, J., Kino, M., *et al.* 2017, *PASJ*, 69, 71
Hodgson, J.A., Rani, B., Lee, S.-S, *et al.* 2018, *MNRAS*, 475, 368
Kim, J.-Y, Trippe, S., Sohn, B.W., *et al.* 2015, *JKAS*, 48, 285
Kim, J.-Y., Krichbaum, T.P., Lu, R., *et al.* 2018, *A&A* in press
Komissarov, S.S., Vlahakis, N., Königl, A., *et al.* 2009, *MNRAS*, 394, 1182
Kovalev, Y.Y., Lister, M.L., Homan, D.C. *et al.* 2007, *ApJ*, 668, L27
Lee, S.-S, *et al.* 2014, *AJ*, 147, 77
Ly, C., Walker, R.C., & Junor, W. 2007, *ApJ*, 660, 200
Mertens. F., Lobanov, A., Walker, R.C., *et al.* 2016, *A&A*, 595, A54
Nagai, H., Suzuki, K., Asada, K., *et al.* 2010, *PASJ*, 62, L11
Nakamura, M., & Asada, K. 2013, *ApJ*, 775, 118
Niinuma, K., Lee, S.-S., Kino, M., *et al.* 2014, *PASJ*, 66, 103
Oh, J., *et al.* 2015, *JKAS*, 48, 299
Suzuki, K., Nagai, H., Kino, M., *et al.* 2012, *ApJ*, 746, 140
Wagner, A.Y., & Bicknell, G.V. 2011, *ApJ*, 728, 29
Wajima, K., *et al.* 2016, in ASP Conf. Ser. 502, Frontiers in Radio Astronomy and FAST Early Sciences Symposium 2015, ed. Qien L. and Li D. (San Francisco, CA: ASP), 81
Walker, R.C., Hardee, P.E., Davies, F.B., *et al.* 2018, *ApJ*, 855, 128
Zhang, Y., An, T., Frey, S., *et al.* 2017, *MNRAS*, 468, 69

Perseus in Sicily: From Black Hole to Cluster Outskirts
Proceedings IAU Symposium No. 342, 2018
K. Asada, E. de Gouveia Dal Pino, M. Giroletti,
H. Nagai & R. Nemmen, eds.

doi:10.1017/S1743921318003915

Molecular gas filamentary structures in galaxy clusters

Francoise Combes

Observatoire de Paris, LERMA, College de France, CNRS, PSL Univ., Sorbonne Univ., F-75014, Paris, France
email: francoise.combes@obspm.fr

Abstract. Recent molecular line observations with ALMA and NOEMA in several Brightest Cluster Galaxies (BCG) have revealed the large-scale filamentary structure at the center of cool core clusters. These filaments extend over 20-100kpc, they are tightly correlated with ionized gas (Hα, [NII]) emission, and have characteristic shapes: either radial and straight, or also showing a U-turn, like a horse-shoe structure. The kinematics is quite regular and laminar, and the derived infall time is much longer than the free-fall time. The filaments extend up to the radius where the cooling time becomes larger than the infall time. Filaments can be perturbed by the sloshing of the BCG in its cluster, and spectacular cooling wakes have been observed. Filaments tend to occur at the border of cavities driven in the X-ray gas by the AGN radio jets. Observations of cool core clusters support the thermal instability scenario, which accounts for the multiphase medium in the upper atmospheres of BCG, where the right balance between heating and cooling is reached, and a chaotic cold gas accretion occurs. Molecular filaments are also seen associated to ram-pressure stripped spiral galaxies in rich galaxy clusters, and in jet-induced star formation, suggesting a very efficient molecular cloud formation even in hostile cluster environments.

Keywords. galaxies: clusters: general – galaxies: ISM – galaxies: jets – galaxies: nuclei – galaxies: peculiar

---

The Perseus cluster, the mascot of our Symposium, is indeed the prototype of cool core clusters, where all physical phenomena can be studied in exquisite detail, due to its proximity. Fabian *et al.* (2003) have published a beautiful X-ray image, showing clearly the cavities sculpted by the radio jets, and their unsharp-masked image revealed multiple ripples, tracing acoustic waves, transferring the AGN energy into the ICM (Intra-Cluster Medium), and contributing to regulate the cooling. Large quantities of molecular gas, gas, $M(H_2) \sim 10^{10}$ $M_\odot$, have been detected with the IRAM-30m telescope (Salomé *et al.* 2006), and there is a good correlation between the CO and Hα filaments, as if the ionized gas was the interface between the dense cold gas and the ICM. Most of the Hα is excited by shocks and cosmic rays, but there are also patches of star formation (Canning *et al.* 2014). The H_2 1-0 S(1) ro-vibrational transition, tracing the shocked and highly excited molecular gas, has been mapped in the near-infrared (2 μm) with WIRCAM at the CFHT by Lim *et al.* (2012). The correspondence between these H_2 filaments and the Hα ones over 85 kpc in length is striking.

1. Molecular observations of cool core clusters

A large number (>50) of cool core clusters have been observed now with ALMA, NOEMA, or IRAM-30m, and large amounts of cold molecular gas have been detected, with always the same features (e.g. Pulido *et al.* 2018). There is a good correlation between Hα and CO emission, and molecular gas masses between 10^9 and 10^{11} $M_\odot$ are well correlated with star formation. The global corresponding depletion time-scale is just below 1 Gyr, somewhat smaller than the usual 2 Gyr value found in average in nearby galaxies (Bigiel *et al.* 2008).

In all resolved studies, where the molecular structures can be distinguished, one strong feature is that the molecular gas (CO) and ionized gas (Hα) are co-spatial. It is difficult to determine whether the filaments are inflowing or outflowing, but statistically there must be both motions. Inflow has been proven in several cases through absorption (see below), and this confirms that molecular gas is fueling the AGN. Outflow is also expected, due to the AGN feedback. In the multiphase gas observed in the central 10-20kpc radii, thermal instabilities create turbulence, clumps, which is necessary to provide more cooling, but certainly makes the kinematics more chaotic. This has also the effect to mix the various phases (cold and hot) and homogeneize abundances, explaining why the hot ICM is observed with relatively high metallicity.

One common characteristics of molecular components detected in cool-core clusters is their low-velocity gradients in filaments, revealing that the filaments are not in free-fall, and also the lack of relaxed structures, that could have settled in a rotating disk in the central BCG. Already in Perseus, Salomé *et al.* (2008, 2011) have shown that the gas velocity in filaments is far from free-fall,1 and that the gas around the center is not in rotation (or only in rare cases, like Hydra-A, cf Hamer *et al.* 2014, or very close to the center, in the central kpc).

Specific examples include the cool-core cluster Abell 1664, where ALMA has detected an H_2 mass of 1.1 $10^{10} M_\odot$ in two velocity components (Russell *et al.* 2014): one $-250 <$ V $<$ 250km/s component around the systemic velocity, and a high velocity one (HVS) at -570km/s (an outflow if in front of the BCG, or outflow otherwise). The structures are clympy and complex, but without strong V-gradient.

In the cluster Abell 1835, ALMA has detected a molecular mass of $M(H_2) = 5\ 10^{10}\ M_\odot$ within 10kpc of the BCG (McNamara *et al.* 2014). The velocity gradient is small, the profile narrow (130km/s) meaning either an almost face-on disk, if in rotation, or inflowing gas to the center. Ther is also a high velocity component, in which it is impossible to distinguish inflow from outflow, some kpc from the center. However, the AGN is not powerful enough to match the energy budget, and there is no bipolar counterpart. Besides the molecular gas is associated to an X-ray bubble, and could be cold gas condensing from the hot ICM.

A very special case is the BCG in Abell 1795, which is sloshing and oscillating in the central potential well of its cluster. McDonald *et al.* (2009) have revealed a long trailing wake in Hα, spanning $\sim$60 kpc in length, as a thin tail behind the BCG, and cospatial with an X-ray tail, with a bubble. The dynamical time-scale of the oscillation is of the same order of the cooling time, i.e. $\sim$300 Myr, and a corresponding molecular gas wake was detected with IRAM interferometer (Salomé & Combes 2004). ALMA results reveal the molecular filaments with exquisite details (Russell *et al.* 2017). As shown in Figure 1, the molecular features are distributed around the cavity sculpted by the radio jet in the ICM gas. The velocity map at the right of the Figure shows no rotation, but small V-gradient, corresponding to slow inflow or outflow.

In some cases, it is possible to distinguish between inflow or outflow, when the molecular gas can be seen in absorption in front of the central AGN continuum source. This is the case in the BCG of Abell 2597, where ALMA has revealed a CO(2-1) absorption spectrum (Tremblay *et al.* 2016, cf Figure 2). The CO emission is centered at the systemic velocity, but the various absorption features are red-shifted only, suggesting inflowing dense clouds, fueling the AGN.

Several cases of such molecular absorptions have been detected now (e.g. Edge *et al.* 2018, in prep) and red-shifted ones are prevailing, although some are blue-shifted, when

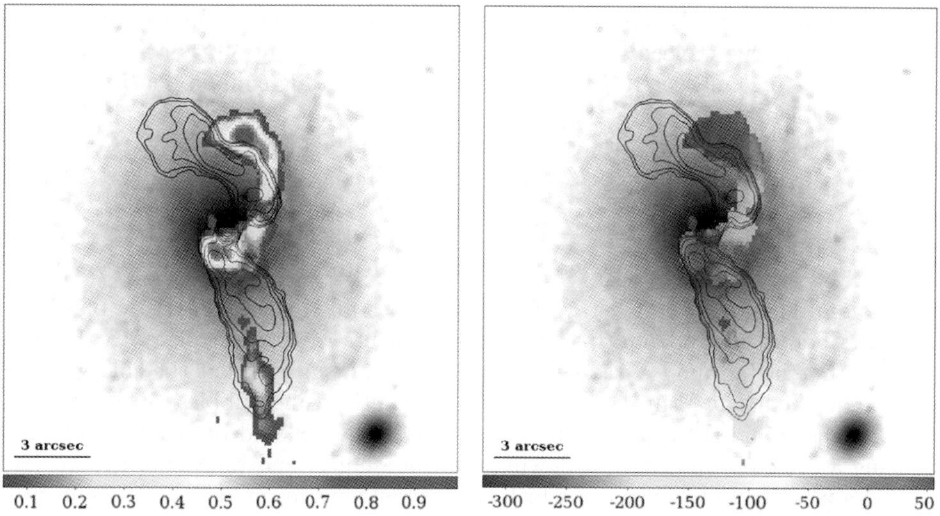

Figure 1. Left: ALMA CO(2-1) map of Abell 1795 (colour scale in Jy km/s), with the VLA 5GHz black contours superposed (from Breugel *et al.* 1984), overlaid on the HST F702W image in gray-scale. **Right:** CO(2-1) velocity map (colour scale from −300 to 50 km/s), overlaid on the HST image, and with the same VLA black contours. From Russell *et al.*, 2017.

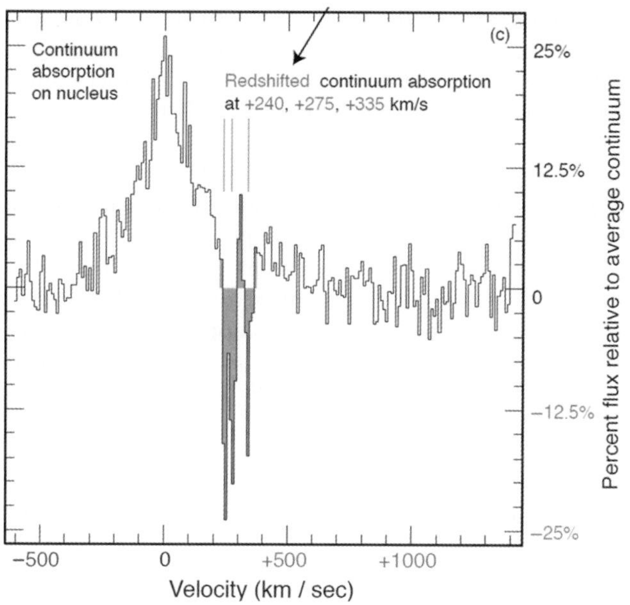

Figure 2. ALMA CO(2-1) spectrum in front of the central continuum source in the galaxy cluster Abell 2597, a cool core cluster. The emission spectrum is seen at the systemic velocity, and then 3 absorbing components are redshifted (highlighted in yellow), indicating inflow of gas towards the nucleus. From Tremblay *et al.* 2016

there is a signature of non-circular rotation, with a sharp edge at terminal velocity and a broad wing. As for Abell 2597, Tremblay *et al.* (2018) detect a molecular mass of 3 10^9 $M_\odot$ within 30kpc, and probably inflow and outflow at large distances, with also deviation of the jet by the molecular clouds. Again there is no evidence for the gas to be in free-fall, but it shows some chaotic motions. This might be explained through ram-pressure forces, if the molecular clouds reach the velocity limit V_l. For an ICM density of $\rho_{ICM} \sim 0.1 - 0.2$ cm^{-3} at R$\sim$ 15-30kpc, the corresponding column density ρ_{ICM} R $\sim$ 10^{22} cm^{-2}, which is similar to the column density in molecular clouds, at GMC scale and below. The order of magnitude of the ram-pressure is then ρ_{ICM} V_l^2 A = M_{cl} g = M_{cl} V_l^2/R, where g is the local acceleration of the cluster.

2. Expectations from theory and simulations

The ubiquitous presence of X-ray cavities, bubbles filled with relativistic plasma, ripples in the hot ICM, and multiple cold gas filaments directed straight to the cluster center, or stretched around X-ray cavities and bubbles, suggest that the cold gas is condensing from the hot ICM in the wakes of the buoyantly rising bubbles. Such a process of hot gas uplifting and then condensing in cold gas was simulated by Revaz *et al.* (2008), producing a cycle of inflowing/outflowing multiphase gas, and McNamara *et al.* (2016) proposed a feedback mechanism in which the low entropy gas at the cluster center (with low t_{cool}), is entrained by the buoyantly rising bubbles, inflated by the AGN radio jets, to an altitude of ~10 kpc, where the cooling time then becomes lower than the infall time t_I, which can be a few times longer than t_{ff}. The free-fall time indeed becomes much longer at high altitude R, being proportional to $(R/g)^{1/2}$, where g is the local gravitational acceleration. In the adiabatic uplift, t_{cool} on the contrary varies very little.

The AGN feedback has been recognized as the main moderator of the huge cooling flows that were previously expected (e.g. Fabian 2012), and now thermal instabilities (TI) (McCourt *et al.* 2012) are shown to be key in allowing the gas to cool far from the center, at R=10-20kpc, as soon as $t_{cool}/t_{ff} \sim 10$. The radio jets inflate bubbles, which create inhomogeneities and trigger further cooling. This has been dubbed Chaotic Cold Accretion (CCA) by Gaspari *et al.* (2011, 2012).

This phenomon is perfectly compatible with the kinematics of the gas observed in cool core clusters. For instance, the Hitomi satellite in Perseus has shown that the gas velocity dispersion $\sigma_{gas} \sim 160$ km/s, at 30-60 kpc (in the Fe XXV-Fe XXVI plasma, Lau *et al.* 2017). This implies a rather quiescent AGN feedback, in spite of cavities, and inhomogeneities. This gentle kinetic feedback corresponds more to the CCA than to violent galaxy cluster mergers. Cool core clusters appear relaxed since 4 Gyr.

Observations and simulations tend to reveal the existence of a floor in the value of the ratio t_{cool} /t_{ff}, from a typical radius of ~10kpc down to the cluster center. While the outer atmosphere, with t_{cool} /$t_{ff} >> 10$ is quite stable, the thermal instability and cooling occurs at a few 10kpc. Cooling is most efficient in the presence of TI because of the strong dependence of the cooling rate on density.

Sharma *et al.* (2012) have shown how local thermal instability (TI) occurs whenever the TI time scale becomes lower than 10 times the free-fall time. Self-regulation ensures a lower-limit in the ratio t_{TI} /t_{ff}, and can explain the floor. Voit *et al.* (2015) have called precipitation the combined process of TI and buoyancy, by analogy with the precipitation of rain drops. This phenomenon is quite similar to what Gaspari *et al.* (2011, 2012) call chaotic cold accretion (CCA) in their simulations. This kind of accretion can be much more efficient than the previous simple model of a Bondi accretion from the hot ICM, and corresponds much better to observations. Gaspari & Sadowski (2017) have tried to unify the widely different scales occuring in groups and clusters, under the self-regulation of AGN feedback, from the ultra-fast outflow (UFO) at scale smaller than a parsec, to kpc scales where thermal instabilities occur, and the outflowing gas comes back, even if launched with the escape velocity.

In a recent generalizing paper, Voit *et al.* (2017) describe in detail the history of ideas in thermal instabilities, and the new important effect brought by the combination with the buoyancy. The latter, intimately linked to AGN feedback, is essential to uplift the gas in the center of cluster, but can be replaced by the sloshing of the BCG at the bottom of the cluster potential well, or by mergers.

That energetic feedback is able to trigger cold gas accretion from a hot atmosphere, is also seen in many other circumstances, in haloes of various masses. There is an analogy with normal spiral galaxies, where the feedback is now supernovae or stellar feedback, inducing a fountain effect. This has been described for the halo of the Milky Way or galaxies such as NGC 891, where cold accretion of HI gas has been observed (e.g. Fraternali & Binney 2008, Marinacci *et al.* 2011). Simulations have shown how some gas from the hot corona could be made cooling, and fueling the gas disk. The cooling time is very large, and without the supernovae feedback, the corona would be stable. Also it is difficult even in flat-entropy coronae to trigger thermal instabilities, in the presence of thermal conduction and buoyancy. However, the fountain effect helps the hot corona to cool: as the fountain gas is moving in the corona, it is stripped off and mixed with the hot coronal gas, and since the metallicity is then high enough, it can radiatively cool. The fountain gas drags a cooling tail in its wake, so that a net fueling of the disk occurs at a few $M_\odot$/yr, about equal to the global star formation rate of the galaxy (Marasco *et al.* 2013), ensuring self-regulation. This mechanism explains the presence of multiphase gas in the atmospere of galaxies.

3. Molecular gas in X-ray groups

Cold gas condensing from the hot X-ray atmospheres is also expected at smaller scale in galaxy groups. One typical example is NGC 5044, at the center of a small group, characterized by multiple and radial Hα filaments, co-spatial with X-ray structures mapped with Chandra. The molecular gas was detected by ALMA, in a series of giant molecular associations (GMA), with masses between 3 10^5 and 10^7 $M_\odot$, and velocity wisths 10-50km/s. There is no rotating disk, but clumps also seen in absorption, only red-shifted (David *et al.* 2014).

Comparable features have been detected in two other groups, NGC 4638 and NGC 5846, by Temi *et al.* (2018). The gas clouds are aligned on Hα features, and delineated by dust lanes. The origin of the cold gas is cooling from the hot medium (which is dust-enhanced due to previous mixing). The clumps or GMA are not bound, their Virial ratio being much larger than 1 (20–200). These observations are compatible with the same CCA mechanism.

4. Origin of the molecules

In these hostile and hot environments, how can dense and cold molecular clouds survive, in the presence of strong shocks due to the AGN feedback and radio jets? Paradoxically, although the energetic incident radiation, or strong jets can destroy the clouds and make them more diffuse, they not only increase the surface of the clouds, but produce Kelvin-Helmholtz instabilities, which trigger the re-formation of molecules downstream (Hopkins & Elvis 2010). Some numerical simulations found that destroyed clouds do not re-condense, and they do not offer enough surface to be entrained (Ferrara & Scanna-pieco 2016). Dust is destroyed by sputtering, and dust is a necessary catalyzer for H_2 molecule formation. New simulations, taking into account chemistry, H_2 formation, and all the dust cycle have shown however that molecules can reform in AGN outflows, with the right order of magnitude (Richings & Giguère 2018).

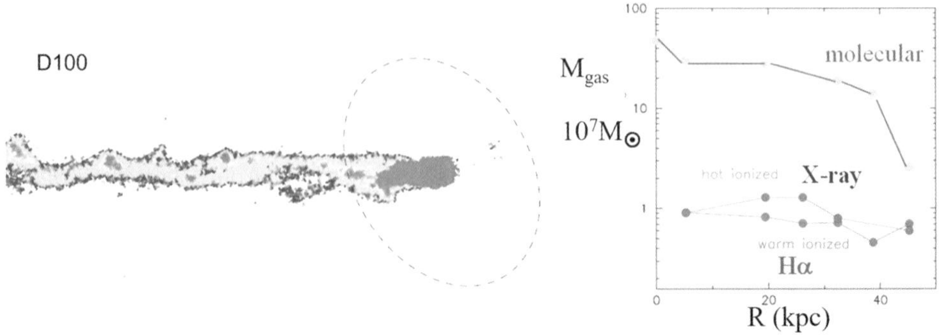

Figure 3. **Left:** Hα image of the D100 ram-pressure tail in Coma, from Yagi *et al.* 2007. The silhouette of the spiral galaxy is indicated with a dashed ellipse. The fraction of the tail shown is 20kpc long. **Right:** Masses of the various gas components, molecular gas, hot X-ray gas and warm Hα gas, in 12" beams (the IRAM CO(2-1) beam) along the tail of D100, until 50kpc from the center. From Jachym *et al.*, 2017

5. Molecular filaments in ram-pressure stripped tails

Other molecular filaments, co-spatial to Hα filaments can be seen in galaxy clusters, this time due to the cluster environment perturbations (tidal and ram-pressure tails). A giant Hα tail, larger than 100 kpc, has been detected in Virgo by Kenney *et al.* (2008). This tail begins at NGC 4438, and turns around M86, and at both extremities, CO emission has been detected (Dasyra *et al.* 2012). This shows that H_2 gas can survive more than 100 Myr in hostile environments, at temperature of 10^7K. Clouds of mass $M(H_2) = 7\ 10^6 M_\odot$ are seen 10kpc NE of M86. These have been formed in situ or come from the NGC 4438 tail.

In rich clusters, the ram-pressure stripping is even more violent, as demonstrated by the ESO137-001 tail in the Norma cluster (Jachym *et al.* 2014). Along 80kpc, a hot X-ray tail is co-spatial to warm Hα gas and molecular gas, emitting in CO. The tail is de-doubled, certainly due to the two sides of the rotating disk. In Coma, the galaxy D100 reveals an even thinner ram-pressure tail, which is the second stage of stripping, when the outer parts have been swept out already (Jachym *et al.* 2017). Figure 3 shows the straight Hα filament (left) and how the mass of the filament is dominated by molecular gas (right).

AGN feedback due to radio jets can also be positive, and trigger star formation in the hot atmosphere of a giant elliptical galaxy. In Centaurus A, a long filament of young star formation aligned along the radio jet is known since a long time. ALMA and APEX have detected large quantities of molecular gas aligned on this Hα filament, just when the radio jet encountered an HI shell, cf Figure 4. On the jet path, all the atomic gas has been transformed in molecules (Salomé *et al.* 2016).

6. Summary

Molecular filaments have now been detected in many cool-core clusters, with common characteristics: they form in the wake of X-ray bubbles, inflated by the AGN radio jets, they are not in free fall, but reveal small velocity gradients. Observations are compatible with their formation by precipitation of cold gas, or chaotic cold accretion (CCA) in a thermally unstable hot medium, as soon as $t_{cool}/t_{ff} \sim 10$. When the entropy profile is relatively shallow in the center of clusters, some low entropy gas can be adiabatially uplifted until this ratio is sufficiently small, and the gas can cool, in a multiphase medium.

This cold gas fuels the central AGN, which can then self-regulate the heating and cooling to the balance observed. Thermal instabilities, combined with buoyancy, and

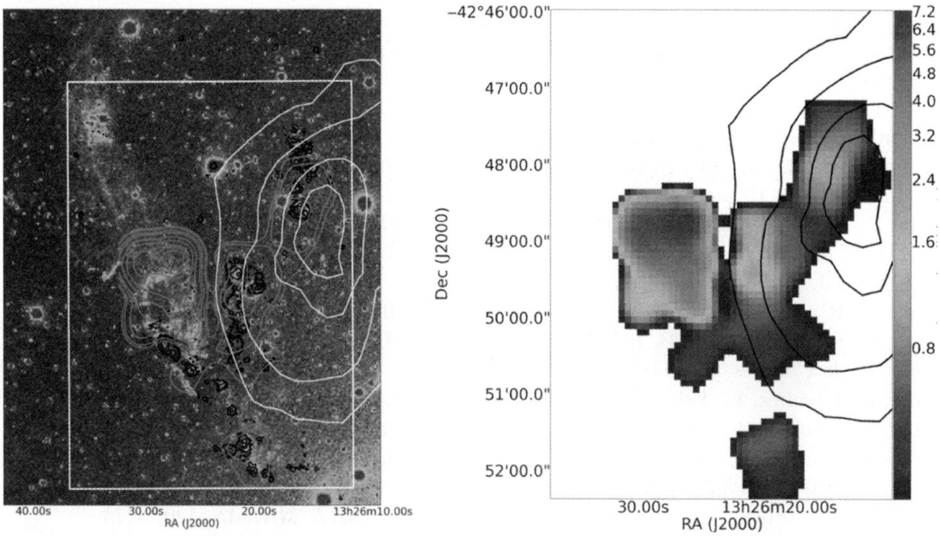

Figure 4. Left: Hα emission of the northern region of Centaurus A with CTIO. The HI emission (VLA; white contours), the CO(2-1)emission (APEX; red contours) and the FUV emission (GALEX; black contours) are overlaid. **Right:** intensity map of the CO(2-1) emission from APEX in K km/s, with the HI emission overlaid in black contours. The white box on the left shows the region of the right panel. From Salomé *et al.* 2006.

AGN feedback can create the conditions of CCA up to 10-20 kpc from the cluster center. The gas in filaments is slowed down from free-fall through ram-pressure forces.

When the BCG is in oscillating motion in the bottom of the potential well, a cooling wake is formed, with X-ray, Hα and CO emission co-spatial in the wake. Molecules can reform easily in the filaments

Molecular filaments are also present in tidal and ram-pressure tails. The cooling of hot coronae can also be triggered by supernovae feedback: fountain gas ejected above the plane of spiral galaxies is then mixing with hot gas, triggering cooling, with a net balance of fueling star formation in the disk. AGN feedback can also be positive, as shown by very clear cases of jet-induced star formation.

References

Bigiel, F., Leroy, A., Walter, F. *et al.*, 2008, *AJ*, 136, 2846
Canning, R. E. A., Ryon, J. E., Gallagher, J. S. *et al.*, 2014, *MNRAS*, 444, 336
Dasyra, K. M., Combes, F., Salomé, P., & Braine, J., 2012, *A&A*, 540, A112
David, L. P., Lim, J., Forman, W. *et al.*, 2014, *ApJ*, 792, 94
Fabian, A. C., Sanders, J. S., Allen, S. W. *et al.*, 2003, *MNRAS*, 344, L43
Fabian, A. C., 2012, *ARA&A*, 50, 455
Ferrara, A., & Scannapieco, E., 2016, *ApJ*, 833, 46
Fraternali, F., & Binney, J. J., 2008, *MNRAS*, 386, 935
Gaspari, M., Melioli, C., Brighenti, F., & D'Ercole, A., 2011 *MNRAS*, 411, 349
Gaspari, M., Ruszkowski, M., & Sharma, P., 2012 *ApJ*, 746, 94
Gaspari, M., & Sadowski, A., 2017, *ApJ*, 837, 149
Hamer, S. L., Edge, A. C., Swinbank, A. M. *et al.*, 2014, *MNRAS*, 437, 862
Hopkins, P. F., & Elvis, M., 2010, *MNRAS*, 401, 7
Jachym, P., Combes, F., Cortese, L. *et al.*, 2014 *ApJ*, 792, 11
Jachym, P., Sun, M., Kenney, J. D. P *et al.*, 2017 *ApJ*, 839, 114
Kenney, J. D. P., Tal, T., Crowl, h. H. *et al.*, 2008, *ApJ*, 687, L69
Lau, E. T., Gaspari, M., Nagai, D., & Coppi, P., 2017, *ApJ*, 849, 54

Lim, J., Ohyama, Y., Chi-Hung, Y. *et al.*, 2012 *ApJ*, 744, 112
Marasco, A., Marinacci, F., & Fraternali, F., 2013, *MNRAS*, 433, 1634
Marinacci, F., Fraternali, F., Nipoti, C. *et al.*, 2011, *MNRAS*, 415, 1534
McCourt, M., Sharma, P., Quataert, E., & Parrish, I. J., 2012, *MNRAS*, 419, 3319
McDonald, M., & Veilleux, S., 2009, *ApJ*, 703, L172
McNamara, B. R., Russell, H. R., Nulsen, P. E. J. *et al.*, 2014, *ApJ*, 785, 44
McNamara, B. R., Russell, H. R., Nulsen, P. E. J. *et al.*, 2016, *ApJ*, 830, 79
Pulido, F. A., McNamara, B. R., Edge, A. C. *et al.*, 2018, *ApJ*, 853, 177
Revaz, Y., Combes, F., & Salomé, 2008, *A&A*, 477, L33
Richings, A. J., & Faucher-Giguère, C.-A., 2018, *MNRAS*, 474, 3673
Russell, H. R., McNamara, B. R., Edge, A. C. *et al.*, 2014, *ApJ*, 784, 78
Russell, H. R., McNamara, B. R., Fabian, A. C *et al.* 2017, *MNRAS*, 472, 4024
Salomé, P., & Combes, F., 2004, *A&A*, 415, L1
Salomé, P., Combes, F., Edge, A. C. *et al.*, 2006, *A&A*, 454, 437
Salomé, P., Combes, F., Revaz, Y. *et al.*, 2008, *A&A*, 484, 317
Salomé, P., Combes, F., Revaz, Y. *et al.*, 2011, *A&A*, 531, A85
Salomé, Q., Salomé, P., Combes, F., & Hamer, S., 2016, *A&A*, 595, A65
Sharma, P., McCourt, M., Quataert, E., & Parrish, I. J., 2012, *MNRAS*, 420, 3174
Temi, P., Amblard, A., Gitti, M. *et al.*, 2018, *ApJ*, 858, 17
Tremblay, G. R., Oonk, J. B. R., Combes, F. *et al.*, 2016, *Nature*, 534, 218
Tremblay, G. R., Combes, F., Oonk, J. B. R. *et al.*, 2018, *ApJ*, sub
van Breugel, W., Heckman, T., & Miley, G., 1984, *ApJ*, 276, 79
Voit, G. M., Donahue, M., Bryan, G. L., & McDonald, M., 2015, *Nature*, 519, 203
Voit, G. M., Meece, G., Li, Y. *et al.*, 2017, *ApJ*, 845, 80
Yagi, M., Komiyama, Y., & Yoshida, M., 2007, *ApJ*, 660, 1209

Perseus in Sicily: From Black Hole to Cluster Outskirts
Proceedings IAU Symposium No. 342, 2018
K. Asada, E. de Gouveia Dal Pino, M. Giroletti,
H. Nagai & R. Nemmen, eds.

doi:10.1017/S1743921318003836

Young radio jets breaking free: molecular and HI outflows in their centers

Raffaella Morganti[1,2], Tom Oosterloo[1,2], Robert Schulz[1], Clive Tadhunter[3] and J. B. Raymond Oonk[1,4]

[1]ASTRON, the Netherlands Institute for Radio Astronomy, Postbus 2, 7990 AA, Dwingeloo, The Netherlands.
email: morganti@astron.nl

[2]Kapteyn Astronomical Institute, University of Groningen, P.O. Box 800, 9700 AV Groningen, The Netherlands

[3] Department of Physics and Astronomy, University of Sheffield, Sheffield, S7 3RH, UK

[4] Leiden Observatory, Leiden University, P.O. Box 9513, 2300 RA Leiden

Abstract. Our view of the central regions of AGN has been enriched by the discovery of fast and massive outflows of H I and molecular gas. Here we present a brief summary of results obtained for young (and restarted) radio AGN. We find that H I outflows tend to be particularly common in this group of objects. This supports the idea that the jet, expanding in a clumpy medium, plays a major role in driving these outflows. The clumpiness of the medium is confirmed by VLBI and ALMA observations. The H I observations reveal that, at least part of the gas, is distributed in clouds with sizes up to a few tens of pc and mass $\sim 10^4$ $M_\odot$. A change of the conditions in the outflow, with an increasing fraction of diffuse components, as the radio jets grow, is suggested by the high resolution H I observations. The molecular gas completes the picture, showing that the radio plasma jet can couple well with the ISM, strongly affecting the kinematics, but also the physical conditions of the molecular gas. This is confirmed by numerical simulations reproducing, to first order, the kinematics of the gas.

Keywords.

1. Introduction

In recent years, a lot of attention has been devoted to understanding the interplay between the energy emitted by an AGN and the surrounding medium. However, given the complexity, the details of this process and the actual impact are still uncertain. For example, the effects of the radio jets on large (cluster) scales are well studied and they represent one of the best examples of feedback in action. Such interactions can also play a role in the central (sub-kpc) regions. This has been known for a long time from studies of outflows of ionised gas, but the contribution in mass of such outflows typically appears to be quite modest. The discovery that radio jets can also drive a more massive outflow of cold gas (H I and cold molecular gas), and the improvement of numerical simulations, has revived the interest for this process and for the impact on galaxy evolution.

Numerical simulations are finding that radio plasma jets couple strongly with the clumpy ISM of the host galaxy (Wagner *et al.* 2012; Mukherjee *et al.* 2018b). The effect of radio jets on galactic scales is found to be particularly strong when a radio jet is young or recently restarted. Furthermore, the simulations show that the impact depends on some key parameters. The jet power, the distribution of the surrounding medium and the orientation at which the jet enters the medium are the most important of such

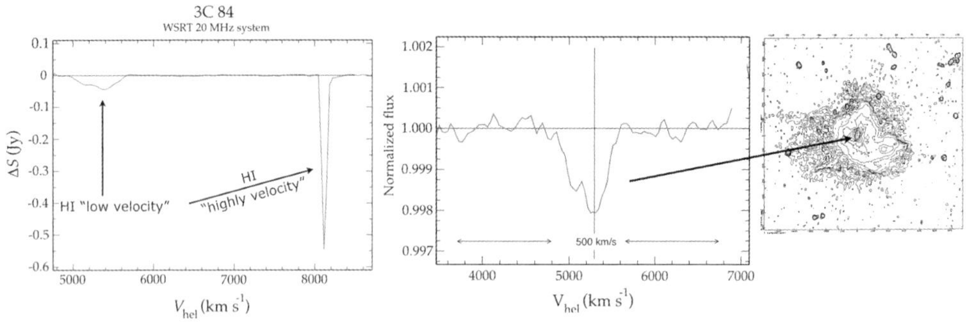

Figure 1. H I absorption profile in 3C 84 (WSRT commissioning data de Bruyn & Oosterloo priv.comm., dashed line marking V_{sys}). Right: continuum image (Sijbring *et al.* (1989)).

parameters (Mukherjee *et al.* 2018b). Last but not least, the simulations show that even low power jets (i.e. with radio luminosity at 1.4 GHz $< 10^{24.5}$ W Hz^{-1}) can play an important role. They are more strongly affected by the interaction with the surrounding medium and can induce more turbulence. This is important because low luminosity radio sources are more numerous, but, because they are often classified as "radio quiet", the effects of their radio jets tend to be neglected (Wylezalek & Morganti 2018).

Here, we show examples of jet-ISM interactions traced by the disturbed kinematics of the gas in young radio sources and discuss what we have learned from them. We follow the impact of the AGN using high spatial resolution observations of H I (using VLBI) and molecular gas (traced by the CO using ALMA).

2. Results from HI outflows and follow-up VLBI

In a pilot H I absorption survey of 248 radio galaxies, at least 5% (15% of H I detections) show fast H I outflows with velocities in the range 500 – 1000 km s^{-1}, (see Geréb *et al.* 2015; Maccagni *et al.* 2017). This is a very conservative lower limit because of the limited depth of the observations. Most importantly, these surveys have shown that the highest detection rate of outflows is seen in *young and restarted radio galaxies*, confirming, as predicted by the simulations, that this is the phase where the radio jet most strongly impacts the surrounding medium.

2.1. *Small detour: how about 3C84?*

H I has been detected in absorption also in 3C 84. This is not too surprising, because this radio galaxy is known to be embedded in cold gas (Lim *et al.* 2008; Nagai *et al.* these proceedings) but provides new insights on the gas. Two H I absorbing systems have been observed against the central region of 3C 84 (Fig. 1). The high-velocity system is likely a chance alignment with a foreground gas-rich galaxy (Momjian *et al.* 2002). Instead, the low-velocity system is broad (~ 500 km s^{-1}) and roughly centred on the systemic velocity. It is likely resulting from a combination of large-scale and of circum-nuclear H I (see discussion in Sijbring *et al.* 1989). However, the profile is asymmetric and mostly blueshifted, suggesting the presence of disturbed kinematics of the gas, perhaps connected to the effect of the radio jet. If confirmed by high spatial resolution observations, it would support the idea that the fast duty cycle of the radio source and the new jet components appearing regularly are continuously disturbing the surrounding ISM.

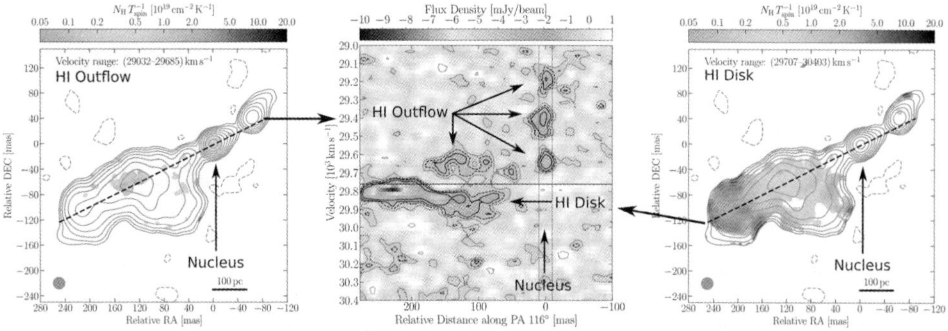

Figure 2. Radio continuum superposed on the H I absorption column density of the H I outflow (left) and disk (right) of the radio galaxy 3C236 obtained with VLBI. Position-velocity plot (center) along the jet position angle showing the outflowing clouds. From Schulz et al. (2018)

2.2. *Do we see evolution of the outflows as the source expands?*

In order to explore how radio jets impact the surrounding medium, we have started a project to trace H I outflows at high spatial resolution using VLBI observations (i.e. with resolution from a few tens to hundred pc). We focus on a small group of young (and re-started) radio galaxies, covering a range in sizes between ~ 100 pc and one kpc. We take the size as a proxy of their age and evolutionary stage.

We find that a clumpy medium is seen in the smaller (i.e. younger) sources, i.e. 4C 12.50 (Morganti *et al.* 2013) and 4C 52.37, an H I outflow discovered by Geréb *et al.* (2015), and located in the inner 100 pc of this source (Schulz et al. in prep). Interestingly, the VLBI observations spatially resolve the outflows and recover all the absorbing flux observed at low resolution. This suggests that the absorption associated with these outflows must be mostly produced by relatively compact structures, easy to detect with VLBI.

The more evolved sources (like 3C 293 and 3C 236) also show evidence of a clumpy medium. Figure 2 illustrates the distribution of H I absorption in the central region of 3C 236 and the outflowing clouds detected in the inner region (Schulz *et al.* 2018). The clouds have a mass of a few $\times 10^4$ $M_\odot$, and are unresolved on VLBI scales (< 40 pc). However, the H I outflows in these sources are only partly recovered by VLBI. This may indicate that the expansion of the jet in the medium changes the structure of the outflows, increasing the fraction of diffuse components. The relevance of these results is that, according to the simulations, a clumpy medium can make the impact of the jet much larger than previously considered: because of the clumpiness of the medium, the jet is meandering through the ISM to find the path of minimum resistance and so creating a cocoon of shocked gas driving the outflow (Wagner *et al.* 2012; Mukherjee *et al.* 2018b).

3. Outflows of molecular gas

The molecular gas is a very good tracer for completing the picture of the impact of a jet on the ISM. The best studied example of a jet-driven outflow is IC 5063. This is also due to the geometry of the system, with the radio jet expanding into the large gas disk. ALMA observations show that in the immediate vicinity of the radio jet, a fast outflow (up to 800 km s^{-1}), is occurring (see Morganti *et al.* 2015 and Fig. 3). The radio jet is also affecting the physical conditions of the molecular gas, as derived from observations of multiple transitions (Oosterloo *et al.* 2017; Fig. 3). The line ratios suggest that the outflowing gas is optically thin with a mass of a few $\times 10^6$ $M_\odot$ and a mass outflow rate $\sim$10 $M_\odot$ yr^{-1}. The kinetic temperatures between 20 – 100 K and the densities between 10^5 and 10^6 cm^{-3}. The pressure in the outflow is about two orders of magnitude

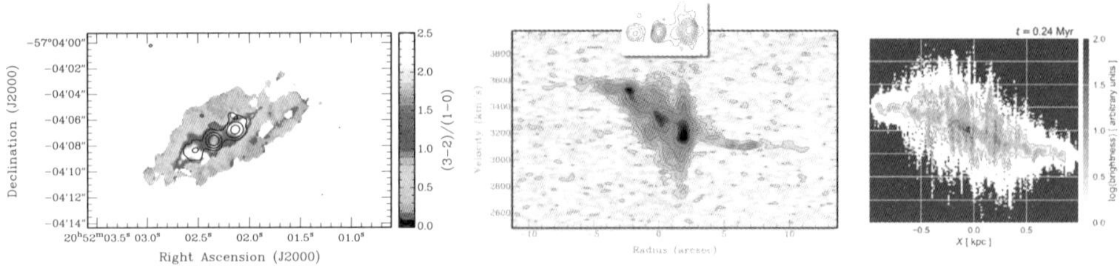

Figure 3. Results from ALMA observations of IC 5063. Left: ratio CO(3-2)/CO(1-0) with superposed the contours of the radio continuum emission (Oosterloo *et al.* 2017). Centre: Position-velocity plot of the CO(2-1), (Morganti *et al.* 2015). The line ratio and kinematics of the gas is clearly disturbed in the region co-spatial with the radio source. Right: position-velocity plot from a simulated data cube of a jet expanding in a gaseous disk (Mukherjee *et al.* 2018a).

higher than in the undisturbed gas and the best fit of line ratios suggests the outflow is very clumpy. Numerical simulations of a jet entering a clumpy medium provide a good description of the observations as seen in Fig. 3 (Mukherjee *et al.* 2018a). Interestingly, as predicted by the simulations (Mukherjee *et al.* 2018b), this impact occurs despite the fact the radio source is low power and that it often would be classified as *radio quiet*. In summary, the radio jet has a significant impact, but only in the inner few kpc and only a small fraction (~0.1% of the total ISM) of the gas will leave the galaxy.

The second object studied is PKS 1549–79, a quasar hosting a strong radio source in the early stages of evolution (Holt *et al.* 2006). The ALMA CO(1-0) data show evidence for both accretion and feedback effects, shown by the presence of a fast, $\sim$2000 $\mathrm{km\,s^{-1}}$, outflow (Oosterloo et al. in prep). The molecular outflow is particularly massive, $\sim$100 $M_{\odot}\ \mathrm{yr}^{-1}$, one of the largest found in radio galaxies. Although the outflow is limited to the region of the inner radio jet (< 200 pc), it is not yet clear whether the jet or the radiation is driving it. If due to the former, this would be in agreement with numerical simulations predicting that more powerful sources (like PKS 1549–79) should drive faster/more massive outflows, compared to e.g. what is seen in IC 5063.

4. Summary of the results so far

H I and molecular outflows are common in young and restarted radio galaxies, suggesting that in this phase the jet is able to remove, reheat or relocate the gas piled up in the centre of the galaxy. Follow-up observations with high spatial resolution confirm, to first order, the predictions of numerical simulations. A clumpy medium is revealed by H I VLBI observations and also derived from observations of multiple molecular transitions in IC 5063. The structure of the H I outflows may change as the radio jet evolves, with a larger fraction of diffuse gas in more extended sources. The main effect of the energy released appears to be the redistribution of the gas (i.e. similar to galactic fountains). Last but not least, a comparison between observations and simulations is now beginning to be possible and is expected to bring new, important insights in this complex process.

References

Geréb, K., Maccagni, F. M., Morganti, R., & Oosterloo, T. A. 2015, *A&A*, 575, A44;
Holt, J., Tadhunter, C., Morganti, R., *et al.* 2006, *MNRAS*, 370, 1633
Lim, J., Ao, Y., & Dinh-V-Trung 2008, *ApJ*, 672, 252;
Maccagni F. M., Morganti R., Oosterloo T. A., Geréb K., & Maddox N., 2017, *A&A*, 604, A43
Momjian, E., Romney, J. D., & Troland, T. H. 2002, *ApJ*, 566, 195
Morganti, R., Fogasy, J., Paragi, Z., Oosterloo, T., & Orienti, M. 2013, Science, 341, 1082
Morganti, R., Oosterloo, T., Oonk, *et al.* 2015, *A&A*, 580, A1

Mukherjee, D., Wagner, A. Y., Bicknell, G. V., *et al.* 2018, *MNRAS*, 476, 80;
Mukherjee, D. *et al.* 2018, *MNRAS* in press, arXiv:1803.08305;
Oosterloo, T., Raymond Oonk, J. B., Morganti, R., *et al.* 2017, *A&A*, 608, A38
Schulz, R., Morganti, R., Nyland, K., *et al.* 2018, *A&A* in press, arXiv:1806.06653
Sijbring, D., de Bruyn, A. G. *et al.* 1989, ESO Conference and Workshop Proceedings, 32, 107
Wylezalek, D., & Morganti, R. 2018, *Nature Astronomy*, 2, 181
Wagner A. Y., Bicknell G. V., & Umemura M. 2012, *ApJ*, 757, 136;

Perseus in Sicily: From Black Hole to Cluster Outskirts
Proceedings IAU Symposium No. 342, 2018
K. Asada, E. de Gouveia Dal Pino, M. Giroletti,
H. Nagai & R. Nemmen, eds.

doi:10.1017/S1743921318007779

Ultra fast outflows, and their connection to accretion and ejection processes in AGNs

Anna Lia Longinotti

CONACyT–Instituto Nacional de Astrofísica, Óptica y Electrónica
Luis E. Erro 1, Sta María Tonantzintla, Puebla, C.P. 72840, Mexico
email: annalia@inaoep.mx

Abstract. The growing evidence for energy-conserving outflows in powerful and luminous AGN supports the idea that high-velocity winds launched from the accretion disc evolve systematically after undergoing a shock with the ambient medium and that they are capable to expel enough mass and energy so as to produce feedback. This talk will give an overview of recent results on AGN ultra fast outflows, with focus on grating X-ray spectra of bright sources. I will review how UFO work, their observational properties and their relation with AGN outflows in other bands, what is their impact on the host galaxies and their role in feedback processes.

Keywords. galaxies: active, hydrodynamics, atomic processes

1. Introduction

Feedback from Active Galactic Nuclei (AGN) is generally thought to be an important ingredient for galaxies evolution. After large amount of gas is accreted during the earliest stage of a quasar life time, the accumulated energy can be released via ejection of powerful outflows driven by the AGN. If the outflow is as strong as 0.5–5% of the Eddington luminosity of the AGN, it has a profound impact on the development of the host galaxy itself. The effect of these winds is to eventually expel the gas that would otherwise be available for forming new stars in the host galaxy therefore providing an effective mechanism of quenching star formation. It is in this sense that we refer to AGN feedback as a mechanism able to regulate the growth of the galaxy and the growth of the central black hole as well (Di Matteo *et al.* 2005, Hopkins *et al.* 2010).

A widely accepted scenario for explaining AGN feedback postulates that a fast wind observable in the X-ray band is launched at accretion disk scale (Faucher-Giguère & Quataert 2012). This highly ionized X-ray gas is currently observed in the form of Ultra Fast Outflows in some AGN spectra (Tombesi *et al.* 2012, Gofford *et al.* 2013). While traveling outward, the impact of the wind with the ISM (inter-stellar medium) gives rise to shock processes (King 2010). After shocking with the gas, deceleration and cooling processes lead to the production of a slower outflow with less ionized lines observable in the optical band and to the formation of a bubble of hot, tenuous gas (e.g. Zubovas & King 2012). As a result of the cooling, the presence of molecular gas outflowing at a much lower velocity is expected. This latest phase is frequently observed in several ULIRGS and Quasars (Cicone *et al.* 2014, Feruglio *et al.* 2010, 2015).

To date, only two cases of ULIRGS are reported where the observed X-ray and molecular phases of the outflow are physically related, IRAS F11119+3257 (Tombesi *et al.* 2015) and Mrk 231 (Feruglio *et al.* 2015). Both results remarkably fit in with the prediction of the energy-conserving outflow model outlined above.

This fascinating picture, among several other variables, relies on the existence and the properties of the nuclear wind, the so-called "Ultra Fast Outflow". The X-ray spectra

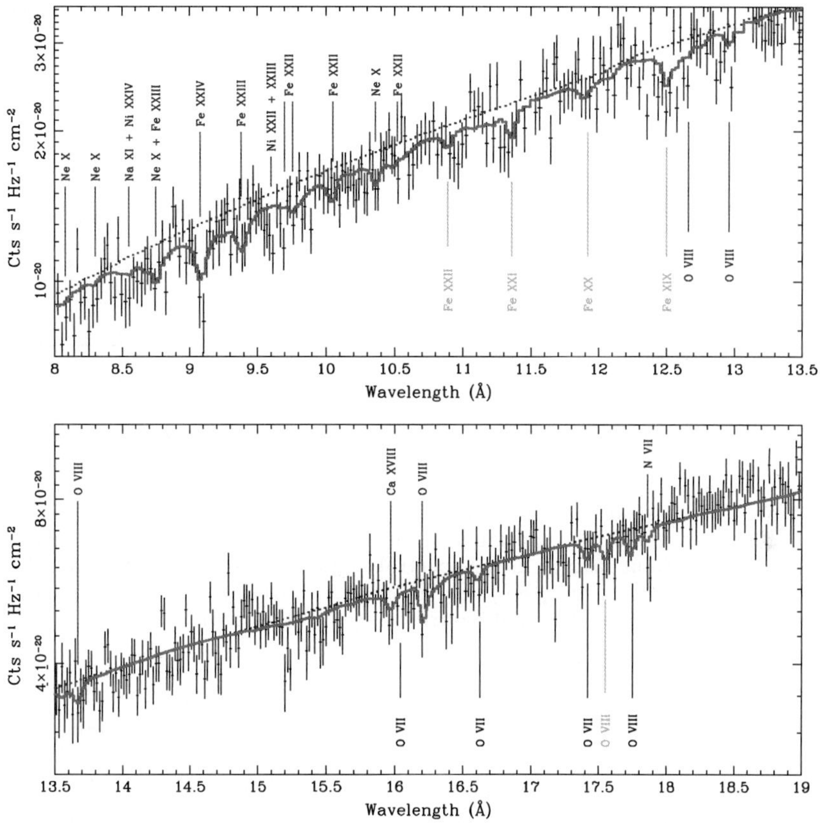

Figure 1. *XMM-Newton*-RGS spectrum of the NLSy1 Galaxy Mrk 1044 fitted by four components of ultra fast outflows (Krongold *et al.* submitted). The fastest component (blue labels) of this system reaches an outflow velocity of ~48,000 km s^{-1} and column density of N$_H$~10$^{23.32}$ cm^{-2}, while the other three are outflowing at 25,000 km s^{-1}.

of some AGNs show signature of gas outflowing at high speed ($v \geq 0.1$ c). This gas is so highly ionized by the nuclear radiation that the only dominant ions left are He-like and Hydrogen-like ionic species. These systems were christened as "Ultra-Fast Outflows" (UFOs) (Tombesi *et al.* 2010) and at the beginning they were observed mainly in the Fe K band at E$\geq$ 7 keV. Several papers reported on UFOs hosted in individual AGNs (Pounds *et al.* 2003, 2011, 2014, Chartas *et al.* 2009, Lanzuisi *et al.* 2012, Nardini *et al.* 2015), and statistical studies show that they are detected in 30-40% of nearby AGN (Tombesi *et al.* 2010, 2012, Gofford *et al.* 2013). The approximate ranges of mass outflow rate (0.01-1M$_\odot$ yr^{-1}) and kinetic energy (10$^{42-45}$ erg s^{-1}) are in good agreement with theoretical predictions for black hole winds (King 2010).

Two launching mechanisms are envisaged for the production of such rapid outflows, and both locate the region of action in the accretion disk of the AGN. One way to produce a fast outflow is through radiation driving (Proga & Kallman 2004, Sim *et al.* 2010), and another one is via magnetic fields in radio-loud sources ("magnetically driven outflows", Fukumura *et al.* 2015).

2. Ultra Fast Outflows in Narrow Line Seyfert 1 Galaxies

In the very recent years, the discovery of fast outflows in the soft X-ray grating spectra of bright Seyfert Galaxies has opened a complementary path to explore the properties

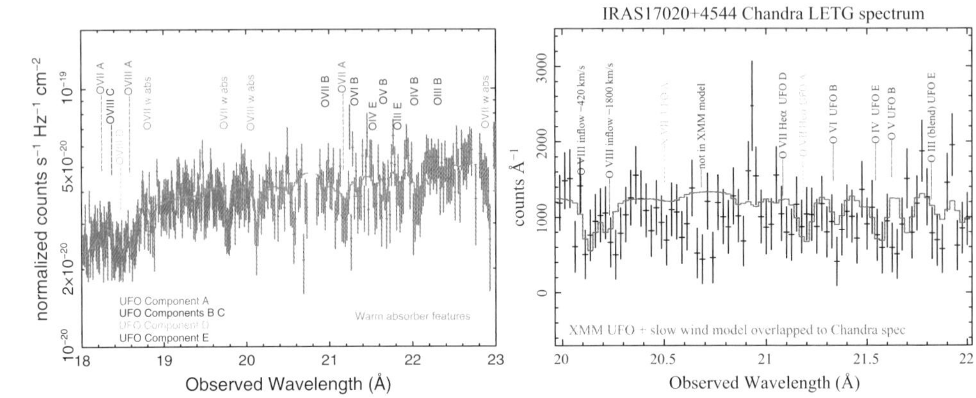

Figure 2. The system of outflowing and inflowing gas in the NLSy1 Galaxy IRAS17020+4544. *Left*: Portion of the *XMM-Newton*-RGS spectrum (observed in 2014) mostly affected by the fast wind absorption. Letters indicate the velocity component of the outflow as described in L15. The label "Warm absorber" marks lines that are part of the slow ionized outflow. *Right*: Close-up of the *Chandra*-LETG spectrum of the source (observed in 2017) with the XMM model overlapped (without fitting) in the spectral region where "slow" inflow and fast outflow are visible (Longinotti *et al.* in prep.). Labels refer to absorption features identified in the *XMM-Newton*-RGS spectrum (L15 and Sanfrutos *et al.* submitted).

of AGN fast winds. Compared to the results based on CCD spectra, the higher detail provided by grating spectroscopy has revealed that soft X-ray fast winds can be made by multiple components of distinct outflowing velocity, ionization state and column density and that they can be massive enough to produce feedback in the host galaxy, e.g. IRAS 17020+4544 (Longinotti *et al.* 2015, L15 throughout) and Mrk 1044 (see Fig 1, Krongold *et al.* submitted). When luminosity variations takes place, the combination of timing and spectroscopy techniques shows that fast X-ray winds respond to flux variations revealing a tight connection between continuum photons emitted in the very inner accretion disc with highly ionized outflowing gas (IRAS 13224-3809 Parker *et al.* 2017; PDS 456 Matzeu *et al.* 2017). With the exception of the luminous QSO PDS 456 (Reeves *et al.* 2016), if we consider the other few cases of fast winds observed in gratings spectra (PG1211+143, Reeves *et al.* 2018; Akn 564, Gupta *et al.* 2013), it is clear that the sources where this phenomenon is detected share the same classification: they are all Narrow Line Seyfert 1 (NLSy1).

This AGN class represents an extreme form of Seyfert activity that is manifested in their peculiar continuum and emission-line properties in almost all bands (see Gallo 2006), their small black hole masses ($M_{BH} \sim 10^{6-7}$ M$\odot$), high Eddington ratios and often, high degree of X-ray variability (Komossa & Xu 2007).

The recent findings of fast X-ray winds in an increasing number of NLSy1 seem to hint to a common behavior of these sources with objects of much higher luminosity (e.g. Chartas *et al.* 2002, Tombesi *et al.* 2015, Nardini *et al.* 2015). Increasing evidence that the velocity of the wind correlates with the source luminosity (i.e. winds are faster at higher luminosities, Pinto *et al.* 2018) strongly favours radiation driving as the launching mechanism for the outflows, leading therefore to the hypothesis that a high accretion rate is the likely driver of fast winds (Matzeu *et al.* 2017).

3. Stratification of the fast outflow: a shocked outflow origin?

We now focus on the peculiar case of the NLSy1 IRAS17020+4544. This source presents the simultaneous presence of a stratified fast outflow (L15) and a multi-layered slower

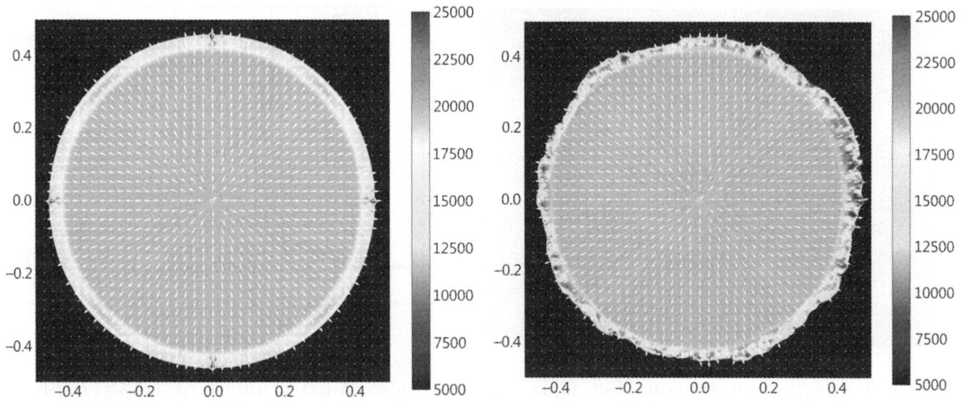

Figure 3. Toy model showing the appearance of a shocked outflow affected by instabilities. The color scale maps the velocity of the ultra fast outflow ejected at 20,000 km s^{-1} that is pushing the shock outward (*Left*). When instabilities are introduced (*Right*), sections of gas with higher velocities start to appear at the border of the shock (simulations to be included in Longinotti *et al.* in prep.)

absorber. Fine quality multi-epoch spectroscopic information available from two *XMM-Newton* observations (obtained in the years 2004 and 2014) and a long look with *Chandra* LETG (obtained in 2017) confirm a surprisingly steady continuum flux and a complex pattern of absorption features (see Fig. 2). While no significant variations are observed in the properties of the fast wind along the 10 years elapsed in between the two *XMM-Newton* observations, the peculiar evolution of the 4 components of the slow absorber (Sanfrutos *et al.* submitted) reveals gas flowing inward (v_{inflow}~1800-3000 km s^{-1}) and outward ($v_{outflow}$~2000-3400 km s^{-1}), thus suggesting a different nature of this wind compared to "standard" warm absorber.

To explain this complex pattern of absorbers we postulate that the coincidence of the fast outflow with slower winds moving in opposite directions may be explained in terms of a "shocked outflow". This model predicts that an initial fast outflow radiatively launched at accretion disc scale with outflow velocity $v_{\rm out} \geq 10^4$ km s^{-1} shocks with the ambient medium (see King & Pounds 2015 for a review). The two shock fronts (reverse and forward) produced by the impact of the wind with gas at an escape velocity lower than the outflow velocity, are separated by a contact discontinuity and whereas the shocked ambient gas could decelerate to velocity of the order of 10^2 km s^{-1}, the wind shock (forward) maintains its high velocity while entraining the ambient gas and pushing it further out (Faucher-Giguère & Quataert 2012).

Analogously to Supernova Remnants (Velazquez *et al.* 1998), fluid instabilities (e.g. Rayleigh-Taylor) are likely to develop at the discontinuity between the two shock fronts due to the difference between the densities of the impacting wind and of the impacted medium. A condition for the Rayleigh-Taylor instability to grow is that the mass of the ISM that is pushed by the discontinuity is higher than the mass of the ejecta (Velazquez *et al.* 1998), which undergoes to a deceleration process that is able to trigger instabilities in the fluid (see Fig. 3). Such instabilities would easily alter the dynamics of the shocked outflow and they may give rise to a re-distribution of the overall velocity field in which slower (and faster) components of the wind may simultaneously cross our line of sight. The effect of the turbulence would then provide a replenishment mechanism that continuously supply new sections/blobs of gas that might as well fall backwards with an opposite direction with respect to the bulk of the outflow.

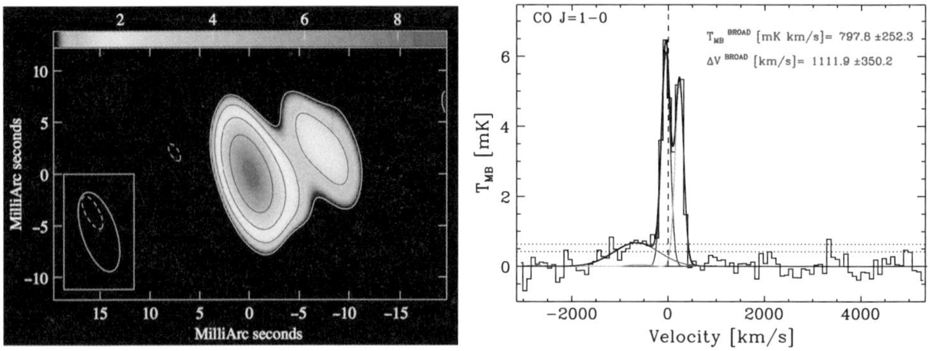

Figure 4. *Left*: Simultaneous 5-8 GHz spectral index VLBA image of IRAS17020+4544 observed in 2014. Contours show total intensity at 8 GHz and colours show the spectral index values (Giroletti *et al.* 2017). *Right*: Spectrum of the CO(1-0) line in IRAS17020+4544 obtained with the LMT telescope in 2017. The double peaked structure is fitted by two narrow Gaussian components (blue and green), and the line wings are fitted by broad Gaussian line (red line). The molecular gas mass is estimated assuming a CO-to-H2 conversion factor appropriate for ULIRGs (Solomon *et al.* 1997). The molecular gas masses estimated in each narrow component are very similar, which is compatible with the presence of a ring of molecular material in the galaxy, while the blue wing of the line is consistent with the presence of molecular gas outflowing at ~660 km s^{-1} (Longinotti *et al.* submitted).

We have developed a toy model for simulating this behavior and testing the viability of this idea. The simulation is based on the GUACHO code (Esquivel & Raga 2013) and a preliminary output is displayed in Fig. 3. An expanding shock is pushed within a turbulent medium by an inner wind with v_{out}=20,000 km s^{-1}. The expansion time is set to 20 yr, the mass of the central object is 10^6 M$\odot$ (i.e. the mass of IRAS17020+4544 and in general of the order of black hole masses in NLSy1 galaxies).

Fig. 3 shows the pattern of velocities developed at the shock front with and without instability. The right panel shows that "plumes" or "fingers" of gas with different velocities are formed as a result of the introduction of instabilities process in the expansion of the shocked outflow. Depending on which "fingers" are intercepted by our line of sight, it is evident that the resulting absorption line spectrum is likely to show a stratified wind consistent with the spectrum observed in IRAS17020+4544 (see Fig. 2). This scenario can easily apply to other NLSy1 sources with the same observational properties.

4. A truly multi-phase outflow in IRAS17020+4544?

Further hints to the presence of a shocked outflow in this source come from radio observations (see Fig. 4) that have revealed an elongated structure on a scale of 10 pc in VLBI images (Giroletti *et al.* 2017). The appealing possibility that such compact jet, which is produced by synchrotron emission, may represent the signature left by the shock of the inner X-ray outflow with the ambient gas has been postulated by several models of galactic outflows (e.g. Zakamska & Greene 2014; Nims *et al.* 2015) in an attempt to link synchrotron emission in radio-quiet sources with galaxy scale molecular outflows.

IRAS17020+4544 is hosted by a barred spiral galaxy with IR luminosity typical of LIRG (L_{FIR}= $1.05\times10^{11} L_{\odot}$) therefore likely to be rich in molecular gas. We have obtained milimetric observations with the Large Millimetric Telescope (LMT, Hughes *et al.* 2010) and found tentative evidence for a galaxy scale outflow traced by CO gas (Fig. 4, Longinotti *et al.* submitted).

Assuming a dynamical timescale of few$\times10^6$ years for the outflow to propagate out of the nucleus at the observed bulk velocity of ~600 km s^{-1}, the spatial scale of the wind is approximately ~0.5-3 kpc. The CO mass in the wind is of the order of 10^8 M⊙ and following Feruglio *et al.* 2015, we find an approximate estimate of the molecular outflow rate of ~100 $M_{\odot}\mathrm{yr}^{-1}$, which is consistent with recent findings reported for Active Galaxies with comparable CO properties (Cicone *et al.* 2014)

The relative proportion of momentum load for the X-ray and molecular outflows in IRAS17020+4544 would provide an additional evidence for the existence of energy conserving winds that propagate through the galaxy after undergoing a momentum boost, supporting the feedback scenario as proposed in earlier works (Feruglio *et al.* 2015).

Admittedly, these numbers suffer from the large uncertainties in the estimates of the mass outflow rate and of the wind spatial extent, therefore we are currently awaiting the outcome of an already performed interferometry observation to constrain the size of the molecular outflow region and its physical properties.

Acknowledgments

The author wishes to thank the organizers of the IAU symposium 342 "Perseus in Sicily: from black hole to cluster outskirts". Financial support from the "Coordinación de Astrofísica" of INAOE Institute is gratefully acknowledged. The author thanks the many colleagues who, through their constant streaming of input and discussion, have actively collaborated to the work presented at this conference: Yair Krongold, Pablo Velazquez, Mario Sanfrutos at UNAM (Mexico City), Olga Vega and Vahram Chavushyan at INAOE (Puebla), Matteo Guainazzi (ESA-ESTEC, Leiden), Marcello Giroletti (INAF-IRA, Bologna), Francesca Panessa (INAF-IAPS, Roma), Elisa Costantini (SRON, Utrecht).

References

Chartas, G., Brandt, W. N., Gallagher, S. C., & Garmire, G. P. 2002, *ApJ*, 579, 169
Chartas, G., Saez, C., Brandt, W. N., Giustini, M., & Garmire, G. P. 2009, *ApJ*, 706, 644
Cicone, C., Maiolino, R., Sturm, E., *et al.* 2014, *A&A*, 562, A21
Di Matteo, T., Springel, V., & Hernquist, L. 2005, *Nature*, 433, 604
Esquivel, A., & Raga, A. C. 2013, *ApJ*, 779, 111
Faucher-Giguère, C.-A., & Quataert, E. 2012, *MNRAS*, 425, 605
Feruglio, C., Maiolino, R., Piconcelli, E., *et al.* 2010, *A&A*, 518, L155
Feruglio, C., Fiore, F., Carniani, S., *et al.* 2015, *A&A*, 583, A99
Fukumura, K., Tombesi, F., Kazanas, D., *et al.* 2015, *ApJ*, 805, 17
Gallo, L. C. 2006, *MNRAS*, 368, 479
Giroletti, M., Panessa, F., Longinotti, A. L., *et al.* 2017, *A&A*, 600, A87
Gofford, J., Reeves, J. N., Tombesi, F., *et al.* 2013, *MNRAS*, 430, 60
Gupta, A., Mathur, S., Krongold, Y., & Nicastro, F. 2013, *ApJ*, 772, 66
Hopkins, P. F., & Elvis, M. 2010, *MNRAS*, 401, 7
Hughes, D. H., Jáuregui Correa, J.-C., Schloerb, F. P., *et al.* 2010, Proceeding of SPIE, 7733, 773312
King, A. R. 2010, *MNRAS*, 402, 1516
King, A., & Pounds, K. 2015, *ARAA*, 53, 115
Komossa, S., & Xu, D. 2007, *ApJL*, 667, L33
Krongold, Y., Nicastro, F., Elvis, M., *et al.* 2007, *ApJ*, 659, 1022
Lanzuisi, G., Giustini, M., Cappi, M., *et al.* 2012, *A&A*, 544, A2
Longinotti, A. L., Krongold, Y., Guainazzi, M., *et al.* 2015, *ApJL*, 813, L39
Matzeu, G. A., Reeves, J. N., Braito, V., *et al.* 2017, *MNRAS*, 472, L15
Nims, J., Quataert, E., & Faucher-Giguère, C.-A. 2015, *MNRAS*, 447, 3612
Nardini, E., Reeves, J. N., Gofford, J., *et al.* 2015, *Science*, 347, 860
Parker, M. L., Pinto, C., Fabian, A. C., *et al.* 2017, *Nature*, 543, 83
Pinto, C., Alston, W., Parker, M. L., *et al.* 2018, *MNRAS*, 476, 1021
Pounds, K. A., Reeves, J. N., King, A. R., *et al.* 2003, *MNRAS*, 345, 705

Pounds, K. A., & Vaughan, S. 2011, *MNRAS*, 413, 1251
Pounds, K. A. 2014, *MNRAS*, 437, 3221
Proga, D., & Kallman, T. R. 2004, *ApJ*, 616, 688
Reeves, J. N., Braito, V., Nardini, E., *et al.* 2016, *ApJ*, 824, 20
Sim, S. A., Proga, D., Miller, L., Long, K. S., & Turner, T. J. 2010, *MNRAS*, 408, 1396
Solomon, P. M., Downes, D., Radford, S. J. E., & Barrett, J. W. 1997, *ApJ*, 478, 144
Tombesi, F., Cappi, M., Reeves, J. N., *et al.* 2010, *A&A*, 521, A57
Tombesi, F., Cappi, M., Reeves, J. N., & Braito, V. 2012, *MNRAS*, 422, L1
Tombesi, F., Melendez, M., Veilleux, S., *et al.* 2015, *Nature*, 519, 436
Velazquez, P. F., Gomez, D. O., Dubner, G. M., de Castro, G. G., & Costa, A. 1998, *A&A*, 334, 1060
Zakamska, N. L., & Greene, J. E. 2014, *MNRAS*, 442, 784
Zubovas, K., & King, A. 2012, *ApJL*, 745, L34

Perseus in Sicily: From Black Hole to Cluster Outskirts
Proceedings IAU Symposium No. 342, 2018
K. Asada, E. de Gouveia Dal Pino, M. Giroletti,
H. Nagai & R. Nemmen, eds.

doi:10.1017/S1743921318003745

AGN feedback and the origin and fate of the hot gas in early-type galaxies

Silvia Pellegrini[1], Luca Ciotti[1], Andrea Negri[2] and Jeremiah P. Ostriker[3]

[1]Department of Physics and Astronomy, University of Bologna, via Gobetti 93/2, 40129 Bologna, Italy
email: silvia.pellegrini@unibo.it

[2]Instituto de Astrofísica de Canarias, calle Vía Láctea, E-38205 La Laguna, Tenerife, Spain

[3]Department of Astronomy, Columbia University, 550 West 120th St, New York, NY 10027, USA

Abstract. We present the results of two-dimensional, grid-type hydrodynamical simulations, with parsec-scale central resolution, for the evolution of the hot gas in isolated early-type galaxies (ETGs). The simulations include a physically self-consistent treatment of the mechanical (from winds) and radiative AGN feedback, and were run for a large set of realistic galaxy models. AGN feedback proves to be very important to maintain massive ETGs in a time-averaged quasi-steady state, keeping the star formation at a low level, and the central black hole mass on observed scaling relations. A comparison with recent determinations of the X-ray properties of ETGs in the local universe shows that, at later epochs, AGN feedback does not dramatically alter the gas content originating in stellar recycled material. Thus, the present-day X-ray luminosity is not a robust diagnostic of the impact of AGN activity, within a scenario where the hot gas mostly originates from the stellar population.

Keywords. ISM: evolution, galaxies: elliptical and lenticular, cD, galaxies: evolution, galaxies: ISM, galaxies: nuclei, X-rays: galaxies, X-rays: ISM

1. Introduction

A few basic facts relate the supermassive black holes (MBHs), the ISM, and the stellar population of ETGs. The first is that ETGs host central MBHs, whose mass M_{BH} scales with the stellar mass, and was mostly built by accretion during bright QSO phases (at least for $M_{BH} \geq 10^8 M_\odot$; Yu & Tremaine 2002). The second is that star formation (SF) stopped at early times, and local ETGs show very low levels of SF, and little (if any) young stellar population (e.g., Kuntschner *et al.* 2010, Davis *et al.* 2014). The third is that ETGs contain a hot ISM, whose X-ray properties have been accurately determined from *Chandra* observations (Kim & Fabbiano 2015; Goulding *et al.* 2016). This ISM is continuously replenished by the collective mass input provided by the stellar population during its normal ageing; this mass input rate $\dot{M}$ and its evolution [$\dot{M}$(t)$\approx 10^{-11} L_B(L_{B\odot})$ t(12 Gyrs)$^{-1.3} M_\odot$/yr, with L_B the galaxy luminosity, for a passively evolving stellar population, after ~ 2 Gyr of age] is accurately known from theory and observations; it corresponds to a significant fraction of the initial stellar mass of the ETG, when integrated over its lifetime. The fourth, final fact is that this continuous mass input develops a flow directed towards the galactic center (e.g., Ciotti *et al.* 1991) that feeds the MBH and causes the "AGN feedback" phenomenon.

Basic questions are then raised: can the MBH accretion energy prevent MBH masses from growing too much, and remain comparable to those built by the end of the quasar phase? How does the accretion energy interact with the galactic ISM? How much of it is absorbed, and with what effects on the ISM? Is the gas displaced from the galactic central regions, and even removed from the galaxy? Does the gas content originating in the stellar population have an X-ray luminosity L_X that agrees with observations? Does SF remain low? To address these questions, we ran high-resolution 2D hydrodynamical simulations with an adapted grid-type code (ZEUS-MP) to study the evolution of the ISM in presence of stellar and AGN feedback. The main results are briefly presented below (see Ciotti *et al.* 2017, hereafter C17, and Pellegrini *et al.* 2018, for more details).

2. The models

The main features of the simulations and of the numerical implementation of the various inputs of mass and energy to the ISM are found in our recent works (Negri *et al.* 2014, 2015, C17; see Ciotti & Ostriker 2012 for a detailed description of the realization of all input sources to the hydrodynamical equations, and of AGN feedback).

The simulations have been run for a large set of axisymmetric two-component (star+dark matter) galaxy models, of shapes ranging from E0 to E7. The Jeans equations provide the internal stellar kinematics, on which the stellar kinematical heating is then based. The stellar density profile follows a deprojected ellipsoidal Sersic law, and the main galactic observables are related in order to lie on the main scaling laws. The spherical, NFW dark matter halo has a dark-to-stellar mass ratio of $M_h/M_\star \simeq 20$, and a dark mass content within the effective radius R_e lower than that in stars, following the results of dynamical modeling of kinematic observations of ETGs. The hot gas originates from stellar mass losses and SN ejecta, and evolves under the action of gravity, of cooling, and of the various energy sources. SF is also included, via a simple scheme shown to reproduce well the Kennicutt-Schmidt relation (Negri et al. 2015).

The models include an accurate and physically self-consistent implementation of AGN feedback, both radiative and mechanical, the latter due to AGN winds. Thanks to the central resolution (5 pc) the fiducial accretion radius is resolved (e.g., Pellegrini 2010, Ciotti & Pellegrini 2017). The heating by AGN feedback results from the mass accretion rate on the MBH directly computed with the high central resolution, from a self-consistent treatment of the mass, energy and momentum balance of the inflowing and outflowing material at the innermost gridpoint, and from radiative and mechanical efficiencies that vary with the mass accretion rate, in agreement with current observational and theoretical findings (C17). The strength of AGN feedback is then not "adjusted", since the heating of the ISM resulting from the accretion process (that is resolved down to the accretion radius) is self-determined.

In the simulations the galaxies start almost empty of ISM, a situation expected to result from the combined effects of the SNe explosions and of the AGN energy injection, that are believed to clear the galaxies from the ISM and 'quench' star formation (e.g., Pellegrini *et al.* 2018 and references therein). The ISM is then replenished by the collective input provided by the stellar population, and its evolution is followed from an age of ~ 2 Gyr (after the main galaxy formation phase), for ~ 10 Gyr. The simulations were run for a large set of representative (isolated) galaxy models, varying their $M_\star$, their shape, internal kinematics, presence of AGN feedback and star formation.

3. Results

Many simulation outputs are considered by C17: maps of the hydrodynamical properties of the gas, surface brightness maps, maps of the SF rate, of the mass in newly formed

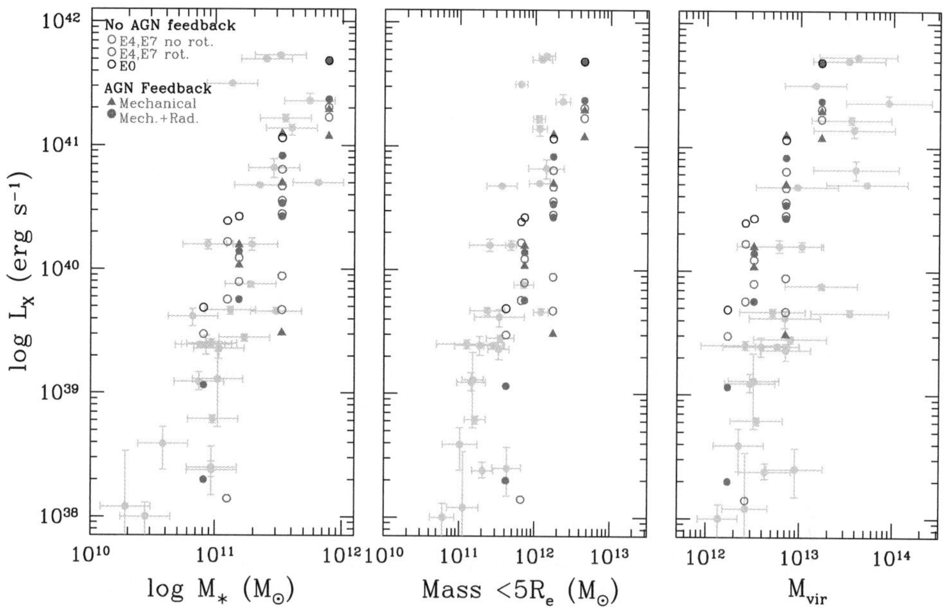

Figure 1. Observed (green) and model X-ray luminosities versus the total stellar mass $M_\star$ (left), the total enclosed mass within $5R_e$ (middle), and the total virial mass M_{vir} (right). Observed L_X values for ETGs are taken from Forbes *et al.* (2017). The models have E0, E4 and E7 shapes; full symbols refer to non-rotating galaxies with AGN feedback (mechanical, or mechanical and radiative); open symbols refer to simulations without AGN feedback, and in this case the E4 and E7 galaxies can be rotating (blue open circles) or not (red open circles). See Pellegrini *et al.* 2018 for more details.

stars, the distribution in age of the new stars, the growth of the MBH mass, the nuclear luminosity, the duty cycle of AGN activity and of the gas. At the nucleus, high and low radiative efficiency accretion periods alternate; a negligible time is spent in the (cold mode) phase of high radiative efficiency, and most of the time is spent in a "quiescent" (hot mode) low nuclear luminosity phase; in an average ETG, the Eddington-scaled mass accretion rate keeps below 10^{-3} for $\simeq 75\%$ of the time, as observed in the local universe (Pellegrini 2010, Gallo *et al.* 2010). The duty-cycle of AGN activity is a few percent. At the end, $M_{\rm BH}$ has grown by a factor of 2–3, instead of a factor of $\simeq 100$, as found without AGN feedback. During nuclear outbursts, cold, inflowing, and hot, outflowing gas phases coexist in the central galactic region; overall, AGN activity has a positive effect for star formation, as already found previously (Ciotti & Ostriker 2007; Nayakshin & Zubovas 2012). At the end of the simulations, in a representative galaxy, roughly half of the total mass loss is recycled into new stars, and just 3% of it is accreted on the MBH, the remainder being ejected from the galaxy.

At the present epoch, L_X, and the emission weighted temperature T_X, agree well with those observed (see Fig. 1). In the figure the models successfully reproduce even the largely different L_X values at the same galaxy mass, thanks to the sensitivity of the gas flow to many galaxy properties (C17). Another important finding is that there is not a significant difference in L_X of models with and without AGN feedback, when a large set of galaxy models is considered. The basic explanation for this finding is that recurrent feedback helps to temporarily displace the gas from the center (out to a radius of the order of 10 kpc), thus L_X is temporarily reduced, even considerably; but feedback does not clear the galaxy from all its gas, it just produces an increase (of the order of 20%–40%) in the ejected mass from the galaxy (C17).

4. Conclusions

We performed simulations of the ISM behavior in ETGs starting from initial conditions created at the epoch of quenching ($z > 2$), when SNe and AGN-driven powerful outflows are supposed to remove the gas residual of the major star formation process, and stop further intense SF. The mass input to the ISM comes from supernovae and stellar winds, takes into account their secular trend, and the contributions of both the ageing and the newly born stars.

We found that after $z \simeq 2$ AGN feedback is very important to maintain massive ETGs in a time-averaged quasi-steady state, keeping low the SF, and the MBH mass; however, it does not dramatically alter the gas content originating in stellar recycled material, that is able to account for a major part of the observed L_X. Thus, the distribution of values for the present-day L_X is not a robust diagnostic of the impact of AGN activity, within a scenario where the hot gas mostly originates from the stellar population.

Evidently, if AGN-driven outflows are to clear the gas out of massive ETGs at early epochs, this requires an efficiency of energy transfer from the AGN to the ISM much larger than found by C17 at low redshift. Similarly, if cosmological infall at later epochs is very important, contributing to a mass within a few R_e much larger than that provided by stars, then again the effects of AGN feedback should be stronger than found by C17, to prevent substantial cosmological accretion of material from outside the ETGs, and then an exceedingly large L_X. Further work including the confining action exerted by an intragroup/intracluster medium in central-dominant ETGs, or cosmological infall, would be helpful to assess more thoroughly the origin of the hot gas, and the role of AGN feedback in ETGs.

References

Ciotti L., D'Ercole A., Pellegrini S., & Renzini A., 1991, *ApJ*, 376, 380
Ciotti, L., & Ostriker, J. P. 2007, *ApJ*, 665, 1038
Ciotti, L., & Ostriker, J. P. 2012, in *Hot Interstellar Matter in Elliptical Galaxies*, Kim D.-W., Pellegrini S., eds., Astrophys. and Space Sci. Library, vol. 378. Springer-Verlag, p. 8
Ciotti, L., Pellegrini, S., Negri, A., & Ostriker, J. P. 2017, *ApJ*, 835, 15 (C17)
Ciotti, L., & Pellegrini, S., 2017, *ApJ*, 848, 29
Davis T. A. *et al.*, 2014, *MNRAS*, 444, 3427
Forbes, D. A., Alabi, A., Romanowsky, A. J., *et al.* 2017, *MNRAS*, 464, L26
Gallo, E., Treu, T., Marshall, P. J., *et al.* 2010, *ApJ*, 714, 25
Goulding, A. D., Greene, J. E., Ma, C.-P., *et al.* 2016, *ApJ*, 826, 167
Kim, D.-W., & Fabbiano, G., 2015, *ApJ*, 812, 127
Kuntschner, H., *et al.* 2010, *MNRAS*, 408, 97
Nayakshin, S., & Zubovas, K. 2012, *MNRAS*, 427, 372
Negri, A., Posacki, S., Pellegrini, S., & Ciotti, L. 2014, *MNRAS*, 445, 1351
Negri, A., Pellegrini, S., & Ciotti, L. 2015, *MNRAS*, 451, 1212
Pellegrini, S. 2010, *ApJ*, 717, 640
Pellegrini, S., Ciotti, L., Negri, A., & Ostriker, J. P. 2018, *ApJ*, 856, 115
Yu, Q., & Tremaine, S. 2002, *MNRAS*, 335, 965

Perseus in Sicily: From Black Hole to Cluster Outskirts
Proceedings IAU Symposium No. 342, 2018
K. Asada, E. de Gouveia Dal Pino, M. Giroletti,
H. Nagai & R. Nemmen, eds.

doi:10.1017/S174392131800371X

Numerical study of active galactic nucleus feedback in an elliptical galaxy with *MACER*

Feng Yuan[1], Jeremiah P. Ostriker[2], DooSoo Yoon[1], Ya-Ping Li[1], Luca Ciotti[3], Zhao-Ming Gan[1], Luis C. Ho[4,5] and Fulai Guo[1]

[1]Shanghai Astronomical Observatory, Chinese Academy of Sciences,
80 Nandan Road, Shanghai 200030, China
email: fyuan@shao.ac.cn

[2]Department of Astronomy, Columbia University, 550 W. 120th Street, New York, USA

[3]Department of Physics and Astronomy, University of Bologna, 40129 Bologna, Italy

[4]Kavli Institute for Astronomy and Astrophysics, Peking University, Beijing 100871, China

[5]Department of Astronomy, School of Physics, Peking University, Beijing 100871, China

Abstract. This paper summarizes our recent works of studying AGN feedback in an isolated elliptical galaxy by performing high-resolution hydrodynamical numerical simulations. Bondi radius is resolved and the mass accretion rate of the black hole is calculated. The most updated AGN physics, namely the discrimination of cold and hot accretion modes and the exact descriptions of the AGN radiation and wind for a given accretion rate are adopted and their interaction with the gas in the host galaxy is calculated. Physical processes such as star formation and SNe feedback are taken into account. Consistent with observation, we find the AGN spends most of the time in the low-luminosity regime. AGN feedback overall suppresses the star formation; but depending on location in the galaxy and time, it can also enhance it. The light curve of specific star formation rate is not synchronous with the AGN light curve. These results put a serious challenge to the observational test of the relation between AGN activity and star formation. We find that wind usually plays a dominant role in controlling the AGN luminosity and star formation, but radiation also cannot be neglected.

Keywords. accretion, accretion disks – black hole physics – galaxies: evolution – galaxies: nuclei

---

1. Introduction

There is growing evidence for the coevolution of central supermassive black holes and their host galaxies (see review by Kormendy & Ho 2013). It is generally believed that active galactic nucleus (AGN) feedback plays an important role in the evolution of galaxies from both observational and theoretical arguments (see review by Fabian 2012; King & Pounds 2015; Naab & Ostriker 2017). The basic scenario of the AGN feedback is illustrated by Figure 1. The outputs from the AGN, namely radiation, wind, and jet, interact with the interstellar medium (ISM) in the host galaxy and change its density and temperature. Consequently, star formation rate will be changed, which changes the evolution of the galaxy. On the other hand, the change of the gas properties will in turn affect the fueling of the AGN. Obviously, the most crucial factor to determining the effects of AGN feedback is the outputs from AGN, which is determined by the value of AGN accretion rate and the AGN outputs for a given accretion rate.

There have been many works on AGN feedback using hydrodynamical numerical simulation. Most of these work focus on very large spatial scales, their resolution is typically

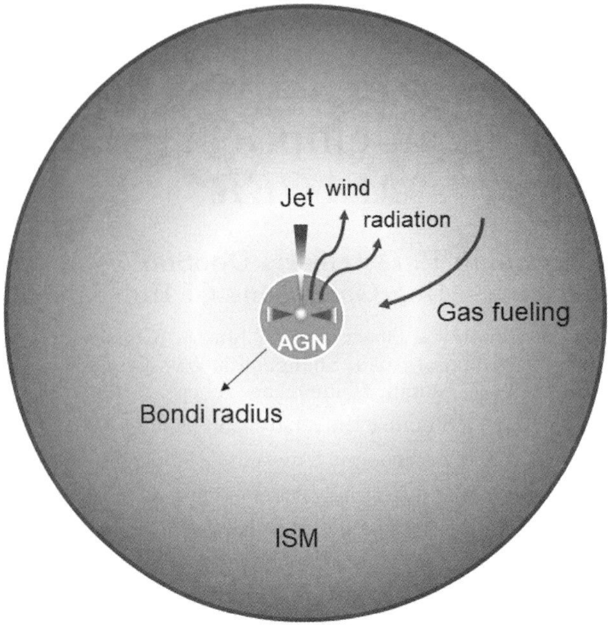

Figure 1. Schematic figure of AGN feedback.

several *kpc* or even larger. Thus it is difficult to resolve the Bondi radius, which is typically several tens of *pc*. In this case, they have to estimate the accretion rate, which may have an uncertainty as large as ~ 300 (Negri & Volonteri 2017; see also Korol *et al.* 2016 for an analytical discussion). Moreover, although after several decades' effort we now have quite robust understanding to accretion physics, these physics unfortunately have not been properly adopted in most current works of AGN feedback. Lastly, the interaction of the AGN outputs and the ISM is also not carefully calculated.

In our works, we study AGN feedback by focusing on a single galaxy using our high-resolution two-dimensional numerical simulations. There are three key features in our model. First, the inner boundary of the simulation domain is ten times smaller than Bondi radius so we can precisely calculate the accretion rate. Second, we adopt the most updated AGN physics. Third, we carefully calculate the interaction between AGN outputs and ISM.

2. AGN accretion physics

Since the inner boundary of our simulation domain ($r_{\rm in}$) is $\sim 0.1 R_{\rm Bondi}$, we do not need to use the Bondi formula, which is an approximation, to calculate the accretion rate. Instead, the accretion rate at $r_{\rm in}$ can be directly calculated by the following precise equation (we use spherical coordinates):

$$\dot{M}(r_{\rm in}) = 2\pi r_{\rm in}^2 \int_0^{\pi} \rho(r_{\rm in}, \theta) \, \min\left[v_r(r_{\rm in}, \theta), 0\right] \sin\theta \, d\theta. \tag{2.1}$$

According to the black hole accretion theory and observations of black hole X-ray binaries (Yuan & Narayan 2014 and references therein), black hole accretion is divided into "cold" and "hot" modes. The boundary between them is $\dot{M} \approx 2\% \dot{M}_{\rm Edd} (\equiv 20\% L_{\rm Edd}/c^2)$ assuming a 10% radiative efficency. In the following, we describe radiation and wind in each mode. Jet is neglected since we assume it deposits very little energy in the galaxy. It is necessary to examine this assumption in future work.

2.1. *Cold accretion mode*

Within $r_{\rm in}$, the accretion still cannot be resolved and must be treated as subgrid physics. The infalling gas initially will freely fall until a small disk is formed. So there will be a time lag and difference of values between the accretion rate in the small disk ($\dot{M}_{\rm BH}$) and $\dot{M}(r_{\rm in})$ (see Yuan *et al.* 2018 for details). The luminosity emitted from the disk is $L_{\rm BH} = \epsilon_{\rm cold}\dot{M}_{\rm BH}c^2$. The corresponding Compton temperature needed to calculate the Compton heating is $T_{\rm C,cold} = 2 \times 10^7$K.

Winds in the cold mode have been widely observed in many sources, so we have good observational constraint on the properties of wind. In our work, we adopt the results compiled in Gofford *et al.* (2015). The mass flux and velocity of wind are described by:

$$\dot{M}_{\rm W,C} = 0.28 \left(\frac{L_{\rm BH}}{10^{45}\,{\rm erg\,s^{-1}}}\right)^{0.85} M_\odot\,{\rm yr}^{-1}, \tag{2.2}$$

$$v_{\rm W,C} = 2.5 \times 10^4 \left(\frac{L_{\rm BH}}{10^{45}\,{\rm erg\,s^{-1}}}\right)^{0.4} {\rm km\,s^{-1}}. \tag{2.3}$$

In Gofford *et al.* (2015), the winds are detected at a distance of $10^{2-4}r_s$ from the black hole, which is roughly consistent with the value of the inner boundary of our simulation.

2.2. *Hot accretion mode*

In the hot accretion mode, the accretion flow consists of an outer truncated thin disk and an inner hot accretion flow. The truncation radius depends on the mass accretion rate (Yuan & Narayan 2014),

$$r_{\rm tr} = 3r_s \left[\frac{2 \times 10^{-2}\dot{M}_{\rm Edd}}{\dot{M}(r_{\rm in})}\right]^2. \tag{2.4}$$

The dynamics and radiation of the hot accretion flow have been intensively studied and are well understood (see review by Yuan & Narayan 2014). Different from the thin disk, the radiative efficiency of a hot accretion flow is a function of accretion rate (Xie & Yuan 2012):

$$\epsilon_{\rm EM,hot}(\dot{M}_{\rm BH}) = \epsilon_0 \left(\frac{\dot{M}_{\rm BH}}{0.1L_{\rm Edd}/c^2}\right)^a, \tag{2.5}$$

the values of ϵ_0 and a are given in Xie & Yuan (2012). Note that the efficiency is comparable to that of the thin disk when the accretion rate is high. The radiation emitted from a hot accretion flow has relatively more hard photons compared to a thin disk; thus, the Compton temperature is higher, $T_{\rm C,hot} \approx 10^8$K (Xie, Yuan & Ho 2017).

For the wind in the hot accretion mode, although we are accumulating more and more observational evidences (e.g., Wang *et al.* 2013; Cheung *et al.* 2016), most of them are indirect evidences, and so it is hard to directly get the observational constraint on the properties of wind. By contrast, we have very good theoretical understanding to the wind from hot mode (see review in Yuan *et al.* 2015; Yuan & Narayan 2014). In particular, the mass flux and velocity of wind have been well studied based on three-dimensional general relativity MHD numerical simulations (Yuan *et al.* 2015),

$$\dot{M}_{\rm W,H} \approx \dot{M}(r_{\rm in}) \left[1 - \left(\frac{3r_s}{r_{\rm tr}}\right)^{0.5}\right], \tag{2.6}$$

$$v_{\rm W,H} \approx (0.2 - 0.4)v_{\rm K}(r_{\rm tr}). \tag{2.7}$$

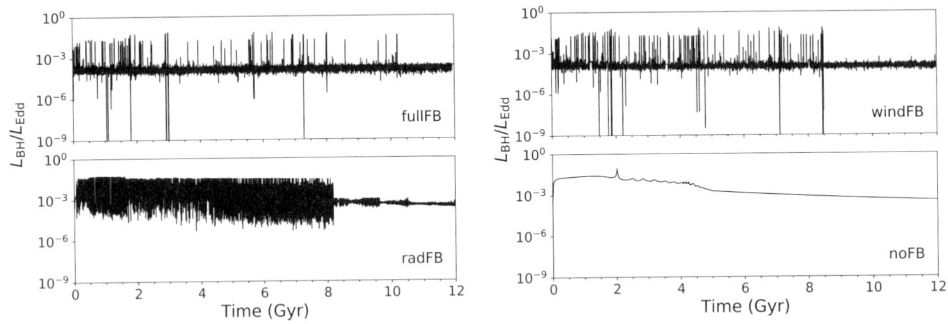

Figure 2. Light curves of AGN luminosity as a function of time for various models. For clarity, we choose the data point so that the two adjacent ones have a relatively large time interval of 2.5 Myr; in this case, some outbursts are filtered out. Taken from Yuan *et al.* (2018).

3. Model

The evolution of the galactic gas flow, with the effect of AGN feedback, is described by the conservation of mass, momentum, and energy (e.g., Ciotti & Ostriker 2012; Ciotti *et al.* 2017; Pellegrini *et al.* 2018). We solve them by the parallel ZEUS code using two-dimensional axisymmetric spherical coordinates. The radial direction of simulation domain covers the range of 2.5 pc – 250 kpc, the finest resolution is at the innermost grid, ~0.3 pc. The wind and radiation are injected at the inner boundary, and their interaction with ISM is calculated. Other physical processes such as the radiative cooling, star formation, and SNe are taken into account. We name our code *MACER* (*M*assive *A*GN *C*ontrolled *E*llipticals *R*esolved).

We have investigated two cases, with the specific angular momentum of the has being low and high in Yuan *et al.* (2018) and Yoon *et al.* (2018), respectively. In this paper, we will take Yuan *et al.* (2018) as an example when we introduce the simulation results.

4. Results

We have simulated several models: noFB (no AGN feedback included), fullFB (both wind and radiation feedback included), windFB (only wind), and radFB (only radiation). The main results are summarized below.

4.1. *AGN light curve*

Figure 2 shows the AGN light curves. For the noFB model, the light curve is featureless; it gradually decreases because of the depletion of gas in star formation. Once feedback is included, the AGN luminosity strongly fluctuates. This is because, when the AGN is luminous, strong radiation and wind will be emitted, pushing the surrounding gas away and heating it. Subsequently, the accretion rate of the AGN will substantially decrease, and the AGN will dim. The surrounding gas will then gradually cool by radiation and become dense, and so the accretion rate will increase. From the figure, we can see that the AGN spends most of its time in the low-luminosity phase (hot mode), with a typical luminosity $L \sim 10^{-4} L_{\rm Edd}$.

The variability amplitudes for the radFB and windFB models are similar, and both of them are similar to the fullFB model. This indicates that wind and radiation feedback cause a similar amplitude of variability. However, in the time-average sense, the AGN luminosity in the radFB model is $10^{-2} L_{\rm Edd}$, almost two orders of magnitude higher than that in the windFB model. The main reason for such a difference is that wind can deposit their momentum much more efficiently than radiation (Yuan *et al.* 2018).

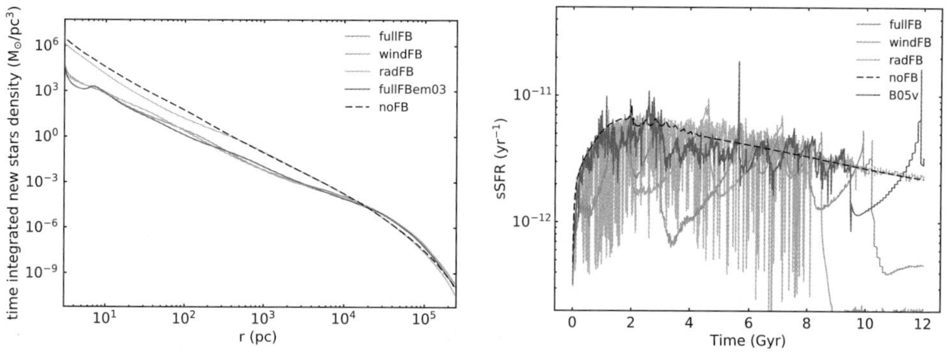

Figure 3. Effects of AGN feedback on star formation for various models. Left panel: Time-integrated mass of newly born stars at a given radius per unit volume. Right panel: Specific star formation rate over time for various models. Taken from Yuan *et al.* (2018).

We can see that in the fullFB model, the baseline AGN luminosity is very similar to that of the windFB model. This indicates that the mass accretion rate of the black hole is controlled by the wind feedback rather than the radiation. However, when we carefully compare the zoom-in plots of the fullFB and windFB models, we find that there are more outbursts in the fullFB model than in the windFB model (Yuan *et al.* 2018). This indicates that feedback by radiation and wind are coupled together, and neither of them can be neglected. At last, the light curve amplitude suddenly decreases after $\sim$ 8 Gyr. The large amplitude is mainly because of the strong "perturbation" of the strong wind in the cold feedback mode. Due to the gradual loss of the gas in the galaxy, after ~ 8 Gyr, the AGN fails to reach the cold accretion mode and the strong wind disappears.

We have also calculated the AGN duration (lifetime) using our simulation data. We find that the typical value is $\sim 10^5$ yr. This is consistent with some observations (e.g., Schawinski *et al.* 2015). We note that the typical lifetime obtained in Gan *et al.* (2014), which has the same model framework but different AGN physics, is almost two orders of magnitude higher. This indicates the impact of specific AGN physics adopted in the feedback model.

4.2. *Mass growth of the black hole*

We have calculated the growth of the black hole mass. When no AGN feedback is included, we confirm (see e.g., Ciotti *et al.* 2017) that the mass can easily reach above $10^{10} M_\odot$, which is obviously too large. In the radFB model, interestingly, we find that the black hole mass becomes even larger. This is because when we include radiation, the star formation, which can deplete some gas in the galaxy, becomes weaker due to the radiative heating. Therefore there will be more gas left to fuel the black hole. In the windFB model, the final black hole mass becomes substantially smaller, $\sim 2 \times 10^9 M_\odot$, only slightly larger than the initial value. This indicates that wind plays a dominant role in suppressing the accretion rate and the black hole mass. Finally, for the fullFB model, the black hole mass is slightly higher than that in the windFB model. The reason is that star formation becomes weaker when radiative feedback is included.

4.3. *Star formation*

The left plot of Figure 3 shows the time-integrated total mass of newly born stars per unit volume as a function of radius. In the radFB model, within several hundred pc, star formation is slightly suppressed compared to noFB model. In the windFB model, we can see that star formation is strongly suppressed all the way up to ~ 20kpc. This is because of the momentum feedback of the wind, i.e., winds push the gas away from the central region beyond ~ 20kpc. We note that the Gemini integral field unit observations by Liu *et al.* (2013) do find the wind can reach that distance. The result of fullFB model is very similar to windFB, indicating that wind plays a dominant role in controlling star formation. At the region of $r \gtrsim 20$kpc, the gas is accumulated there, so star formation is slightly enhanced.

The right plot of Fig. 3 shows the evolution of specific star formation rate (sSFR), i.e., the star formation rate normalized by the stellar mass of the galaxy. We find three important results. First, by comparing the windFB and fullFB models, we can see that their general patterns are similar, but their light curves are not synchronous with each other. There is an obvious offset between them, and the "amplitudes" of the fullFB model are also larger. This indicates that, although in the time-integrated sense the wind seems to be much more important than radiation in suppressing star formation, radiation also plays a very important role. The wind and radiation couple together in affecting star formation. Second, by comparing the fullFB and noFB models, we can see that the sSFR in fullFB model is in general suppressed compared to the noFB model, but occasionally the sSFR can also be enhanced. Third, by comparing the light curve of sSFR of the fullFB model with the AGN light curve shown in Fig. 2, we find that they are not synchronous with each other. This is partly because the timescale of star formation episodes ($\tau_{SF} \gtrsim 100$Myr) is much longer than the timescale of AGN activity ($\tau_{\rm AGN} \lesssim 1$Myr) (e.g., Harrison 2017). The last two results put a serious challenge to the observational test of the effect of AGN feedback on star formation, and this may explain why the conclusion reached by observational studies of the relation between AGN and star formation is so diverse (see review by Xue 2017).

4.4. *AGN duty cycle*

From the AGN light curve, we can calculate the AGN duty cycle, which is defined as the percentage of the total time of AGN spent below or above a given Eddington ratio. From the light curve shown in Fig. 2, we can see that the AGN must spend most of the time in the low-luminosity regime, i.e, the hot accretion (feedback) mode. In fact, we find that AGN spends over 80% of its evolution time with Eddington ratios below 2×10^{-4}. We have compared the simulation result with the observational data which are available for low-redshift sources and general consistency is found. These results suggest the potential importance of feedback effect by low-luminosity AGNs.

Observations show that, although AGNs spend most of their time in the low-luminosity AGN phase, they emit most of their energy during the high-luminosity phase. To compare with this observation, we have calculated the percentage of the total energy emitted above or below a given Eddington ratio. Unfortunately, we find that the AGN only emits 6% of the entire energy at the Eddington ratio above 0.02, which is not consistent with observation. What is the reason? Note that in our current model we only adopt one value for each model parameter and have not done any parameter survey. Our ongoing work of studying the effect of different model parameter seems to indicate that this discrepancy with observation can be solved by adjusting the values of some parameters (Yao *et al.* in preparation).

References

Cheung, E., Bundy, K., Cappellari, M., *et al.* 2016, *Natur*, 533, 504
Ciotti, L., Pellegrini, S., Negri, A., & Ostriker, J. P. 2017, *ApJ*, 835, 15

Ciotti, L., & Ostriker, J. P. 2012, in Hot Interstellar Matter in Elliptical Galaxies, Astrophysics and Space Science Library, Vol. 378 (Berlin: Springer-Verlag), 83
Fabian, A. C. 2012, *ARA&A*, 50, 455
Gan, Z., Yuan, F., Ostriker, J. P., Ciotti, L, Novak, G. S., 2014, *ApJ*, 789, 150
Gofford, J., Reeves, J. N., McLaughlin, D. E., *et al.* 2015, *MNRAS*, 451, 4169
Harrison, C. M. 2017, *NatAs*, 1, 0165
King, A., & Pounds, K. 2015, *ARA&A*, 53, 115
Kormendy, J., & Ho, L. C. 2013, *ARA&A*, 51, 511
Korol, V., Ciotti, L. & Pellegrini, S. 2016, *MNRAS*, 460, 1188
Liu, G., Zakamska, N. L., Greene, J. E., Nesvadba, N. P. H., & Liu, X. 2013, *MNRAS*, 436, 2576
Naab, T. & Ostriker, J. P. 2017, *ARA&A*, 55, 59
Negri, A., & Volonteri, M. 2017, *MNRAS*, 467, 3475
Pellegrini, S., Ciotti, L., Negri, A., & Ostriker, J. P. 2018, *ApJ*, 856, 115
Schawinski, K., Koss,M., Berney, S., & Sartori, L. F. 2015, *MNRAS*, 451, 2517
Wang, Q. D., Nowak, M. A., Markoff, S. B., *et al.* 2013, *Sci*, 341, 981
Xie, F., & Yuan, F, 2012, *MNRAS*, 427, 1580
Xie, F., Yuan, F., & Ho, L. C. 2017, *ApJ*, 844, 42
Xue, Y. Q. 2017, *NewAR*, 79, 59
Yoon, D., Yuan, F., Gan, Z. M. *et al.* 2018, *ApJ*, in press (arXiv:1803.03675)
Yuan, F., & Narayan, R. 2014, *ARA&A*, 52, 529
Yuan, F., Gan, Z., Narayan, R. Sadowski, A., Bu, D., Bai, X. 2015, *ApJ*, 804, 101
Yuan, F., Yoon, D., Li, Y., Gan, Z., Ho, L. C., Guo, F. 2018, *ApJ*, 857, 121

Perseus in Sicily: From Black Hole to Cluster Outskirts
Proceedings IAU Symposium No. 342, 2018
K. Asada, E. de Gouveia Dal Pino, M. Giroletti,
H. Nagai & R. Nemmen, eds.

doi:10.1017/S1743921318007329

Prodigious and Continuous Formation of Super Star Clusters from Cooled Intracluster Gas

Jeremy Lim[1], Emily Wong[1], Youichi Ohyama[2], Tom Broadhurst[3] and Elinor Medezinski[4]

[1]Department of Physics, The University of Hong Kong, Pokfulam Road, Hong Kong
email: jjlim@hku.hk, emily79@connect@hku.hk

[2]Academia Sinica, Institute of Astronomy and Astrophysics, Taipei, Taiwan

[3]Department of Theoretical Physics, University of Basque Country UPV/EHU, Bilbao, Spain

[4]Department of Astrophysical Sciences, Princeton University, Princeton, NJ 08544, USA

Abstract. Globular clusters (GCs) — compact and massive star clusters found ubiquitously around galaxies — are believed to be ancient relics (ages $\gtrsim$10 Gyr) from the early formative phase of galaxies, although their physical origin remains widely debated. The most numerous GC populations are hosted by giant elliptical galaxies, where they can exhibit a broad dispersion in colour interpreted as a wide spread in metallicity. Here, we show that many thousands of similarly compact and massive super star clusters have formed at an approximately steady rate over, at least, the past $\sim$1 Gyr around the nearby giant elliptical galaxy, NGC 1275, at the centre of the Perseus cluster. The number distribution of these young star clusters appears to exhibit a similar dependence in luminosity and mass as the even more numerous but older GCs around NGC 1275. In just a few Gyr, these super star clusters will evolve to become indistinguishable in broadband optical colours from the older GCs, and their spread in age add to the dispersion in colour of these GCs. The spatial distribution of the super star clusters resembles the filamentary network of multiphase gas in the cluster core, implying that they formed from molecular gas amassed from cooling of the hot intracluster gas. The sustained formation of super star clusters from cooled intracluster gas constitutes a previously unrecognised but prodigious source of GCs over cosmic timescales, and contributes to both their enormous numbers and broad colour dispersion in giant elliptical galaxies.

Keywords. Super Star Clusters, Globular Clusters, Intracluster Gas, Giant Elliptical Galaxy

---

1. Introduction

Images taken with the Advanced Camera for Surveys (ACS) on the NASA/ESA Hubble Space Telescope (HST) are able to clearly separate, and even resolve, individual star clusters in NGC 1275. We have identified all the star clusters towards this galaxy from images taken in three filters: F435W in the blue (B), F550M in the middle of the visual band (V), and F625W in the red (R). Previous studies of NGC 1275 with the HST have been restricted to either small fields of view spanning the innermost region of this galaxy (Holtzman *et al.* 1992; Carlson *et al.* 1998), or those that employ the same images as we do but only identify star clusters at the outskirts of the galaxy (Canning *et al.* 2010; Canning *et al.* 2014). Although these studies revealed numerous compact and massive star clusters around NGC 1275, they did not establish the full properties of these star clusters (total number, overall spatial distribution, full age spread, luminosity or mass function, and formation rate over time), nor their relationship with other galaxies along the line

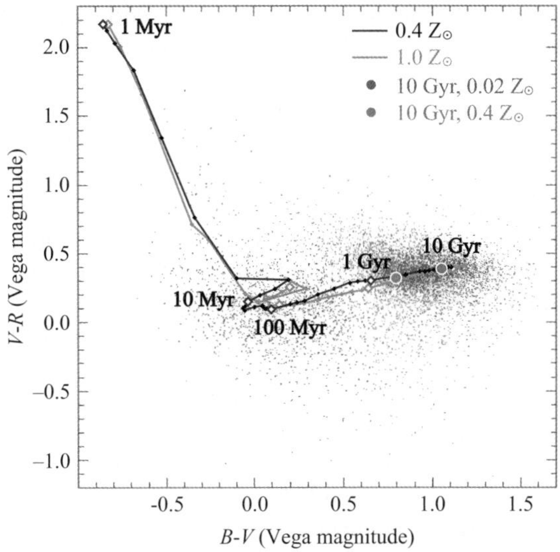

Figure 1. Colour-colour diagram of star clusters belonging to NGC 1275.

of sight. In our work, we identify star clusters throughout the $3'.4 \times 4'.0$ (73.4 kpc $\times$ 86.4 kpc) field of view common to all three filters, including the inner region of NGC 1275 where the light distribution is complex making a complete census of star clusters in this region challenging. As demonstrated below, in this way we are able to separate out star clusters associated with other galaxies along the line of sight, and determine in full the properties of those belonging to NGC 1275.

2. Results

Fig. 1 shows a colour-colour diagram of the star clusters detected in all three filters that do not include those bounded by two galaxies along the line of sight towards NGC 1275. Each point in this figure denotes an individual star cluster. Their ages can be inferred from theoretical evolutionary tracks for coeval stellar populations, provided their metallicity is known. The black line in Fig. 1 show evolutionary tracks for star clusters having a metallicity of $0.4Z_{\odot}$, corresponding to that of the intracluster gas (Werner *et al.* 2013), and the grey line to the Solar metallicity of $1.0Z_{\odot}$. These tracks are much more sensitive to age than metallicity over the parameter range plotted. The colour of a star cluster becomes redder with age as bluer, more massive, stars evolve away from the main sequence, resulting in generally increasing values in both $(B-V)$ and $(V-R)$ with age. The youngest star clusters having ages $\lesssim$6 Myr are bluest in $(B-V)$, but relatively red in $(V-R)$ owing to Hα+[NII] emission from leftover gas photoionized by massive stars that adds to the stellar continuum in the R filter. A striking feature of this colour-colour diagram is the very broad and continuous range in $(B-V)$ spanned by the star clusters, covering the full range of ages spanned by the stellar evolutionary tracks from the present epoch back in time to the early Universe.

3. Analyses

Before assigning ages to individual star clusters, we first have to separate out GCs, which individually can have a very different metallicity than the values adopted in the two evolutionary tracks plotted in Fig. 1. GCs generally exhibit a bimodal colour distribution, indicating distinct metal-poor (bluer) and metal-rich (redder) populations that differ by about a factor of ten in metallicity (Brodie *et al.* 2006; Harris 2010). In giant elliptical galaxies, however, the colour distribution of GCs can be continuous rather than bimodal (Harris *et al.* 2017); if GCs constitute a single ancient population ($\gtrsim$10 Gyr old), their metallicities would then span the enormous range $\sim 0.005 Z_\odot$–$1.0 Z_\odot$ (Harris *et al.* 2017). Even then, GCs are commonly divided into two populations having average metallicities that differ by an order of magnitude. In Fig. 1, the colour of 10-Gyr star clusters having a metallicity of $0.02 Z_\odot$ representative of the metal-poor GC population is indicated by a blue filled circle, and those having a metallicity $0.4 Z_\odot$ representative of the metal-rich GC population is indicated by a red filled circle. There is a clear concentration of star clusters in these regions of the colour-colour diagram, as would be expected for GCs belonging to NGC 1275.

The star clusters plotted in Fig. 1 can be separated into two groups based on whether they have an isotropic or anisotropic spatial distribution. Those coded blue in this figure, hereafter referred to as blue star clusters (BSCs), have a highly anisotropic spatial distribution. In stark contrast, those coded red in Fig. 1 have an isotropic distribution that decreases radially outwards in surface density, as is characteristic of GCs. There is an appreciable concentration of star clusters along the dusty spiral arms of a foreground infalling galaxy known as the High Velocity System (HVS), indicating that some of the star clusters towards the HVS belong to this galaxy. Those in the central innermost region also has been omitted from Fig. 1, as they likely belong to yet another spiral galaxy lying along the line of sight towards NGC 1275. The BSCs in this region are unusually bright (Holtzman *et al.* 1992), and furthermore have a different luminosity and mass function compared with the remaining BSCs. Away from the HVS and central spiral, the vast majority if not all of the BSCs must belong to NGC 1275, as is evident by their spatial concentrations in patterns largely related to the morphology of the gaseous nebula in NGC 1275.

We find the BSCs and GCs to exhibit a similar dependence in their luminosity or mass functions. Furthermore, in a previous study of over two-hundred BSCs in NGC 1275, their radial light profiles have been found to resemble GCs (Carlson *et al.* 2001). At a given luminosity or mass interval, the BSCs over the age range $\sim$100-900 Myr are approximately one-tenth or one-hundredth (strict lower limits, given those bounded by the HVS have been excluded), respectively, as numerous as the GCs over the sky area searched. The GCs in NGC 1275 have maximal masses of $\sim 10^7$ $M_\odot$, similar to the most massive GCs found in other giant elliptical galaxies (Harris *et al.* 2014). The BSCs have maximal masses of $\sim 10^6$ $M_\odot$, about an order of magnitude higher than the typical GC mass at the peak in their number distribution (Brodie *et al.* 2006; Harris 2010), and two orders of magnitude higher than the maximal masses of open star clusters in our Galaxy. Across the three age intervals plotted for the BSCs, their mass function is approximately constant, indicating an approximately steady formation rate over the past $\sim$1 Gyr. Over the mass range $\sim 5 \times 10^4 - 1 \times 10^6$ $M_\odot$ for the BSCs spanning ages $\sim$100-900 Myr, their formation rate is 1.2 Myr^{-1} at a total mass rate of 1×10^5 $M_\odot$ Myr^{-1}; i.e., at a typical rate of one star cluster every 1 million years having a mass close to that of GCs at the peak in their number distribution. These values are lower limits as some bounded by the HVS likely belong to NGC 1275, and furthermore we do not include the even more numerous BSCs detected in only one or two filters and likely having lower individual masses. Note that we cannot rule out the possibility of significant temporal modulations in the formation rate of BSCs on relatively short timescales; indeed, the mass formation rate of the youngest BSCs (ages $\lesssim$50 Myr) has been estimated to be about an order of magnitude larger (Canning *et al.* 2014).

4. Interpretation

The close spatial relationship between the BSCs and the gaseous nebula in NGC 1275 indicates that these star clusters formed from the nebular gas (Canning *et al.* 2010; Canning *et al.* 2014). Indeed, the atomic gas nebula in this galaxy is accompanied by an exact counterpart in molecular gas (Lim *et al.* 2012; Salome *et al.* 2006), the reservoir for star formation. NGC 1275 is the nearest, but by no means atypical, example of a cluster central giant elliptical galaxy possessing multiphase gas and exhibiting recent star formation. Such features are found preferentially if not exclusively among the central giant elliptical galaxies of relaxed clusters (Cavagnolo *et al.* 2008; Werner *et al.* 2014), where the intracluster gas (at temperatures $\sim 10^{7-8}$ K) at the cluster core is especially dense and therefore experiences vigorous radiative cooling.

About 25 years ago, the discovery of relatively large populations of compact and massive but young star clusters in nearby galaxies experiencing a brief period of highly elevated star formation (starbursts) due to interactions or mergers suggested that some ancient GCs may have formed through such events (e.g., Whitmore *et al.* 1993; Whitemore & Schweizer 1995) In such episodes, the available reservoir of cool gas is rapidly consumed by star formation and dispersed by stellar winds, supernovae, or AGN activity. In cluster central galaxies, however, cooled intracluster gas can be continuously or intermittently replenished, depending on AGN activity in the cluster central galaxy as well as global disturbances due to cluster-cluster mergers. The sustained formation of progenitor GCs from cooled intracluster gas can help explain not just the unusually large populations of GCs in cluster central galaxies, but their resulting wide distribution in ages (and perhaps also metallicities owing the enrichment of the intracluster gas) and therefore colours can contribute to the broad dispersion in colours of presumed ancient GCs in these galaxies.

References

Brodie, J. P. & Strader, J. 2006, *ARAA*, 44, 193

Canning, R. E. A., Fabian, A. C., Johnstone, R. M., *et al.* 2010, *MNRAS*, 405, 115

Canning, R. E. A, Ryon, J. E., Gallagher, J. S., *et al.* 2014, *MNRAS*, 444, 336

Carlson, M. N., Holtzman, J. A., Watson, A. M., *et al.* 1998, *AJ*, 115, 1778

Carlson, M. N., & Holtzman, J. A. 2001, *PASP*, 113, 1522

Cavagnolo, K. W., Donahue, M., Voit, G. M., & Sun, M. 2008, *ApJ* (Letters), 683, L107

Harris, W. E. 2010, *Phil. Trans. R. Soc*, 368, 889

Harris, W. E., Morningstar, W., Gnedin, O. Y., *et al.* 2014, *ApJ*, 797, 128

Harris, W. E., Ciccone, S. M., Eadia, G. M., *et al.* 2017, *ApJ*, 835, 101

Holtzman, J. A., Faber, S. M., Shaya, E. J., *et al.* 1992, *AJ*, 103, 691

Lim, J., Ohyama, Y., Chi-Hung, Y., *et al.*, Dinh-V-Trung, & Shiang-Yu, W. 2012, *ApJ*, 744, 112

Salomé, P., Combes, F., Edge, A. C., *et al.* 2006, *AAP*, 454, 437

Werner, N., Urban, O., Simionescu, A., & Allen, S. W. 2012, *Nature* 502, 656

Werner, N., Oonk, J. B. R, Sun, M., *et al.* 2014, *ApR* 439 2291

Whitmore, B. C., Schweizer, F., Leitherer, X. *et al.* 2006, *AJ*, 106, 1354

Whitmore, B. C. & Schweizer, F. 1998, *AJ*, 109, 960

Perseus in Sicily: From Black Hole to Cluster Outskirts
Proceedings IAU Symposium No. 342, 2018
K. Asada, E. de Gouveia Dal Pino, M. Giroletti,
H. Nagai & R. Nemmen, eds.

doi:10.1017/S174392131800844X

Characterizing the Outburst of the Supermassive Black Hole in M87

William Forman[1], Eugene Churazov[2,3], Sebastian Heinz[4], Christine Jones[1], Paul Nulsen[1], Ralph Kraft[1], Scott Randall[1] and Alexey Vikhlinin[1,3]

[1]Harvard-Smithsonian Center for Astrophysics, 60 Garden St., Cambridge, MA, USA
email: wforman@cfa.harvard.edu

[2]MPA, Karl-Schwarzschild Strasse 1, 85748 Garching bei Munchen, Germany

[3]Space Research Institute, 84/32 Profsoyuznaya Str, Moscow, Russia

[4]University of Wisconsin, 4506 Sterling Hall, Madison, WI 53706 USA

Abstract. M87, in the Virgo cluster, allows us to study the interaction of a supermassive black hole (SMBH) with its hot gaseous atmosphere. Deep Chandra observations reveal a nearly circular shock front with a Mach number of 1.2 and a radius of 13 kpc which is driven by a central cavity inflated by an SMBH outburst began 12 million years ago. An outburst with an energy of $\sim 5 \times 57$ ergs and a duration of ~ 2 Myrs provides a good match to all the constraints. For an outburst repetition rate of about 12 Myrs (the outburst age), the outburst energy is sufficient to balance the radiative cooling of the gas. The outburst duration in M87 argues for a "gentle" (long duration) outburst that does not generate strong shocks and where much of the outburst energy is deposited in the cavities that then transfer energy to the surrounding gas as they buoyantly rise.

Keywords. supermassive black hole, cool core, AGN outburst

1. Chronicling the outburst history of the supermassive black hole in M87

Galaxies, groups and clusters form a family of dark matter halos. At the center of each of these halos that are filled with hot, X-ray emitting plasma, especially those referred to as cool core systems where the gas cooling time is short compared to the age of the system, there is generally a single massive galaxy with a supermassive black hole (SMBH) in the core. Fig. 1 shows three examples of hot gaseous atmospheres ranging from galaxy/group scales to the very massive cluster scale. In these examples, each X-ray atmosphere exibits large cavities, which were produced by energetic outbursts from the central SMBH. Often, the only way to derive the total energy release from the SMBH is through the effects of the SMBH on its environment (e.g. the production of cavities and shocks in the surrounding X-ray emitting gas). While the accretion onto the SMBH appears to be radiatively inefficient at the present epoch, the mechanical output from these black hole outbursts is sufficient to inflate cavities with total energies ranging from 10^{56} ergs for individual galaxies to as high as 10^{62} ergs for the largest cavities seen in massive clusters (e.g. Finoguenov & Jones 2001; Fabian *et al.* 2003; McNamara *et al.* 2005; Churazov *et al.* 2002).

In this contribution, we focus on the X-ray and radio observations of M87 based on the work described in Churazov *et al.* (2001) and Forman *et al.* (2005, 2007, 2017) which provide more details. M87 lies at a distance of 16 Mpc and hosts a SMBH with a

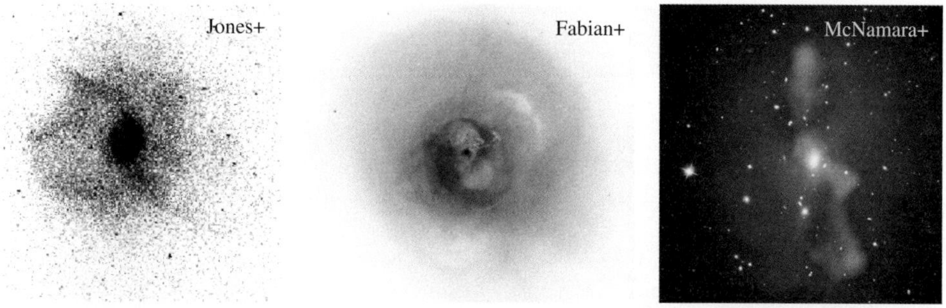

Figure 1. Early type galaxies, groups and clusters form a family of dark matter halos. As the halo mass increases, so does the baryon fraction, gas temperature, and X-ray luminosity. Many systems have strongly peaked X-ray surface brightness profiles that imply short gas cooling times and that are centered on a bright central galaxy (BCG). In the systems shown (from left to right, NGC4636, Perseus/NGC1275, MS0735), each X-ray atmosphere exhibits large cavities, which were produced by energetic outbursts from the central SMBH of the BCG.

mass of $3 - 6 \times 10^9$ $M_\odot$ (e.g. Ford *et al.* 1994; Walsh *et al.* 2013). The X-ray and radio observations of M87 chronicle the outbursts from the SMBH over the past 150 Myr (see Fig. 2 and 3). Fig. 2 (left panel) shows the broad band X-ray image of M87 including the jet (extending $20''$ from the nucleus to the NW) which is filling the central cavity with relativistic plasma clearly seen as the very dark (saturated) region in the VLA radio image (Fig. 2, right panel). The VLA image also shows a pair of arms extending up to $5'$ to the east and southwest. The eastern arm appears as a torus atop a stem (a "mushroom cloud") and represents a buoyant bubble of plasma that has risen about 20 kpc over the past 40-70 Myrs (Owen *et al.* 2000; Churazov *et al.* 2001). Only a twisted filamentary arm remains of the corresponding plasma bubble to the SW. Corresponding X-ray filamentary arms of cool gas, uplifted by the buoyant plasma bubbles, are seen in the Chandra image (left panel). On the largest scales (extending to almost 40 kpc to the north and south), two faint disk-like radio features are probably the remnants of the oldest outbursts from M87 (~ 100 Myrs old; labeled as "outer radio lobe"). The hard band X-ray image (center panel) shows a shock at a radius of $2.8'$ (13 kpc; see also Fig. 4) which is represented in the radio image as a black circle. This shock was produced by a prominent outburst that occurred approximately 12 Myr ago. A surface brightness image in the hard energy band is approximately proportional to the square of the pressure projected on the sky for gas temperatures of 1 - 3 keV (see Churazov *et al.* 2016; Arévalo *et al.* 2016). Images in this band show direct evidence for outbursts as over-pressured, bright regions. The cool filamentary arms, so prominent in the softer band (left panel), are not seen in the hard band and are, hence, in pressure equilibrium with the surrounding plasma.

We use deep Chandra observations of M87, along with a model for the shock observed at a radius of 13 kpc that is being driven by the expansion of the cavity as it is being inflated with relativistic radio-emitting plasma by the SMBH, to derive the properties of the AGN outburst that created the shock. The age of the outburst, giving rise to the shock, is about 12×10^6 years. With the Chandra and radio observations, we can chronicle outbursts from the M87 supermassive black hole over the past 150 Myr.

The 13 kpc shock in M87 was the first classical shock found in the hot gaseous atmosphere around a central cluster galaxy, where both the gas density and gas temperature jumps could be accurately measured. The gas temperature and density jumps are both consistent with the Rankine-Hugoniot shock jump conditions. Fig. 4 shows the projected and deprojected X-ray surface brightness profiles for M87 in the Chandra "hard" (3.5 - 7.5 kev) band (in red) and "soft" (1.2 - 2.5 keV) energy band (in blue).

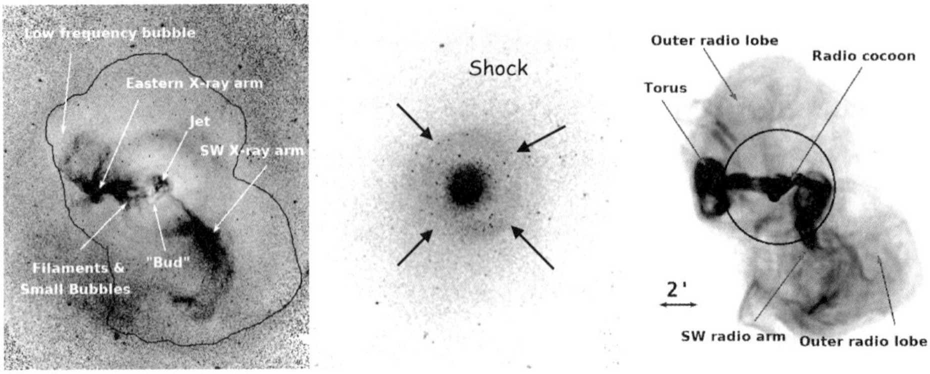

Figure 2. The Chandra broad (left) and hard (center) band images and the 90 cm VLA (right) image, matched in scale, document the major outbursts from M87 (for details see Churazov *et al.* 2001; Forman *et al.* 2007, 2017; Owen *et al.* 2000). (left) Chandra broad band image (0.5-2.5 keV) divided by the average radial profile to show faint surface brightness features. A contour of the faintest surface brightness regions of the radio emission is included. (center) Arrows denote the shock location in the hard band (3.5-7.5 keV) image. (right) The VLA image shows the prominent arms, torus, and outer lobes. The circle shows the location of the X-ray observed shock.

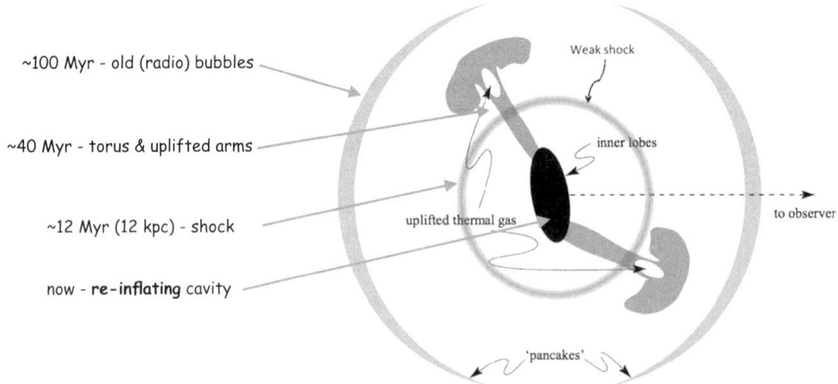

Figure 3. Schematic of M87 activity over the past 100 Myr as chronicled in the radio and X-ray observations. The sketch (adapted from Churazov *et al.* (2001)) shows the oldest outbursts as flattened pancakes, high in the atmosphere, the mushroom shaped tori (one now disrupted), the X-ray shock, and the central cavity that drove the 12 Myr old shock and is now re-pressurizing.

To characterize the outburst properties, simulations were carried out that capture the key physics. The radio plasma, ejected by the jet from the SMBH, inflates a central cavity. The inner radio lobes act as a piston that displaces the hot X-ray gas and drives a shock. By varying the outburst energy and timescale (see Fig. 5), we constrain the outburst parameters and derive the outburst energy, the outburst age, and the outburst duration. Our fiducial model that best characterizes the M87 outburst has an age of 12 Myrs, a total outburst energy of 5.5×10^{57} ergs, and a duration of ~ 2 Myr. At the current time, the shock is expanding at Mach 1.2. By characterizing the outburst, we are also able to derive the energy partition of the outburst. The buoyant bubbles contain the largest fraction of the outburst energy. The initial shock contains $\sim 30\%$ of the outburst energy of which 10% is deposited in shocked gas by the present epoch. The remaining shock energy is carried to large radii as the shock continues to weaken. About 80% of the outburst energy is available for heating the gas in the X-ray bright core.

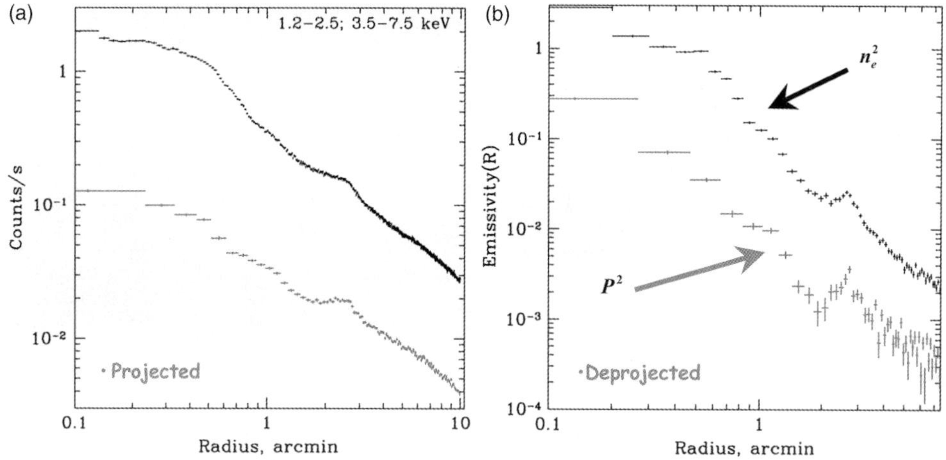

Figure 4. (a) The azimuthally averaged (projected) radial profiles to the north of M87 in the 1.2-2.5 keV (upper curve; mostly sensitive to density) and 3.5-7.5 keV (lower curve; sensitive to pressure) bands; these show a sharp edge at $0.60'$ (most distinctly seen in the hard band as a decrease between the fourth and fifth data points), a moderate flattening of the profile at about $1'$, and a strong excess at $2' - 3'$. (b) The deprojected emissivity profiles, derived from the azimuthally averaged surface brightness profiles in the two energy bands 1.2 - 2.5 keV (upper curve) and 3.5 - 7.5 keV (lower curve). The deprojection shows the very pronounced feature at $2' - 3'$, which represents the 13 kpc shock. In addition, for a spherical shock one expects a rarefaction region trailing the shock front characterized by a density and temperature decrease below the upstream values. Such a rarefaction region is probably seen in panel b at $r \sim 1.5' - 2'$, where the emissivity shows a clear depression.

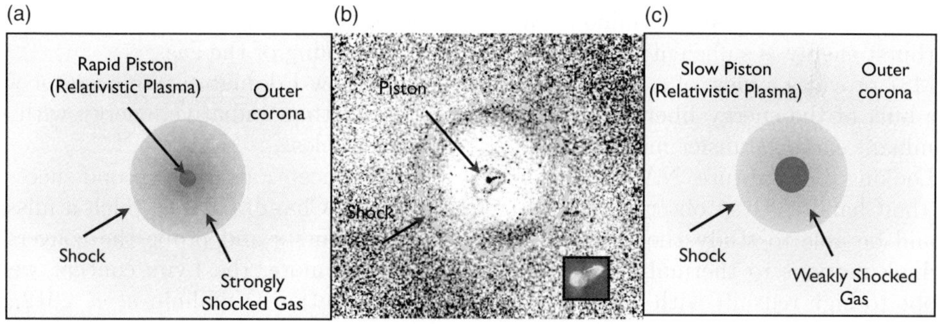

Figure 5. Two shock scenarios **(a - left)** A strong, short duration outburst ($\Delta t = 0.1 \times 10^6$ yrs) drives a strong shock. At present, the region interior to the shock (located at the observed 13 kpc radius) would encircle a central strongly shocked, hot, low density atmosphere. **(b - center)** X-ray image of M87 divided by the average radial profile shows the central piston, that powered the M87 shock, and the jet (6 cm radio image is inset). The dashed blue circle (labeled "Shock") indicates the outer edge of the shock, seen as the bright ring of emission. **(c - right)** A longer duration outburst, ($\Delta t = 2.2 \times 10^6$ yrs; the fiducial model) provides the same strength shock at 13 kpc, but only weakly shocked gas interior to the shock location and a larger central plasma-filled, piston. The observed data are not consistent with a short duration outburst.

Fig. 5 shows a cartoon version of two outbursts – one with a short duration and a second with long duration – which characterize the quantitative analysis. A powerful, short duration outburst, in the left panel ($\Delta t = 0.1 \times 10^6$ yrs), drives a strong shock into the surrounding atmosphere. At the present time, the region interior to the shock (located at the observed 13 kpc radius) would enclose a central hot, strongly shocked, low density atmosphere. The right panel pictures a longer duration outburst, ($\Delta t = 2.2 \times 10^6$ yrs;

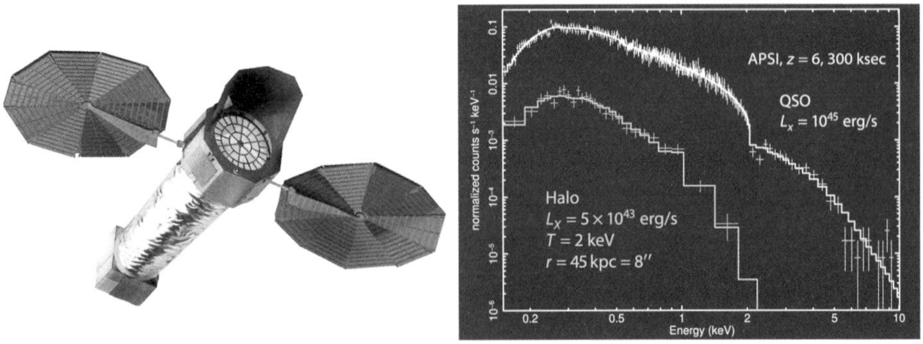

Figure 6. (left) An artist's concept of the Lynx mission that would provide 30 times Chandra's effective area with comparable angular resolution. (right) At redshift $z \sim 6$, Lynx would be able to resolve the bright AGN from its surrounding hot atmosphere and provide a high quality spectrum of the hot atmosphere with a 300 ks observation. With sub-arcsecond angular resolution, Lynx would be able to follow the evolution of hot atmospheres, the dark matter halos in which they are confined, and the AGN, associated with the SMBH at the galaxy nuclei, from high redshift to the present.

referred to as the fiducial model) which provides the same magnitude shock at 13 kpc, but only weakly shocked gas interior to the shock location and a larger central, plasma-filled piston. The central panel shows the actual outburst. The observations do not show a strongly shocked region interior to the shock and the radius of the central cavity matches that observed for our fiducial model with a few Myr duration outburst.

During the subsequent evolution, within the framework of the model, as the buoyant bubble rises, much of its energy is transferred to the ambient cluster gas. For an outburst repetition rate of about 12 million years (the age of the current outburst), 80% of the outburst energy is sufficient to balance the radiative cooling of the gas.

M87 provides a view of a "typical" outburst from a low-Eddington rate accretor with the bulk of the energy liberated as mechanical, rather than radiative, energy with the dominant energy transfer mediated by the buoyant bubbles.

Looking to the future, NASA is studying the Lynx concept, a sub-arcsecond successor to the Chandra X-ray observatory but having 30 times Chandra's area. Such a mission would be able to study the detailed properties of outbursts and probe the conversion of bulk motions to thermal energy (see Fig. 6). Furthermore, the Lynx concept would probe to high redshift with sub-arcsecond angular resolution (Vikhlinin *et al.* 2012, see section 3.1 and Fig. 3) and could fully test models of galaxy evolution and the impact of the SMBHs by studying the AGN and the hot gaseous atmospheres in which they lie. It could trace the evolution of both the AGN and the surrounding hot gaseous atmosphere by deriving their properties (luminosity, temperature, density profile, and abundance) from redshifts of $z \sim 6$ (or higher) to the present.

References

Arévalo, P., Churazov, E., Zhuravleva, I., Forman, W. R., & Jones, C. 2016, *ApJ*, 818, 14
Churazov, E., Forman, W., Jones, C., & Böhringer, H. 2000, *A&A*, 356, 788
Churazov, E., Brüggen, M., Kaiser, C. R., Böhringer, H., & Forman, W. 2001, *ApJ*, 554, 261
Churazov, E., Sunyaev, R., Forman, W., & Böhringer, H. 2002, *MNRAS*, 332, 729
Churazov, E., Sazonov, S., Sunyaev, R., *et al.* 2005, *MNRAS*, 363, L91
Churazov, E., Arevalo, P., Forman, W., *et al.* 2016, *MNRAS*, 463, 1057
Fabian, A. C., Sanders, J. S., Allen, S. W., *et al.* 2003, *MNRAS*, 344, L43
Finoguenov, A., & Jones, C. 2001, *ApJ*, 547, L107
Ford, H. C., Harms, R. J., Tsvetanov, Z. I., *et al.* 1994, *ApJ*, 435, L27

Forman, W., Nulsen, P., Heinz, S., *et al.* 2005, *ApJ*, 635, 894
Forman, W., Jones, C., Churazov, E., *et al.* 2007, *ApJ*, 665, 1057
Forman, W., Churazov, E., Jones, C., *et al.* 2017, *ApJ*, 844, 122
McNamara, B. R., Nulsen, P. E. J., Wise, M. W., *et al.* 2005, *Nature*, 433, 45
Owen, F. N., Eilek, J. A., & Kassim, N. E. 2000, *ApJ*, 543, 611
Vikhlinin, A., Reid, P., Tananbaum, H., *et al.* 2012, *Proc. SPIE*, 8443, 844316
Walsh, J. L., Barth, A. J., Ho, L. C., & Sarzi, M. 2013, *ApJ*, 770, 86

Perseus in Sicily: From Black Hole to Cluster Outskirts
Proceedings IAU Symposium No. 342, 2018
K. Asada, E. de Gouveia Dal Pino, M. Giroletti,
H. Nagai & R. Nemmen, eds.

doi:10.1017/S1743921318007366

X-ray probing of NGC 1275 nuclear region with Hitomi, Swift, and Suzaku

Yasushi Fukazawa

Department of Physics, Hiroshima University, 1-3-1 Kagamiyama, Higashi-Hiroshima, Japan 739-8526
email: fukazawa@astro.hiroshima-u.ac.jp

Abstract. NGC 1275 has been known as a ppint-like X-ray source with a continuum and a Fe-K line. Unlike radio and GeV/TeV gamma-ray emissions, origin of X-ray emission is not yet understood; is it a jet emission like blazars or an accretion corona emission like Seyfert galaxies. X-ray emission is important to determine the SED of jet emission to constrain jet parameters and also understand the relation between accretion and jet. Here we report a recent X-ray probing of NGC 1275 nuclear region with Hitomi/SXS, Swift/XRT, and Suzaku/XIS. Hitomi/SXS gave the first opportunity to measure a Fe-K line of AGNs with several eV resolution. The line center is consistent with the neutral iron emission, and the width is constrained to be 500-1600 km/s (FWHM). This ruled out the origin of broad line region and inner accretion disk. A low-covering-fraction molecular torus or a rotating molecular disk around pc scales, illuminated by accretion corona emission, is suggested as a possible origin. For the continuum emission, Suzaku/XIS monitor observations revealed that the X-ray flux has gradually increased as the GeV gamma-ray flux. Swift/XRT showed a several-days flux increase, associated with the GeV gamma-ray flare. These results on the continuum emission suggests a contribution of jet emission to the X-ray emission. Based on the combined results of Fe-K line and continuum, we discuss some scenarios for X-ray emitting region.

---

1. Results and discussion

NGC 1275 is an interesting AGN since it shows a sign of emissions from jet and disl/corona in the multi-wavelength band. Radio and GeV/TeV gamma-rays have been detected from NGC 1275 (Abdo *et al.* 2009), while emission from disk/corona are suggested in the X-ray and optical band (Yamazaki *et al.* 2013). Here we reported a brief summary of recent X-ray studies of NGC 1275, based on the Hitomi/SXS (Hitomi collaboration 2018), Suzaku and Swift/XRT (Fukazawa *et al.* 2018), and discussed the origin of X-ray emission, refering to Fukazawa *et al.* (2016).

In the X-ray band, a point-like source has been detected from Einstein/HRI, ROSAT/HRI, Chandra, and XMM-Newton (Fabian *et al.* 2015). XMM-Newton detected a Fe-K (6.4 keV) line (Churazov *et al.* 2003). Fe-K 6.4 keV line is typical for Seyfert galaxies, while a low Eddington luminosity ratio of 10^{-4} to 10^{-3} gives a doubt whether a normal stanrard accretion disk and torus exists or not. What is an origin of Fe-K line; does it come from torus or others? Precise spectroscopy of Fe-K 6.4 keV line is very important to constrain the origin by measuring doppler shift, width, ionization, and so on. X-ray CCD cannot distinguish between 2000 km s^{-1} and several 100 km s -1. Grating spectroscopy with Chandra/HETG gave a hint of BLR origin of Fe-K 6.4 keV line (Shu *et al.* (2011)). On the other hand, Fe-K variability time scale of a few years suggests a torus origin (Fukazawa *et al.* 2016).

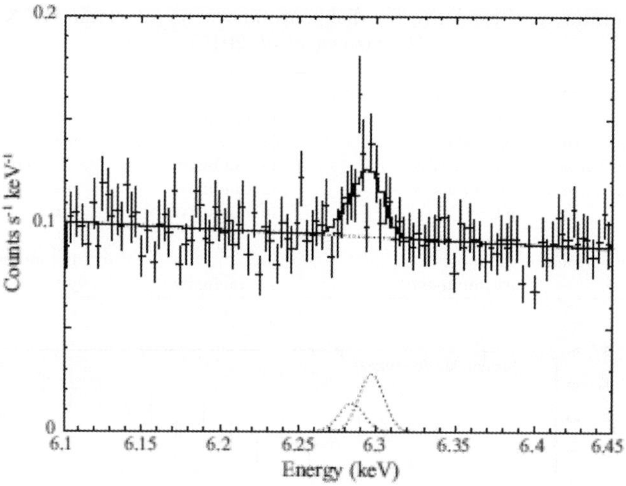

Figure 1. Fe-Kα line of NGC 1275 detected with Hitomi/SXS (Hitomi Collaboration 2018).

Hitomi/SXS observed NGC 1275 during an initial operation observing the Perseus cluster on Feb. 25–27th and Mar. 4–6th, 2016, with an exposure of 240 ks. SXI (CCD) took only a small amount of data, and HXI and SGD have not yet operated. Hitomi/SXS results on Fe-K 6.4 keV line of NGC 1275 has been reported by Hitomi Collaboration (2018).

Hitomi/SXS gave us the first fine spectroscopy of AGN Fe-K line with an energy resolution of 4.9 eV. A resolved line is simple; just Kα_1 and Kα_2, indicating a neutral iron as origin. The center energy is well determined to be 0.01700$\pm$0.00063 in redshift for the neutral Fe-K line, consistent with the latest optical redshift (0.017284) of parent galaxy. The line width is for the first time constrained to be 500–1600 km s^{-1} (FWHM), ruling out the BLR (2750 km s^{-1}) origin. Equivalent width (EW) is just 10 eV for the observed continuum which includes both of AGN and cluster emissions. By PSF photometry technique, we extracted an AGN continuum flux, and an EW of 25 eV was obtained against the AGN conitnuum.

So far several candidates of origin of Fe-K line has been suggested; accretion disk, BLR, and torus. For NGC 1275, molecular disk and molecular clouds have been found (Salomé *et al.* 2006; Scharwächter *et al.* 2013) and they are also a candidate of Fe-K emitter. Hitomi/SXS spectoscopy ruled out an accretion disk and a BLR as origon. Hitomi/SXS also could constrain an extent of Fe-K line emission to be < 42 arcsec (< 17 kpc), and Chandra X-ray data did not show a Fe-K line at 4–30 arcsrc region (1.6–12 kpc) from the center. Based on the Monte Carlo simulation, a Fe-K line flux from the molecular clouds is estimated to be too weak to explain the observed flux. So, torus or molecular disk remains as an origon. A Fe-K line width is consistent with [Fe II] line width of 380–1000 km s^{-1} from the molecular disk (Scharwächter *et al.* 2013), and this prefers an origin of molecular disk. A small EW of Fe-K against 100–200 eV of typical Seyfert galaxies (Fukazawa *et al.* (2011)) indicates a small column density of torus if the torus is an origin.

Then, how about other radio galaxies are on the X-ray properties? Fukazawa *et al.* (2014) reported Suzaku obseravtions of GeV gamma-ray emitting radio galaxies. Some radio galaxies show a Fe-K line; Cen A, 3C 120, 3C 111, and NGC 1275, while other galaxies do not; M87, NGC 6251, PKS 0625-354, and 3C 78. Jet emission is likely to be dominant in the X-ray band for the latter galaxies. Fe-K line of NGC 1275 suggests a disk/corona emission in the X-ray band, since jet emission is so beamed that only a

Table 1. Summary of evidence for disk/corona versus jet origin for X-ray emission (Fukazawa *et al.* 2015)

Source	Fe-K line	X-ray spectral index	X-ray variability	[O III] line	T ype [ref.]
3C 78	jet	jet	inconclusive	jet	LERG
NGC 1275	disk/corona	inconclusive	jet	disk/corona	HERG/LERG
3C 111	disk/corona	inconclusive	inconclusive	disk/corona	HERG
3C 120	disk/corona	inconclusive	inconclusive	disk/corona	HERG
PKS 0625-354	jet	jet	inconclusive	jet	LERG
M 87	jet	jet	jet	jet	LERG
Cen A	disk/corona	inconclusive	jet	inconclusive	HERG
NGC 6251	jet	inconclusive	inconclusive	jet	LERG

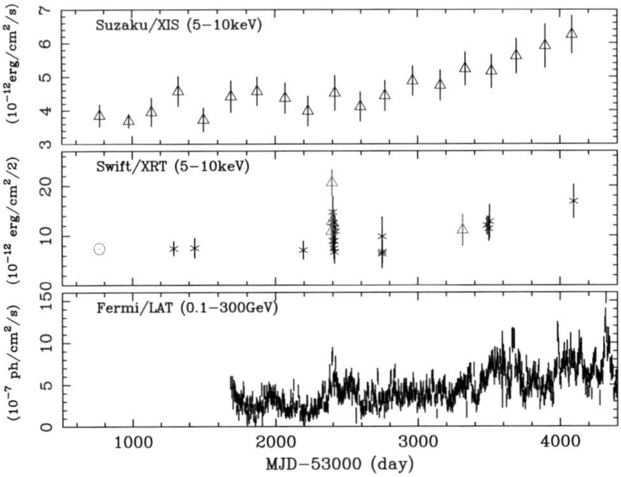

Figure 2. Fermi GeV gamma-ray, Suzaku X-ray, and Swift/XRT X-ray light curves of NGC 1275 (Fukazawa *et al.* 2018).

small fraction of Fe-K emission is predicted. On ther other hand, GeV/TeV gamma-ray emission has been detected and broad-band SED including X-ray band can be represented by jet SSC model. Then, does X-ray continuum come from disk/corona or jet?

Fukazawa *et al.* (2018) reported that Suzaku X-ray monitoring observations every half year shows a recent gradual flux increase as well as GeV gamma-ray, and X-ray and GeV flux correlate well with each other. This long-term gradual X-ray and GeV gamam-ray flux increase is likely to come from the radio C3 spot (Nagai *et al.* (2016)), without any change of SED shape. Swift/XRT observed this galaxy several times, and a short-term flare was detected at the GeV gamma-ray flare in 2010, where the GeV spectrum became hard, suggesting a freshly accelerated electrons. This indicates that a jet emission also exists in the X-ray band of NGC 1275.

Fukazawa et al. (2014) discussed about X-ray emission properties of radio galaxies in terms of Fe-K line, X-ray spectral index, X-ray variability, [O III] line, and optical type. For low excitation radio galaxies (LERG), which has a low accretion rate, X-ray emission is likely to be a jet-origin. For high excitation radio galaxies (HERG), X-ray emission is likely to be a disk/corona-origin. NGC 1275 is an intermediate type between LERG and HERG, and this is consistent with that X-ray emission is contributed from both disk/corona and jet.

In the near future, XARM observations will give us opportunity to measure a variability of Fe-K line and 10–20 keV continuum. Also Athena can provide more on Fe-K line spectroscopy with a good S/N and a large EW avoiding a cluster emission. These are important to constrain the Fe-K origin and also probe a central engine of NGC 1275.

References

Abdo, A. A., Ackermann, M., & Ajello, M., *et al.* 2009, *ApJ*, 699, 31
Churazov, E., Forman, W., Jones, C., & Böhringer, H. 2003, *ApJ*, 590, 225
Fabian, A. C., Walker, S. A., Pinto, C., Russell, H. R., & Edge, A. C. 2015, *MNRAS*, 451, 3061
Fukazawa, Y., Hiragi, K., Mizuno, M., *et al.* 2011, *ApJ*, 727, 19
Fukazawa, Y., Finke, J., Stawarz, Ł., *et al.* 2015, *ApJ*, 798, 74
Fukazawa, Y., Furui, S., Hayashi, K., *et al.* 2016, *ApJ*, 821, 15
Fukazawa, Y., Shiki, K., Tanaka, Y., *et al.* 2018, *ApJ*, 855, 93
Hitomi Collaboration, Aharonian, F., Akamatsu, H., *et al.* 2018, *PASJ*, 70, 13
Nagai, H., Chida, H., Kino, M., *et al.* 2016, *Astronomische Nachrichten*, 337, 69
Salomé, P., Combes, F., Edge, A. C., *et al.* 2006, *A&A*, 454, 437
Scharwächter, J., McGregor, P. J., Dopita, M. A., & Beck, T. L. 2013, *MNRAS*, 429, 2315 1
Shu, X. W., Yaqoob, T., & Wang, J. X. 2011, *ApJ*, 738, 147
Yamazaki, S., Fukazawa, Y., Sasada, M., *et al.* 2013, *PASJ*, 65, 30

Perseus in Sicily: From Black Hole to Cluster Outskirts
Proceedings IAU Symposium No. 342, 2018
K. Asada, E. de Gouveia Dal Pino, M. Giroletti,
H. Nagai & R. Nemmen, eds.

doi:10.1017/S1743921318005537

Life-cycles & Energetics of Radio-Loud AGN

V. H. Mahatma, M. J. Hardcastle, W. L. Williams + LOFAR Surveys team

Centre for Astrophysics Research, School of Physics, Astronomy and Mathematics, University of Hertfordshire, College Lane, Hatfield AL10 9AB, UK
email: v.mahatma2@herts.ac.uk

Abstract. Radio jets are the large-scale and extragalactic footprints of accretion onto supermassive black holes, and are suggested to be the key ingredient controlling the galaxy stellar mass function. Of particular importance is their jet power - the time-averaged energetic feedback into their environment. Hence, the dynamics, energetics and life-cycles of radio-loud AGN (RLAGN) must be understood in order to build a qualitative and quantitative picture of their impact over cosmic time. Here, we present a study of the spectral age of two powerful, cluster-center radio galaxies, and compare with an analytic model to robustly determine their jet powers. We also present some recent LOFAR observations of the different phases of RLAGN activity, namely the remnant and subsequent restarting phases, which are key to understanding the dynamics of RLAGN over their total lifetime.

Keywords. galaxies: jets, galaxies: nuclei, radio continuum: galaxies

1. Introduction

Radio galaxies or RLAGN drive powerful jets into their surrounding environment. This can have profound implications for cosmic evolution: the well known AGN feedback mechanism (e.g. Croton *et al.* 2006; Wyle *et al.* 2016) attempts to account for the shape of the galaxy stellar mass function – there is a steep decline of galaxies with increasing stellar mass. This can naturally be explained if heating by AGN outflows suppresses star-formation in the most massive galaxies (e.g. McNamara & Nulsen 2007), while this mechanism can also similarly explain the lack of cold gas at the centres of the thermal intra-cluster medium (ICM) in galaxy clusters (Smith *et al.* 2002; Mingo *et al.* 2014). It is therefore important to understand and quantify the energetics of radio galaxies over their lifetime.

The jet power, the time-averaged energetic output of the radio source, is therefore a crucial ingredient of galaxy and cluster evolution models. The determination of the jet power, however, requires an accurate estimate of the age of the radio source, which historically has been difficult to obtain. An estimation of the dynamical age of the source requires some model of the lobe advance speed, which itself is a time-dependent value since the radio jets are expected to decelerate on large scales due to their dense environments. On the other hand, the physical process of spectral ageing (discussed below), has been invoked to determine the age of the radio source, although traditionally the spectral age of a given radio galaxy almost always seems to be underestimated relative to its dynamical age (e.g. Eilek 1996; Harwood *et al.* 2013; Harwood *et al.* 2015).

With future wide area radio surveys (e.g. the LOFAR Two-Metre Sky Survey, ASKAP, MeerKAT and eventually, the SKA itself) on the horizon, it is essential that methods and techniques used to determine jet powers are available, in order to quantify the energetic impact of the population of RLAGN. Low-redshift studies of powerful sources are the necessary precursors for these future population-based studies.

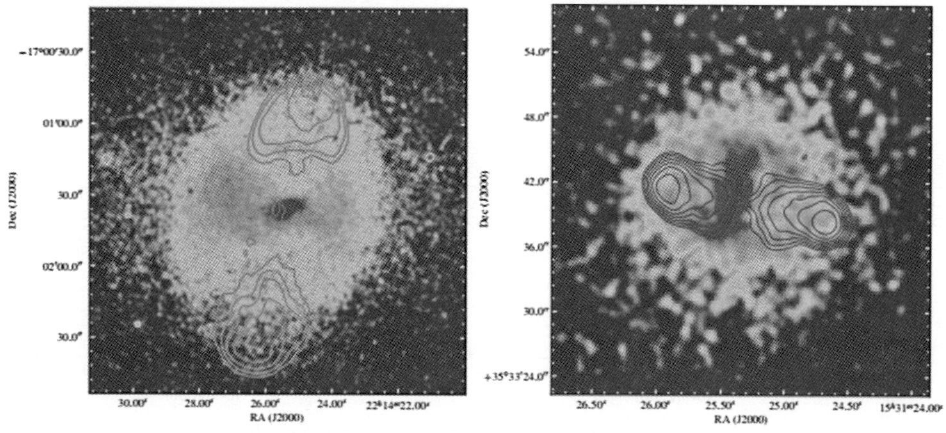

Figure 1. X-ray *Chandra* observations of the cluster environments of 3C444 and 3C320, with contours from archival observations of the radio source overlaid.

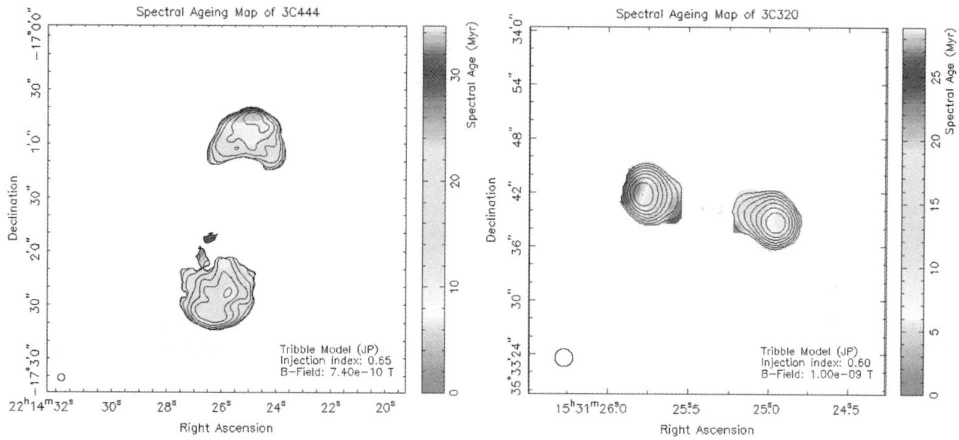

Figure 2. Spectral age maps of 3C444 and 3C320 produced by BRATS Harwood *et al.* (2013).

2. Spectral ageing

Spectral ageing is the radiative cooling process by which the particles in the lobes of radio galaxies emit synchrotron emission. This produces a distinct and characteristic (age-dependent) spectrum at GHz radio frequencies, and can be used to determine the age of the population of electrons in the lobe producing the emission.

3C320 and 3C444, two powerful radio galaxies sitting in rich cluster media emitting hot X-ray gas (see Figure 1), were observed with the VLA in its broad-band setup at multiple array configurations, and at 1.4 GHz and 6 GHz, producing well sampled data sets with a broad frequency coverage enabling robust measurements of their lobe spectra. Using the `BRATS` software package (Harwood *et al.* 2013), which fits contemporary models of ageing spectra to the observed spectra on a pixel by pixel basis, we determined the spectral ages of 3C320 and 3C444 and, for comparison purposes, their dynamical ages based on a model of the lobe advance speed using X-ray Chandra data. Their spectral and dynamical ages agree relatively well, based on certain assumptions regarding the lobe magnetic field strengths and X-ray shock dynamics, contrary to previous studies on the discrepancy of spectral and dynamical ages.

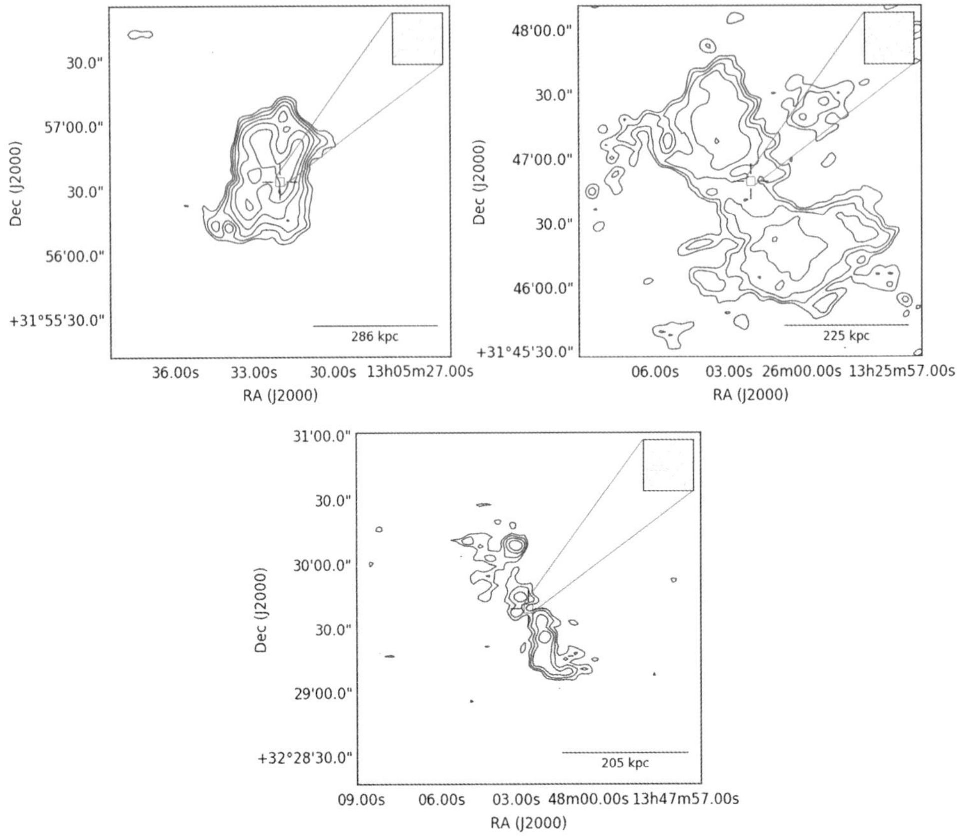

Figure 3. LOFAR observations of candidate remnants in the Herschel-ATLAS field (Mahatma *et al.* 2018) (blue contours) overlaid on follow-up VLA observations showing a lack of a radio core – see zoom-in.

To test the spectral ageing model in determining accurate ages, we use the model of Hardcastle (2018) – a simulation based analytic model of the evolution of radio galaxies. Using physical source parameters based on our observations, we produced a model for the time-evolution of 3C320 against physical source parameters such as radio luminosity and source size, as a function of jet power. We obtained self-consistent dynamical models that agree with our radio and X-ray measurements, and they show that the spectral age/dynamical age discrepancy increases as the lobe magnetic field goes towards the equipartition value. This shows that the lobes of radio galaxies are not in equipartition, agreeing with many previous studies. Our spectral ages are therefore robust, and give accurate determinations of jet powers of radio galaxies.

Of further importance are the jet powers of the population of RLAGN. Hardcastle *et al.* (2019) inferred the jet kinetic luminosity function, essentially the number density of sources as a function of jet power, for the ($\sim 23,000$) objects selected as RLAGN from the LOfar Two-metre Sky Survey (LoTSS: Shimwell *et al.*, submitted, Duncan *et al.* 2019, Williams *et al.* 2019) of the HETDEX field. Their result that RLAGN as a population, with jet powers inferred from their radio luminosities and an assumption of their environment, produce sufficient energy to overcome radiative cooling of the hot, X-ray emitting gas in the centers of clusters is an important requirement of AGN feedback models. Even with these advances, the energetic impact of the other evolutionary phases

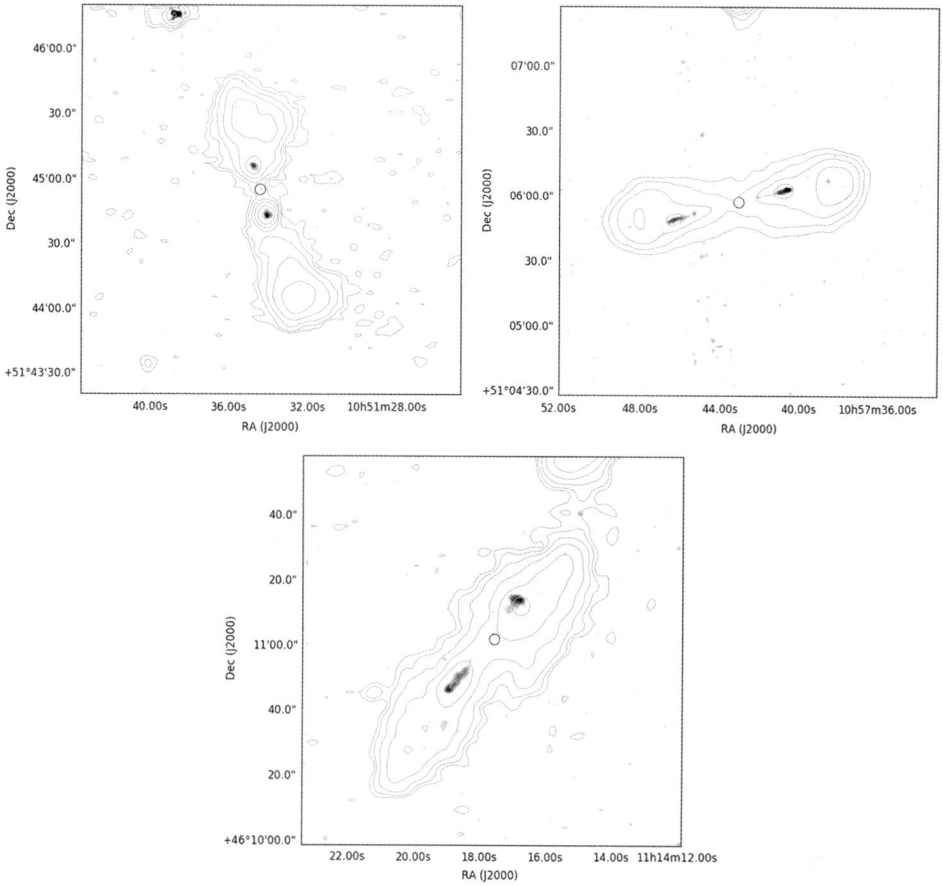

Figure 4. LOFAR observations of DDRGs in the HETDEX field (Mahatma *et al.*, submitted) (orange contours) overlaid on follow-up VLA observations (gray-scale) confirming the presence of inner jets.

of RLAGN are far less understood, namely the remnant (switched off) and restarting phases, observations of which have been severely limited in the past.

3. Remnant and Restarting Radio-Loud AGN

With the current best low-frequency radio telescope at ~ 5 arcsec resolution capable of wide-area observations, LOFAR, RLAGN are beginning to be observed down to low radio luminosities ($L_{150} \sim 10^{24}$ W Hz^{-1}) (Hardcastle *et al.* 2019). Studies of such low surface brightness sources have given the opportunity for large-sample studies of the remnant and subsequent restarting phases of radio galaxies – crucial to our understanding of the overall impact of RLAGN.

LOFAR observations in *Herschel*-ATLAS field (Hardcastle *et al.* 2016) have enabled candidate remnant RLAGN to be captured, and sampled on a systematic level (Mahatma *et al.* 2018), a few of which are shown in Figure 3. Crucially, the remnant fractions of ~ 10 per cent found by Brienza *et al.* (2017), Godfrey *et al.* (2017) and Mahatma *et al.* (2018) imply a rapid fading process for the synchrotron plasma left behind by the previously active AGN.

RLAGN, particularly for the class of low-excitation radio galaxies (LERGs: e.g. Hardcastle *et al.* 2007), are expected to restart some time after the beginning of the remnant phase. The class of double-double radio galaxies (DDRGs: Schoenmakers *et al.* 2000) are the most clearly visible restarting sources found in radio observations, showing two pairs of radio lobes – an inner pair denoting the restarted active jets, and an outer pair denoting the remnant lobes from the previous activity. One of the outstanding questions surrounding the nature of DDRGs is what causes the jets to switch to a remnant and subsequent restarting phase: if the host galaxies of DDRGs are significantly different in physical appearance, such as in their integrated stellar emission or colour, which may be caused by galaxy mergers, then a host galay dichotomy between DDRGs and single-cycle RLAGN should exist. With the first data release of LoTSS, compromising of LOFAR observations in the HETDEX field at 6 arcsec resolution with a median sensitivity of 70μJy beam^{-1}, we visually identified a sample of 40 DDRGs, with follow-up VLA observations at 1.3 arcsec, which confirmed the presence of inner edge-brightened jets typical of the restarted jets of DDRGs, and hence allowed us to confirm or reject DDRGs in our sample. With a final robust sample of 33 DDRGs (a few of which are shown in Figure 4), we compared their host galaxy properties with those of the RLAGN sample of Hardcastle *et al.* (2019), matched in radio luminosity. We find that the host galaxy rest-frame absolute magnitudes and colours (optical and near infra-red K_s, r and $WISE$ bands) for the DDRGs are indistinguishable compared to RLAGN. This confirms, particularly for LERGs which perhaps dominate our selection samples, that restarted activity is not driven by significant changes in the host galaxy, and is therefore a normal process in the lifetime of powerful RLAGN.

These findings have constrained useful information on the nature of RLAGN and their various phases of evolution. With the upcoming data release of LoTSS, and the advent of future wide-area sensitive radio surveys such as the SKA, high fidelity observations of radio galaxies will routinely be available, enabling further insights into the energetics and lifetimes of RLAGN.

References

Brienza, M., Godfrey, L., Morganti, R., Prandoni, I., Harwood, J., Mahony, E. K., Hardcastle, M. J., Murgia, M., Röttgering, H. J. A., Shimwell, T. W., & Shulevski, A. 2017, *Astronomy & Astrophysics*, 606, A98

Croton, D. J., Springel, V., White, S. D. M., De Lucia, G., Frenk, C. S., Gao, L., Jenkins, A., Kauffmann, G., Navarro, J. F., & Yoshida, N. 2006, *MNRAS*, 365, 11–28

Duncan *et al.* 2019, https://doi.org/10.1051/0004-6361/201833562

Eilek, J. A. 1996, Astronomical Society of the Pacific Conference Series, 100, 281

Godfrey, L. E. H., Morganti, R., & Brienza, M. 2017, *MNRAS*, 471, 891

Hardcastle, M. J., Evans, D. A., & Croston, J. H. 2007, *MNRAS*, 376, 1849

Hardcastle, M. J. 2018, *MNRAS*, 475, 2768

Hardcastle *et al.* 2019, https://doi.org/10.1051/0004-6361/201833893

Harwood, J. J., Hardcastle, M. J., Croston, J. H., & Goodger, J. L., 2013, *MNRAS*, 435, 3353

Harwood, J. J., Hardcastle, M. J., & Croston, J. H. 2015, *MNRAS*, 454, 3403

Mahatma, V. H., Hardcastle, M. J., Williams, W. W. *et al.* 2018, *MNRAS*, 475, 4557

McNamara, B. R., & Nulsen, P. E. J. 2007, *Annual review of Astronomy and Astrophysics*, 45, 117

Mingo, B., Hardcastle, M. J., Croston, J. H., Dicken, D., Evans, D. A., Morganti, R., & Tadhunter, C. 2014, *MNRAS*, 440, 269

Schoenmakers, A. P., de Bruyn, A. G., Röttgering, H. J. A., van der Laan, H., & Kaiser, C. R. 2000, *MNRAS*, 315, 371–380

Smith, D. A., Wilson, A. S., Arnaud, K. A., Terashima, Y., & Young, A. J. 2002, *ApJ*, 565, 195

Williams *et al.* 2019, https://doi.org/10.1051/0004-6361/201833564

Wylezalek, D., & Zakamska, N. L. 2016, *MNRAS*, 461, 3724–3739

Perseus in Sicily: From Black Hole to Cluster Outskirts
Proceedings IAU Symposium No. 342, 2018
K. Asada, E. de Gouveia Dal Pino, M. Giroletti,
H. Nagai & R. Nemmen, eds.

doi:10.1017/S1743921318005756

Deep Chandra observations of the core of the Perseus cluster

Jeremy S. Sanders

Max-Planck-Institut für extraterrestrische Physik,
Giessenbachstr. 1, 85748 Garching, Germany
email: jsanders@mpe.mpg.de

Abstract. The Perseus cluster is the X-ray brightest cluster in the sky and with deep Chandra observations we are able to map its central structure on very short spatial scales. In addition, the high quality of X-ray data allows detailed spatially-resolved spectroscopy. In this paper I review what these deep observations have told us about AGN feedback in clusters, sloshing and instabilities, and the metallicity distribution.

Keywords. galaxies: clusters: individual (Perseus cluster, Abell 426)

---

1. Introduction

The Perseus cluster is the X-ray-brightest galaxy cluster in the sky. The central galaxy, NGC 1275 contains the well-known radio source 3C 84. It is also the prototypical 'cool core' cluster, showing short central cooling times. Given its proximity ($z \sim 0.018$) and brightness, the cluster is an ideal location to study the interaction of an active galactic nucleus (AGN) with its surroundings.

The Chandra X-ray observatory (Weisskopf *et al.* 2002) was launched in 1999. With its sub-arcsecond spatial resolution it has revealed an increasing amount of detail in a series of increasingly deep observations of Perseus. The initial findings in 24 ks of data were reported in Fabian *et al.* (2000), with deeper observations totalling ~ 200 ks in Fabian *et al.* (2003a), and very deep observations totalling ~ 900 ks in Fabian *et al.* (2006). This was later followed by 500 ks of wider-field observations making use of the Chandra ACIS-I array (Fabian *et al.* 2011).

2. The central cavities and shock

Galaxy clusters are filled with a hot X-ray emitting plasma, the intracluster medium (ICM), which dominates the baryonic component. Using ROSAT, Böhringer *et al.* (1993) observed that there were two central cavities in the X-ray emission. These cavities were coincident with the radio lobes or bubbles of the central radio source 3C 84. As the jets from the nucleus impact with the ICM, the bubbles are formed. The thermal ICM is displaced by the material in the bubbles, leading to cavities in the X-ray emission. Fabian *et al.* (2000) revealed these cavities with Chandra observations in much greater detail. Since then, they have been discovered in the majority of nearby clusters with short central cooling times (for a review see Fabian 2012). This is important because in many of these systems the ICM would be rapidly cooling in the centre, but evidence for this is not seen in their X-ray spectra or at other wavelengths. Energetically, the enthalpy of the material in the bubbles ($4PV$ if relativistic; Churazov *et al.* 2002) could be available to replace the energy lost by cooling. Indeed, if the heating power (calculated

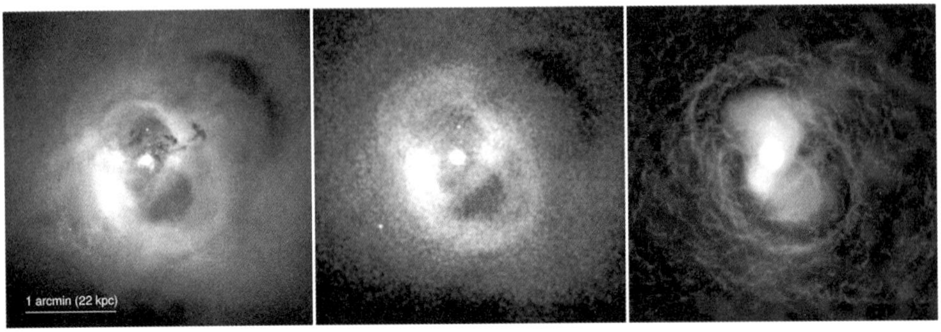

Figure 1. (Left panel) Full band image of the centre of Perseus. (Middle panel) Pressure sensitive, 3.5 to 8 keV image. (Right panel) Edge-filtered 3.5 to 8 keV image in red, with 330 MHz radio overlayed in cyan. North is to the top and east is to the left in images unless otherwise noted.

from the enthalpy and assuming some timescale) is compared to the cooling luminosity for a sample of clusters and groups, there is good agreement.

Fig. 1 (left panel) shows the Chandra image of the centre of the cluster. Either side of the nucleus are the cavities in the X-ray emission. In this image are also soft filaments (discussed in Section 4) and the shadow of a galaxy along the line of sight to the galaxy cluster blocking out soft X-rays (Gillmon *et al.* 2004; Sanders & Fabian 2007). The cavities can be seen more clearly in a hard X-ray image which is more sensitive to X-ray thermal pressure rather than density variations (Fig. 1 centre panel). We have applied an edge filtering technique (Gaussian Gradient Magnitude or GGM) to filter X-ray images to be sensitive to gradients in the X-ray surface brightness (Sanders *et al.* 2016). The pressure-sensitive edge-filtered image is shown in Fig. 1 (right panel), overplotting an image of the 330 MHz radio emission from the radio bubbles.

Fabian *et al.* (2000) found that the material in the bright X-ray emitting rims of the cavities was around 3 keV in projected temperature, cooler than the surrounding 4 keV ICM. Beyond the bright cavity rims, Fabian *et al.* (2003a) identified the thick shell around the northern cavity as a weak shock with a Mach number of around 1.5. There is also likely a second shock around the southern cavity, although the projected cooler material makes this harder see (Fig. 1 centre and right panels show the second southern shock more clearly). The inflation of the central cavities by the radio jet is therefore shocking the surrounding ICM. Although there is a clear density and pressure jump, surprisingly there was not the expected temperature jump across the edge (Graham *et al.* 2008). Possible explanations for the lack of temperature jump include mixing of cooler filamentary gas at the shock (the filaments abruptly end at the shock front). If the material in the bubbles is relativistic, it can release its $4PV$ enthalpy into the ICM, where V is the cavity volume (estimated from the area on the sky) and P is the pressure of the material (inferred from the pressure of the surrounding X-ray emitting gas assuming it is close to pressure equilibrium). Graham *et al.* (2008) measured $3.5PV$ excess energy in the shocked region, close to the cavity enthalpy.

To the north-west of the image is visible a third cavity, previously seen in Einstein observatory images (Fabian *et al.* 1981; Branduardi-Raymont *et al.* 1981). There is another cavity outside the image to the south. These outer cavities are seen in low frequency emission (Fabian *et al.* 2002). These bubbles were likely formed in an earlier outburst of the central nucleus and have buoyantly risen in the gravitational potential of the cluster.

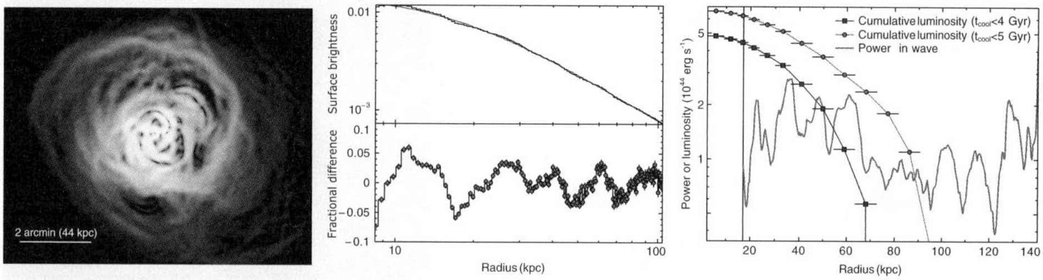

Figure 2. (Left panel) GGM-filtered X-ray image of Perseus, showing gradients on 4 arcsec (1.5 kpc) scales. (Centre panel) Surface brightness profile along an east-southeast sector (Sanders & Fabian 2007) and residuals to a smooth model. (Right panel) Sound wave power as a function of radius, if they are generating the observed surface brightness fluctuations, compared to the cumulative cooling luminosity in regions where the cooling time is < 4 or < 5 Gyr, from Sanders & Fabian (2007).

3. Sound waves and feedback

Fabian *et al.* (2003a) discovered a series of surface brightness ripples around the core of the cluster (Fig. 2 left and centre panels). An explanation is that these could be sound waves generated by the central AGN feedback. If this is the case, then if they can be dissipated in the ICM then they could provide a mechanism by which the AGN could heat the core of the cluster in a distributed fashion. Sound waves are seen in simulations of AGN feedback (e.g. Ruszkowski *et al.* 2004; Sijacki & Springel 2006).

Sanders & Fabian (2007) calculated the energy in the ripples assuming that they are spherical sound waves (Fig. 2 right panel). The sound waves can carry several 10^{44} erg s^{-1} , a substantial fraction of the energy lost by X-ray radiation. Combined with the from weak shocks in the centre, the combination could plausibly heat the cooling region. The amplitude of the ripples has an e-folding decay length of ~ 50 kpc, similar to the radius of the cooling region.

Alternatively, turbulence may be the mechanism by which heating is distributed over the core of the cluster. If the density fluctuations are used to infer turbulence via simulations (Zhuravleva *et al.* 2014), assuming that the fluctuations do not have another origin, there is a good correspondence between the inferred heating rate and the radiative cooling rate in radial bins in Perseus.

If turbulence is responsible it has been noted that the responsible gravity waves travel slowly radially and there is not a great deal of energy in turbulence in Perseus as seen by Hitomi (Fabian *et al.* 2017). The speed of sound is much faster than the gravity wave radial velocity and does not have this problem. Gravity waves would be required to be created at each radius, by cavities for example, to have a close heating-cooling balance. Future deep observations with XRISM and Athena (Croston *et al.* 2013) will tell us in detail what the velocity structure of the gas is in Perseus and other systems, allowing us to know which heating mechanisms are the most important.

4. The filamentary nebula

The core of the cluster contains a well-known filamentary nebula (Kent & Sargent 1979) spanning out to 70 kpc in radius. These filaments are also seen in soft X-rays by Chandra (Fig. 3; Fabian *et al.* 2003b). The morphology of the filaments behind the outer cavity to the north-west suggests that they are tracing the velocity structure of the intracluster medium, as also suggested by their velocity structure (Hatch 2006), although not all the filaments appear to be dragged out by bubbles (Gendron-Marsolais *et al.* 2018). It has been suggested that particle heating may be responsible for the emission of these filaments (Ferland *et al.* 2009).

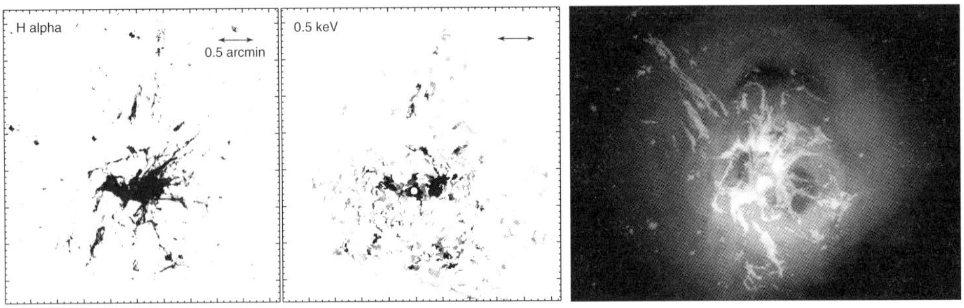

Figure 3. (Left and centre panels) Comparison of the Hα nebula with the X-ray emitting gas of a temperature ~ 0.5 keV (Sanders & Fabian 2007). The X-ray map was made by spectral fitting of gas in regions with a model made up of multiple temperature components. (Right panel) Chandra X-ray image of the cluster in blue with an Hα image of the filaments from Conselice *et al.* (2001) in red, rotated so that the outer bubble is upwards.

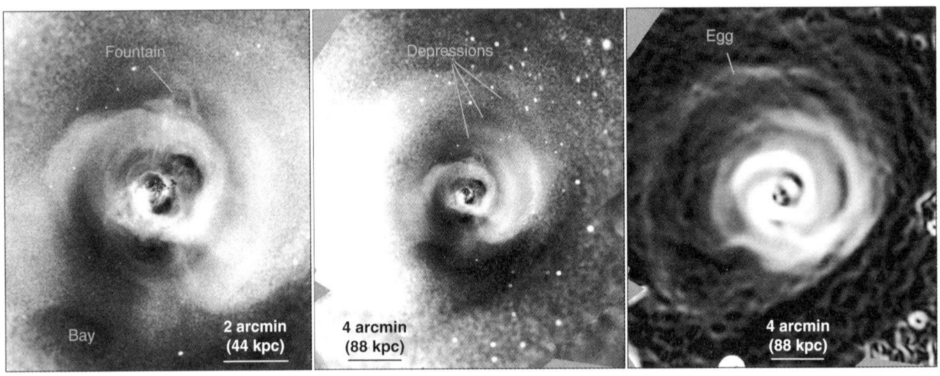

Figure 4. (Left panel) Adaptively-smoothed image of the cluster, showing the fractional difference from the average surface brightness at each radius. (Centre panel) Zoomed-out image of the fractional difference from each radius. (Right panel) GGM-filtered log X-ray image, showing the gradient on scales of 16 arcsec.

5. Sloshing and instabilities

Fig. 4 (left and centre panels) show the fractional difference from the radial average of adaptively smoothed Chandra images of the cluster. In the centre we see a spiral structure. This is cooler and denser gas within the intracluster medium. In addition, there is a strong enhancement on surface brightness 8 arcmin to the east of the cluster core, and 4 arcmin to the west of the cluster core. On larger scales there is a east-west asymmetry in the X-ray surface brightness (Churazov *et al.* 2003).

Spiral structures and alternating enhancements and decrements in surface brightness and temperature are characteristic of gas sloshing within the potential well of a cluster (for a review see Markevitch & Vikhlinin 2007). These regions are separated by contact discontinuities where the pressure is continuous, but the temperature and temperature jump.

Several such features are seen in Perseus. These edges in surface brightness and temperature are seen out to 700 kpc radius (Simionescu *et al.* 2012; Urban *et al.* 2014). There is a chain of galaxies in the western side of the cluster which may be responsible for the generation of the sloshing (Churazov *et al.* 2003).

The concave-shaped Bay region to the south could be an ancient cavity (e.g. Fabian *et al.* 2011), although there is no associated low frequency radio emission. Another possibility might be that it is a 50 kpc roll generated by Kelvin-Helmholtz instabilities

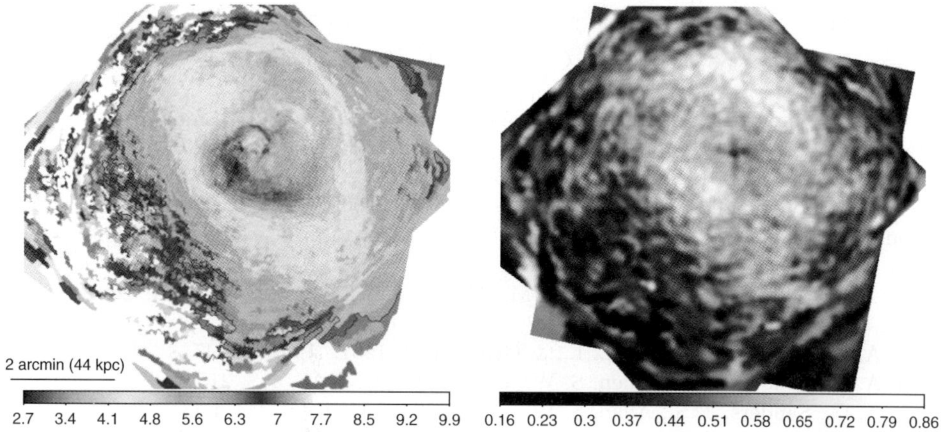

Figure 5. Temperature and metallicity maps, created by spectral fitting spectra from regions containing 10^4 counts (Sanders & Fabian 2007). The maps have been smoothed by Gaussians of 2 and 8 arcsec, respectively. The units of temperature are keV and the metallicity $Z_\odot$.

(Walker *et al.* 2017). If so, comparisons with simulations suggest that a magnetic field pressure of $\beta = 200$ is required to reproduce this feature.

6. Larger scale structures

Also on large scales to the north (Fig. 4) are several depressions in surface brightness (Sanders & Fabian 2007). The largest of these to the north (known as the northern trough; Fabian *et al.* 2011) is a 10 per cent drop in surface brightness and lies along the main jet axis to the north and is also along the axis of the long Hα filament. It is plausible that the trough may be the cavity associated with bubbles which dragged out this filament. If so, then it would be equivalent in energy (estimated using $4PV$ enthalpy) of 11 of the bubbles in the centre ($\sim 10^{60}$ erg). Alternatively, it might be the accumulation of several episodes of feedback. The radio mini halo of the cluster also extends in this northern direction to the radius of the northern trough (Fabian *et al.* 2011).

The entire core of the cluster is surrounded by an egg-shaped region of several hundred kpc radius (Fig. 4 right panel). There is, in particular, a very sharp gradient to the south-west of the nucleus. The edge of the radio mini halo also located here. It is unclear what the origin of this structure is, as a cold front would not be expected to be on all sides of the cluster.

7. The metallicity distribution

The centre of the cluster shows a strong metal enhancement, though the metallicity declines in the innermost 20 kpc (Fig. 5). This does not appear to be a bias due to inadequate modelling of the spectrum (Sanders & Fabian 2007). There are also blobs of high metallicty gas (e.g. 40 kpc to the south-west of the cluster core). These have scales of ~ 5 kpc and total $\sim 3 \times 10^4\ M_\odot$ of excess iron.

One explanation is that these are high metallicity material which is dragged out by a rising cavity in the cluster. The spatial connection between the cavities and metals in Perseus, is however, not that clear. In a sample of systems a connection is found between the radius of the metals and jet power (Kirkpatrick & McNamara 2015).

The presence of these blobs shows they must be able to survive within the ICM for reasonable timescales, also providing additional constrains on the motions of the gas (e.g. Rebusco *et al.* 2005; Graham *et al.* 2006).

References

Böhringer, H., Voges, W., Fabian, A. C., Edge, A. C., & Neumann, D. M. 1993, *MNRAS*, 264, L25
Branduardi-Raymont, G., Fabricant, D., Feigelson, E., *et al.* 1981, *ApJ*, 248, 55
Churazov, E., Forman, W., Jones, C., & Böhringer, H. 2003, *ApJ*, 590, 225
Churazov, E., Sunyaev, R., Forman, W., & Böhringer, H. 2002, *MNRAS*, 332, 729
Conselice, C. J., Gallagher, III, J. S., & Wyse, R. F. G. 2001, *AJ*, 122, 2281
Croston, J. H., Sanders, J. S., Heinz, S., *et al.* 2013, ArXiv e-prints, arXiv:1306.2323
Fabian, A. C. 2012, *ARA&A*, 50, 455
Fabian, A. C., Celotti, A., Blundell, K. M., Kassim, N. E., & Perley, R. A. 2002, *MNRAS*, 331, 369
Fabian, A. C., Hu, E. M., Cowie, L. L., & Grindlay, J. 1981, *ApJ*, 248, 47
Fabian, A. C., Sanders, J. S., Allen, S. W., *et al.* 2011, *MNRAS*, 418, 2154
Fabian, A. C., Sanders, J. S., Allen, S. W., *et al.* 2003a, *MNRAS*, 344, L43
Fabian, A. C., Sanders, J. S., Crawford, C. S., *et al.* 2003b, *MNRAS*, 344, L48
Fabian, A. C., Sanders, J. S., Ettori, S., *et al.* 2000, *MNRAS*, 318, L65
Fabian, A. C., Sanders, J. S., Taylor, G. B., *et al.* 2006, *MNRAS*, 366, 417
Fabian, A. C., Walker, S. A., Russell, H. R., *et al.* 2017, *MNRAS*, 464, L1
Ferland, G. J., Fabian, A. C., Hatch, N. A., *et al.* 2009, *MNRAS*, 392, 1475
Gendron-Marsolais, M., Hlavacek-Larrondo, J., Martin, T. B., *et al.* 2018, *MNRAS*, 479, L28
Gillmon, K., Sanders, J. S., & Fabian, A. C. 2004, *MNRAS*, 348, 159
Graham, J., Fabian, A. C., Sanders, J. S., & Morris, R. G. 2006, *MNRAS*, 368, 1369
Graham, J., Sanders, J. S., & Fabian, A. C. 2008, *MNRAS*, 386, 278
Hatch, N. A., Crawford, C. S., Johnstone, R. M., & Fabian, A. C. 2006, *MNRAS*, 367, 433
Kent, S. M. & Sargent, W. L. W. 1979, *ApJ*, 230, 667
Kirkpatrick, C. C. & McNamara, B. R. 2015, *MNRAS*, 452, 4361
Markevitch, M. & Vikhlinin, A. 2007, *Phys. Rep.*, 443, 1
Rebusco, P., Churazov, E., Böhringer, H., & Forman, W. 2005, *MNRAS*, 359, 1041
Ruszkowski, M., Brüggen, M., & Begelman, M. C. 2004, *ApJ*, 611, 158
Sanders, J. S. & Fabian, A. C. 2007, *MNRAS*, 381, 1381
Sanders, J. S., Fabian, A. C., Russell, H. R., Walker, S. A., & Blundell, K. M. 2016, *MNRAS*, 460, 1898
Sijacki, D. & Springel, V. 2006, *MNRAS*, 366, 397
Simionescu, A., Werner, N., Urban, O., *et al.* 2012, *ApJ*, 757, 182
Urban, O., Simionescu, A., Werner, N., *et al.* 2014, *MNRAS*, 437, 3939
Walker, S. A., Hlavacek-Larrondo, J., Gendron-Marsolais, M., *et al.* 2017, *MNRAS*, 468, 2506
Weisskopf, M. C., Brinkman, B., Canizares, C., *et al.* 2002, *PASP*, 114, 1
Zhuravleva, I., Churazov, E., Schekochihin, A. A., *et al.* 2014, *Nature*, 515, 85

Perseus in Sicily: From Black Hole to Cluster Outskirts
Proceedings IAU Symposium No. 342, 2018
K. Asada, E. de Gouveia Dal Pino, M. Giroletti,
H. Nagai & R. Nemmen, eds.

doi:10.1017/S1743921318005185

Close-up view of an ongoing merger between the NGC 4839 group and the Coma cluster

Natalia Lyskova[1,2], Eugene Churazov[2,3], Congyao Zhang[3], William Forman[4] and Christine Jones[4]

[1]National Research University Higher School of Economics, Myasnitskaya str. 20, Moscow 101000, Russia

[2]Space Research Institute (IKI), Profsoyuznaya str. 84/32, Moscow 117997, Russia
email: natalya.lyskova@gmail.com

[3]Max Planck Institute for Astrophysics, Karl-Schwarzschild-Strasse 1, 85741 Garching, Germany

[4]Harvard-Smithsonian Center for Astrophysics, 60 Garden Street, Cambridge, MA 02138, USA

Abstract. We analyse archival *XMM-Newton* observations of the massive galaxy group NGC 4839 falling into the Coma cluster core, which reveal a complex morphology for the merger. By comparing high quality X-ray maps of the merging subcluster with SPH simulations, we propose an infall scenario which qualitatively reproduces the observed structure of the NGC 4839 tail.

Keywords. galaxies: clusters: individual (Abell 1656 (Coma), NGC 4839 group), X-rays: galaxies: clusters

1. Introduction

According to the current concordance cosmological model, galaxy clusters form hierarchically through rare dramatic merger events and more gentle continuous accretion of smaller objects. The Coma cluster of galaxies, as one of the nearest massive galaxy clusters, provides an excellent laboratory for studies of structure formation processes. Optical and X-ray observations have revealed a wealth of substructures in Coma (Vikhlinin, Forman & Jones 1997; Adami *et al.* 2005; Neumann *et al.* 2003). One of the most prominent is the NGC 4839 group of galaxies. It lies in the cluster outskirts (~ 1 Mpc in projection) South-West from the cluster center. The X-ray image exhibits an edge-like structure at the head of the galaxy group and an elongated tail of ram pressure stripped gas extending toward the South-West (see Fig. 1, upper right panel), i.e. opposite to the direction of the Coma center. The NGC 4839 group appears to be falling onto the Coma cluster core, and the infall direction coincides with a filament connecting the Coma cluster with Abell 1367. It has been debated for a long time (i) whether this infall is radial (Colless & Dunn 1996) or non-radial (Biviano *et al.* 1996; Adami *et al.* 2005), and (ii) whether we observe the first passage (Colless & Dunn 1996; Neumann *et al.* 2001; Ogrean & Brüggen 2013; Akamatsu *et al.* 2013) or if the group has already passed the Coma cluster center (Burns *et al.* 1994).

Here, we analyse available *XMM-Newton* observations and compare X-ray maps with SPH simulations to constrain the 3D geometry and a trajectory of the infalling group. More details can be found in Lyskova *et al.* (2019). All our results are scaled to a flat ΛCDM cosmology with $\Omega_m = 0.3$, $\Omega_\Lambda = 0.7$, and a Hubble constant $H_0 = 70$ km s^{-1} Mpc^{-1}, which implies the linear scale of 27.98 kpc arcmin^{-1} at Coma's redshift $z = 0.0231$ (NED database).

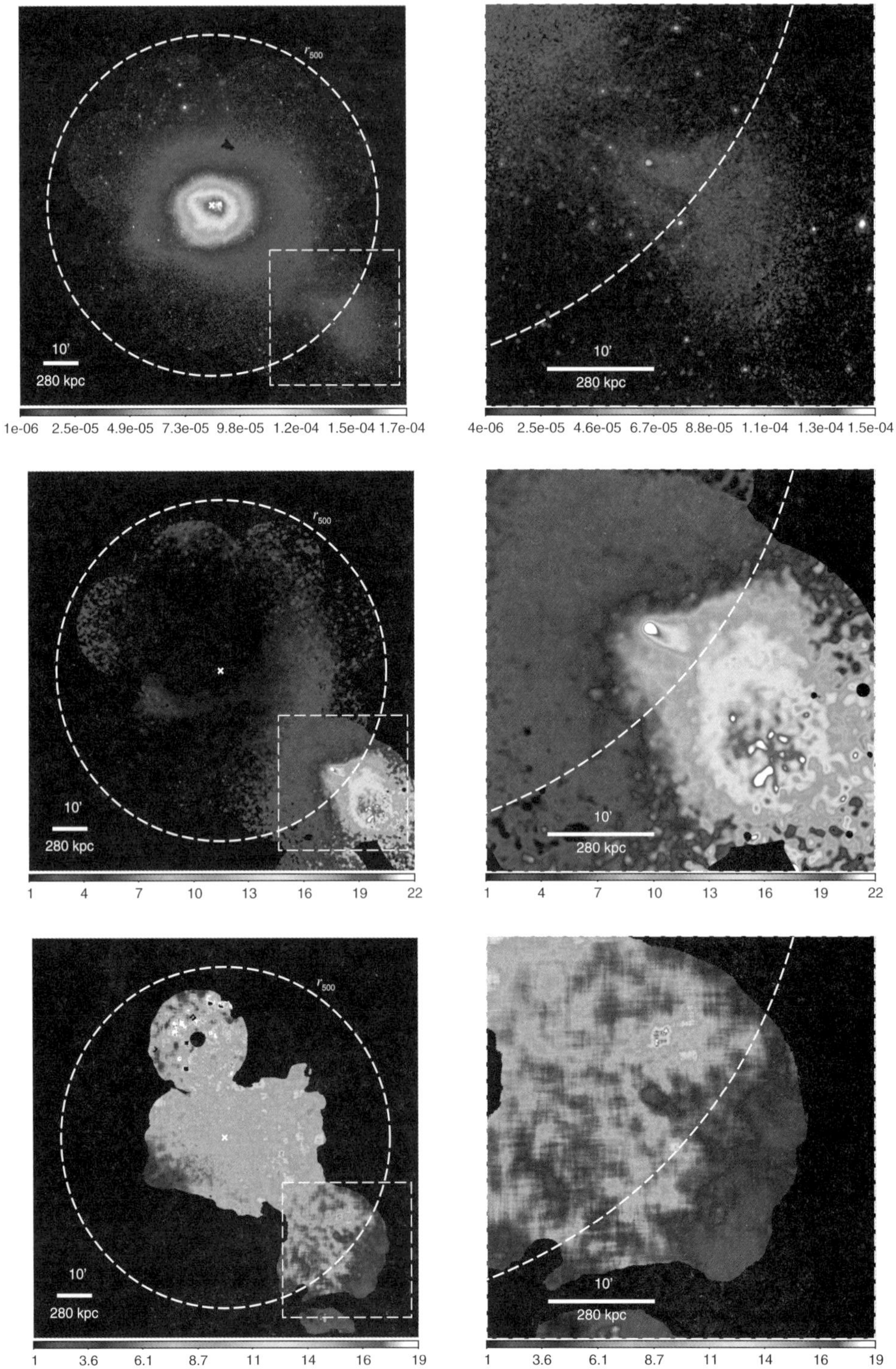

Figure 1. *Upper row: XMM-Newton* image of the Coma Cluster (left panel) and the NGC 4839 group (right panel) in the 0.5–2.5 keV energy band. *Middle row:* Surface brightness excess over the best β-model. *Bottom row:* Projected temperature map. The white dashed circle marks the radius of $r_{500} = 47$ arcmin (Planck Collaboration *et al.* 2013).

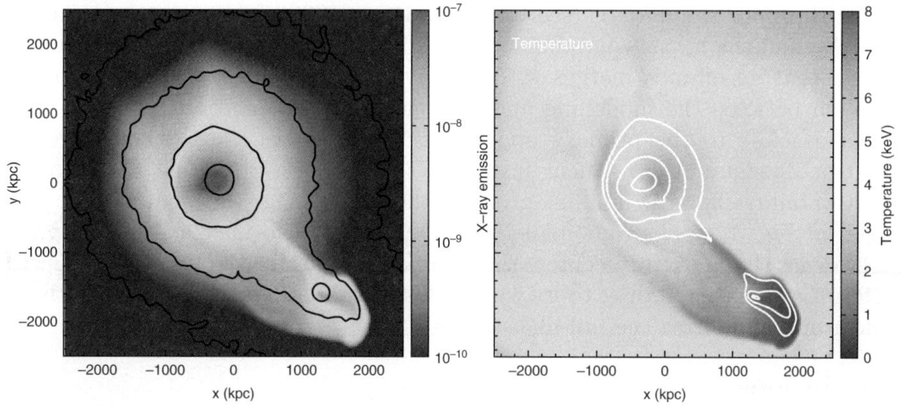

Figure 2. X-ray surface brightness (left panel) and X-ray weighted temperature (right panel) images of our 'best-fit' model at the moment just after the primary apocentric passage of the group. The black and white contours represent the mass surface density and the X-ray surface brightness distributions, respectively. Our model shows a good match to the X-ray morphology of the NGC 4839 group (e.g., its asymmetric tail).

2. The Coma Cluster and the NGC 4839 group

We have analysed publicly available *XMM-Newton* data of Coma and constructed the background-subtracted, exposure- and vignetting-corrected image in the 0.5-2.5 keV energy band (Fig. 1, upper left panel). The zoom-in view of the NGC 4839 group is shown in the upper right panel of Fig. 1.

To describe the surface brightness profile of the Coma cluster, we extracted X-ray counts from within circular annuli centred at (RA, DEC) = (194.94458, 27.95). We exclude the sector containing the NGC 4839 group. The measured distribution is well fit with the one-dimensional β-model (Cavaliere & Fusco-Femiano 1978) with the core radius $r_c = 9.9$ arcmin $\simeq 277$ kpc and $\beta = 0.71$. To remove the global cluster emission and highlight the surface brightness deviations from the smooth underlying model, we divide the X-ray image (Fig. 1, upper row) by the best-fitting β-model. The result is shown in the middle row of Fig. 1. In addition to the extended X-ray emission around and behind (SW) the NGC 4839 group, there is a prominent X-ray excess in the West, elongated in the North-South direction, and a filamentary shaped excess (though not very prominent in this image) to the East of the cluster centre, elongated in the East-West direction (Vikhlinin, Forman & Jones 1997; Neumann *et al.* 2003).

To construct a (projected) temperature map of the Coma cluster, we follow a technique described in Churazov *et al.* (1996). We model the ICM as a single thermal component with the thin-plasma emission code APEC (Smith *et al.* 2001) absorbed by the Galactic hydrogen column density $N_H = 9 \times 10^{19}$ cm^{-2} (Dickey & Lockman 1990). The resulting temperature map is shown in Fig. 1 (bottom row).

3. Comparison with simulations

We performed smoothed-particle hydrodynamic (SPH) simulations† to investigate the merging scenario of the NGC 4839 group. We model the Coma cluster and the group both as spherical objects consisting of a dark matter halo and a gas halo (see Zhang, Yu & Lu 2014, 2015 for a detailed description of the simulation method). We fix the mass of the Coma cluster $M = 1.2 \times 10^{15} M_{\odot}$ and the mass ratio between the main cluster and the

† We adopt the Gadget-2 code throughout this work (Springel *et al.* 2001).

group $\xi = 10$ in our simulations (see Colless & Dunn 1996), but vary the initial relative velocity (V) and the impact parameter (P) to find a 'best-fit' merging configuration.

We find that an off-axis merger with $V = 500$ km s^{-1} and $P = 2000$ kpc provides a good match between the modelling and the real X-ray observations (see Fig. 1), when the group is close to the primary apocentre. Fig. 2 shows the simulation results for our 'best-fit' model at the moment just after the apocentric passage of the group. An asymmetric tail trailing the infalling group in the X-ray image (left panel) is similar with that seen in Fig. 1, upper right panel. This structure is formed in the following way. While crossing the apocentre center, the group slows down, which however contributes a significant gravitational drag to its wake. This overshooting wake subsequently sweeps over the group and forms the tail-like structure.

4. Conclusions

The majority of recent studies (e.g., Ogrean & Brüggen 2013; Akamatsu *et al.* 2013) argue that the NGC 4839 group is most likely to come from the south-west direction along the filament connecting Coma with Abell 1367 and that it has just started to penetrate the Coma ICM. Our analysis of *XMM-Newton* data and comparison of obtained X-ray maps with SPH simulations suggest a possible alternative scenario. Indeed, before the pericenter passage, the tail of the subcluster appears symmetric, while X-ray data reveal a complex structure and an asymmetric shape for the tail of the NGC 4839 group. The group may be caught at the moment just after the apocentric passage. In this merger scenario, rotation of the gas core forms the observed morphology of the tail.

Acknowledgments

This work was supported by the Russian Science Foundation (grant 14-22-00271). NL is thankful to the organizers of the IAU symposium 342 for providing financial support.

References

Adami, C., Biviano, A., Durret, F., & Mazure, A. 2005, *A&A*, 443, 17

Akamatsu, H., Inoue, S., Sato, T., Matsusita, K., Ishisaki, Y., & Sarazin, C. L. 2013, *PASJ*, 65, 89

Biviano, A., Durret, F., Gerbal, D., Le Fevre, O., Lobo, C., Mazure, A., & Slezak, E. 1996, *A&A*, 311, 95

Burns, J. O., Roettiger, K., Ledlow, M., & Klypin, A. 1994, *ApJ*, 427, L87

Cavaliere, A. & Fusco-Femiano, R. 1978, *A&A*, 70, 677

Churazov, E., Gilfanov, M., Forman, W., & Jones, C. 1996, *ApJ*, 471, 673

Colless, M. & Dunn, A. M. 1996, *ApJ*, 458, 435

Dickey, J. M. & Lockman, F. J. 1990, *ARA&A*, 28, 215

Lyskova, N. *et al.* 2019, *MNRAS*, 485, 2922

Neumann, D. M. *et al.* 2001, *A&A*, 365, L74

Neumann, D. M., Lumb, D. H., Pratt, G. W., & Briel, U. G. 2003, *A&A*, 400, 811

Ogrean, G. A. & Brüggen M. 2013, *MNRAS*, 433, 1701

Planck Collaboration *et al.* 2013, *A&A*, 554, A140

Springel, V., Yoshida, N., & White, S. D. M. 2001, *NewA*, 6, 79

Smith, R. K., Brickhouse, N. S., Liedahl, D. A., & Raymond, J. C. 2001, *ApJ*, 556, L91

Vikhlinin, A., Forman, W., & Jones, C. 1997, *ApJ*, 474, L7

Zhang, C., Yu, Q., & Lu, Y. 2014, *ApJ*, 796, 138

Zhang, C., Yu, Q., & Lu, Y. 2015, *ApJ*, 813, 129

Perseus in Sicily: From Black Hole to Cluster Outskirts
Proceedings IAU Symposium No. 342, 2018
K. Asada, E. de Gouveia Dal Pino, M. Giroletti,
H. Nagai & R. Nemmen, eds.

doi:10.1017/S1743921318007020

Thermal and non-thermal connection in radio mini-halos

A. Ignesti[1,2], G. Brunetti[2], M. Gitti[1,2] and S. Giacintucci[3]

[1]DIFA, University of Bologna, Via Gobetti 93/2, 40129 Bologna, Italy
email: alessandro.ignesti2@unibo.it

[2]IRA INAF, Via Gobetti 101, 40129 Bologna, Italy

[3]U.S. NRL, 4555 Overlook Avenue SW, Code 7213, Washington, DC 20375, USA

Abstract. Several cool-core clusters are known to host a radio mini-halo, a diffuse, steep-spectrum radio source located in their cores, thus probing the presence of non-thermal components as magnetic field and relativistic particles on scales not directly influenced by the central AGN. The nature of the mechanism that produces a population of radio-emitting relativistic particles on the scale of hundreds of kiloparsecs is still unclear. At the same time, it is still debated if the central AGN may play a role in the formation of mini-halos by providing the seed of the relativistic particles. We aim to investigate these open issues by studying the connection between thermal and non-thermal components of the intra-cluster medium. We performed a point-to-point analysis of the radio and the X-ray surface brightness of a compilation of mini-halos. We find that mini-halos have super-linear scalings between radio and X-rays, with radio brightness declining more steeply than the X-ray brightness. This trend is opposite to that generally observed in giant radio halos, thus marking a possible difference in the physics of the two radio sources. Finally, using the scalings between radio and X-rays and assuming a hadronic origin of mini-halos we derive constraints on the magnetic field in the core of the hosting clusters.

Keywords. radiation mechanisms: nonthermal, radiation mechanisms: thermal, galaxies: clusters: general

---

1. Introduction

About 70% of relaxed galaxy clusters show cool cores with peaked X-ray surface brightness where the intra-cluster medium (ICM) is efficiently radiating away its thermal energy (e.g., Hudson *et al.* 2010). Those clusters are characterized by the presence of luminous active galactic nuclei (AGN) at their centre (e.g., Sun 2009) and magnetic field of the order of 10 μG (e.g., Carilli & Taylor 2002). Most of them also show diffuse radio emission in their centre in form of ą radio mini-halos (MHs), diffuse radio sources with a radius $R_{\rm MH} < 0.2R_{500} \simeq 100$ kpc (e.g., Giacintucci *et al.* 2017) and steep spectral index ($\alpha > 1$ with $S \propto \nu^{-\alpha}$). The main open problem regarding the origin of MHs is the slow-diffusion problem, which arises from the fact that the diffusion time required to the cosmic rays electrons (CRe) to propagate up to the scale of the radio emission is longer ($> 10^9$ yr) than their radiative time ($\leqslant 10^8$ yr) (e.g., Brunetti & Jones 2014). There are two possible solutions to this problem. One is the leptonic scenario, where the CRe injected by the central AGN are re-accelerated by ICM turbulence (e.g., Gitti *et al.* 2002). The other is the hadronic scenario, where CRe are produced by collisions between cosmic ray protons (CRp) and thermal protons (e.g., Pfrommer & Enßlin 2004). Contrary to the case of giant radio halos, current γ-ray observations do not provide stringent constraints to the hadronic scenario (e.g., Brunetti & Jones 2014; Ahnen *et al.* 2016). In both models the central AGN may play an important role in the origin of MHs as source of CRs.

Table 1. Physical properties of the sample and results of the Monte Carlo *ptp* analysis.

Cluster name	ID	RA$_{J2000}$ [h,m,s]	DEC$_{J2000}$ [deg,',"]	z	R$^{\dagger}_{500}$ [Mpc]	M$^{\dagger}_{500}$ [10^{14} $M_{\odot}$]	Radio freq. [GHz]	$k^{\ddagger}_{\rm MC}$
2A0335	S1(a,b)	03 38 44.4	+09 56 34	0.035	0.92	$2.3^{+0.2}_{-0.3}$	1.4, 5.5	1.2, 1.1
RBS 797	S2	09 47 00.2	+76 23 44	0.345	1.16	$6.3^{+0.6}_{-0.7}$	1.4	1.3
Abell 3444	S3(a,b)	10 23 54.8	−27 17 09	0.254	1.27	$7.6^{+0.5}_{-0.6}$	0.6, 1.4	1.2, 1.2
MS 1455	S4	14 57 15.1	+22 20 34	0.258	0.98	$3.5^{+0.4}_{-0.4}$	0.6	0.9
RXC J1504	S5	15 04 05.4	−02 47 54	0.215	0.98	$7.0^{+0.6}_{-0.6}$	0.3	2.0
RX J1532	S6	15 32 53.8	+30 20 58	0.345	1.04	$4.7^{+0.6}_{-0.6}$	1.4	1.1
RX J1720	S7	17 20 12.6	+26 37 23	0.164	1.24	$6.3^{+0.4}_{-0.4}$	0.6	1.2

Notes. † Radius and total mass at a mean overdensity of 500 with respect to the cosmological critical density at redshift z (e.g., Giacintucci *et al.* 2017); ‡ The dispersions are $\leqslant 10\%$.

2. Thermal and non-thermal connection

Radio MHs represent valuable probes to investigate the properties of cool cores, because they link the thermal and non-thermal components of the ICM. On the one hand, the X-ray emission of the ICM is produced via thermal bremmstrahlung, whose bolometric emissivity is $j_{\rm X} \propto n_{\rm th}^2 T^{1/2}$, where $n_{\rm th}$ and T are the particle density and the temperature of the ICM. On the other hand, the radio synchrotron emissivity depends on the magnetic field and on the energy distribution of CRe. For the hadronic model the injected energy rate of CRe depends on the densities of CRp and thermal protons (e.g., Pfrommer & Enßlin 2004), whereas for the leptonic model the CRe turbulent heating depends on the turbulent energy flux and turbulent dissipation scales (e.g., Brunetti & Lazarian 2016). Thus, both the radio and X-ray emissivities depend on the density of the thermal plasma, inducing possible scalings between the radio surface brightness, $I_{\rm R}$, and the X-ray surface brightness, $I_{\rm X}$, in the form:

$$I_{\rm R} \propto I_{\rm X}^k \tag{2.1}$$

where the index k can be used to study the non-thermal components.

Feretti *et al.* (2001) and Govoni *et al.* (2001) analyzed the scalings between radio and X-ray brightness in giant radio halos by looking at the point-to-point (*ptp*) correlations. They analyzed five radio halos, finding a sub-linear scaling ($k \leqslant 1$). In our work we want to estimate the $I_{\rm R}$-$I_{\rm X}$ scaling for a sample of MHs to study the ICM properties in the cool core.

3. Sample selection and Monte Carlo point-to-point analysis

We select our sample of MHs by collecting from literature radio maps with a resolution suitable for the *ptp* analysis. For these clusters we then reduced archival Chandra observations to obtain the $I_{\rm X}$ images (0.5-2.0 keV). We report in Tab. 1 the properties of our sample and the frequencies of the radio maps that we used.

The *ptp* analysis is a straightforward method to estimate the $I_{\rm R}$-$I_{\rm X}$ connection by fitting the values of $I_{\rm R}$ and $I_{\rm X}$ that are measured in the cells of a mesh tailored on the radio diffuse emission (see Fig. 1). However, different *ptp* analysis performed with different meshes may produce different estimates of k for the same MH. In order to properly account for this bias related to the arbitrary choice of a particular mesh, we introduce the Monte Carlo *ptp* analysis that relies on repeated random sampling of the diffuse radio and X-ray emission. Basically, we perform 1000 cycles of *ptp* analysis with a randomly-generated mesh for each cycle. Each mesh is modeled to sample only the radio emission over a given $I_{\rm R}$ threshold (we used the 3σ level), while excluding the emission not related to the MH (AGN, radio-filled cavities or field sources), and produces a different estimate of k ($k_{\rm SM}$). Therefore, we end up with a distribution of values of k, instead of a

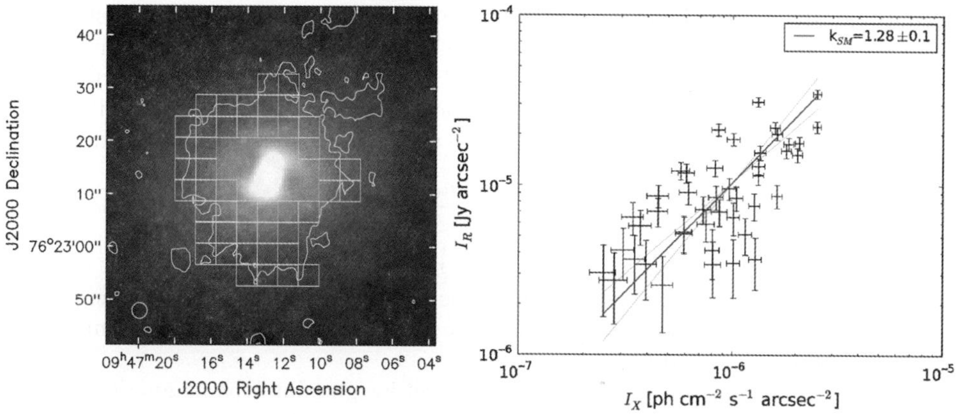

Figure 1. *Left:* Chandra 0.5-2.0 keV X-ray image of RBS 797 with the 3σ contour (white) of the 1.4 GHz radio emission (1σ=10 μJy beam^{-1}, Doria *et al.* 2012) and a sampling mesh (green); *Right:* $I_{\rm R}$ vs $I_{\rm X}$ measured in the cells of the mesh. The index of the best-fit power-law (blue) is reported in the legend.

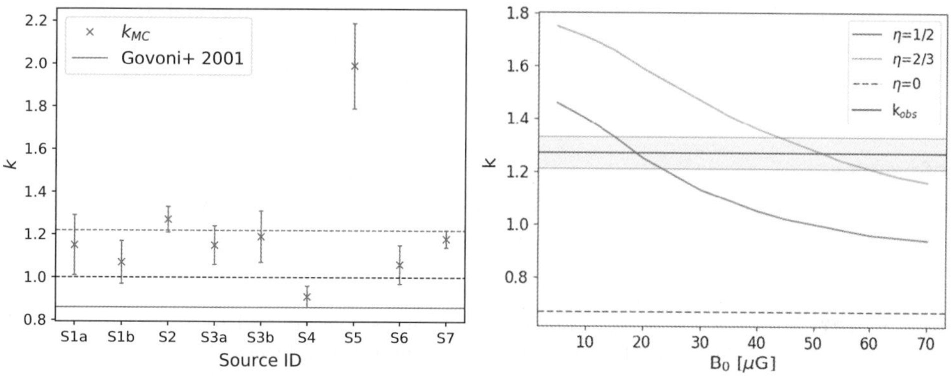

Figure 2. *Left:* Results of the Monte Carlo *ptp* analysis. The source IDs are reported in Tab. 1. The horizontal lines point out the average value of the scalings that we measured for MHs (red), the k=1 level (black) and the mean value observed for radio halos (green). *Right:* Comparison between the observed (green) and the theoretical estimates of k for RBS 797.

single estimate, and our final estimate for the scaling, $k_{\rm MC}$, is given by the mean of this distribution.

4. Results and implications

The results of the Monte Carlo *ptp* analysis are reported in Tab. 1. In Fig. 1 (right panel) we show the scaling observed in RBS 797 as an example. We observe that MHs have super-linear $I_{\rm R}$-$I_{\rm X}$ scaling ($k>1$). This is opposite to what was observed for giant radio halos ($k\leqslant 1$), thus this difference may indicate an intrinsic physical difference between these objects (Fig. 2, left panel).

The $I_{\rm R}$-$I_{\rm X}$ scaling may allow us to constrain the intra-cluster magnetic field. We assumed an hadronic model for the origin of CRe, in this case the radio emissivity is:

$$j_{\rm R} \propto \epsilon_{\rm CRp} n_{\rm th} \frac{B^{\alpha+1}}{B^2 + B_{\rm CMB}^2} \tag{4.2}$$

where $B_{\rm CMB} = 3.25(1+z)^2$ μG is the equivalent magnetic field of the cosmic microwave background and $\epsilon_{\rm CRp}$ is the CRp energy density. In our model we assume that these CRp are continuously injected by the central AGN and diffuse in the core of the cluster. Under stationary conditions, it is $\epsilon_{\rm CRp} \propto Q_{\rm p}/r$ where r is the distance from the central source and $Q_{\rm p}$ is the CRp injected luminosity (Blasi & Colafrancesco 1999). We assume the isothermal β-model (Cavaliere & Fusco-Femiano 1976) to describe the radial distribution of $n_{\rm th}$ inside the cool core and we parametrize the intra-cluster magnetic field as:

$$B(r) = B_0 \left[\frac{n_{\rm th}(r)}{n_0}\right]^{\eta} \tag{4.3}$$

where B_0 and n_0 are the central values of the magnetic field and the number density of thermal particles and η describes the scaling between the ICM density and the magnetic field. For the X-ray emissivity we assume $j_{\rm X} \propto n_{\rm th}^2$, that is the thermal bremmstrahlung emissivity in the case of isothermal core.

Therefore, the scaling $I_{\rm R}$-$I_{\rm X}$ may allow us to get insights into the magnetic field of the ICM and the CR luminosity of the AGN by constraining the free parameters of our model B_0, η and $Q_{\rm p}$. We integrated numerically the emissivity models to obtain a theoretical estimate of k as function of B_0 and η that we compared with the observed scaling. We tested the equipartion ($\eta = 1/2$), the isotropically compressed ($\eta = 2/3$, Tribble 1993) and the uniform ($\eta = 0$) scalings for the magnetic field. Here we report the preliminary results for RBS 797 in Fig.2, right panel. We found that the observed $I_{\rm R}$-$I_{\rm X}$ scaling is reproduced with $B_0 \simeq 20\ \mu G$ for $\eta = 1/2$ and $B_0 \simeq 50\ \mu G$ for $\eta = 2/3$, whereas the uniform magnetic field can not generate a super-linear scaling. Results for the whole sample and the calculation of the γ-ray expectations will be presented in Ignesti et al., in preparation.

References

Ahnen, M. L., Ansoldi, S., Antonelli, L. A., *et al.* 2016, Astronomy and Astrophysics, 589, A33

Blasi, P. & Colafrancesco, S. 1999, Astroparticle Physics, 12, 169

Brunetti, G. & Jones, T. W. 2014, International Journal of Modern Physics D, 23, 30007

Brunetti, G. & Lazarian, A. 2016, Monthly Notices of the RAS, 458, 2584

Carilli, C. L. & Taylor, G. B. 2002, Annual Review of Astronomy and Astrophysics, 40, 319

Cavaliere, A. & Fusco-Femiano, R. 1976, Astronomy and Astrophysics, 49, 137

Doria, A., Gitti, M., Ettori, S., *et al.* 2012, Astrophysical Journal, 753, 47

Feretti, L., Fusco-Femiano, R., Giovannini, G., & Govoni, F. 2001, Astronomy and Astrophysics, 373, 106

Giacintucci, S., Markevitch, M., Cassano, R., *et al.* 2017, Astrophysical Journal, 841, 71

Gitti, M., Brunetti, G., & Setti, G. 2002, Astronomy and Astrophysics, 386, 456

Govoni, F., Enßlin, T. A., Feretti, L., & Giovannini, G. 2001, Astronomy and Astrophysics, 369, 441

Hudson, D. S., Mittal, R., Reiprich, T. H., *et al.* 2010, Astronomy and Astrophysics, 513, A37

Pfrommer, C. & Enßlin, T. A. 2004, Astronomy and Astrophysics, 413, 17

Sun, M. 2009, Astrophysical Journal, 704, 1586

Tribble, P. C. 1993, Monthly Notices of the RAS, 263, 31

Perseus in Sicily: From Black Hole to Cluster Outskirts
Proceedings IAU Symposium No. 342, 2018
K. Asada, E. de Gouveia Dal Pino, M. Giroletti,
H. Nagai & R. Nemmen, eds.

doi:10.1017/S174392131800515X

Perseus – A Huge Reservoir of Dark Matter investigated with MAGIC

Michele Doro[1,2], Joaquim Palacio[3], Javier Rico[3] and Monica Vazquez Acosta[4]

[1]Department of Physics and Astronomy, University of Padova, I-35131 Padova, Italy
email: michele.doro@unipd.it

[2]Istituto Nazionale Fisica Nucleare sez. Padova, I-35131 Padova, Italy

[3]Institut de Fìsica dAltes Energies (IFAE), The Barcelona Institute of Science and Technology (BIST), E-08193 Bellaterra (Barcelona), Spain

[4]Inst. de Astrofisica de Canarias, E-38200 La Laguna, and Universidad de La Laguna, Dpto. Astrofisica, E-38206 La Laguna, Tenerife, Spain

Abstract. Galaxy clusters are excellent targets for high energy astrophysics with gamma rays. Not only they may host active galaxies, but they are often expected to provide signatures of accelerations of electrons and protons to PeV energies. Furthermore, according to ΛCDM scenario, they should be embedded in an extremely massive dark matter halo, the largest halo expected. In this report, we summarize the recently published MAGIC lower limits on the decaying dark matter lifetime using 202 h of selected high quality data taken on the Perseus galaxy cluster, in a 5-year long campaign.

Keywords. Keyword1, keyword2, keyword3, etc.

1. Dark Matter in galaxy cluster

In the current consensus theory called ΛCDM, the gravitational imbalance observed especially in the velocity distribution of stars in galaxies, is coped with the supposition that a new massive particle (or a set of them) populates the region around the galaxies in the form of halos. This new particle is generically known as dark matter (DM). There are no satisfactory Standard Model particles that can be valid candidate for DM. The main search is for weakly interacting massive particles (WIMPs), that are particle in the GeV-TeV mass range, with cross section of the order of the eletroweak one. There are plenty of theories producing valid DM candidates that need to be tested. Gamma rays provide a powerful probe, because, if the DM has some sort of interaction with the standard model (otherwise we would probably never see it), it would be through heavy elementary particles such as gauge bosons or quarks, that soon after production (via DM annihilation or decay) can give rise to an ample broad spectrum photon emission peaked in the MeV-GeV range (Bertone *et al.* (2004)). The highest fluxes for annihilating DM models are where DM is strongly concentrated, such as at the center of closeby satellite galaxies of the Milky Way or at its center. In the case of galaxy clusters, their large distance fosters a reduction of the annihilation signal, however, in case of decaying DM, the expected emission, connected to the total cluster mass, would still sufficiently high to allow for detection in some cases (Sanchez-Conde *et al.* (2011)).

Perseus is a galaxy clusters located at 77 Mpc from the Earth, with a total mass of the order of $10^{15} M_\odot$ mostly in the form of DM. It was subject of a multi-year campaign with the MAGIC telescopes started in 2012 and still ongoing, that so far has gathered more

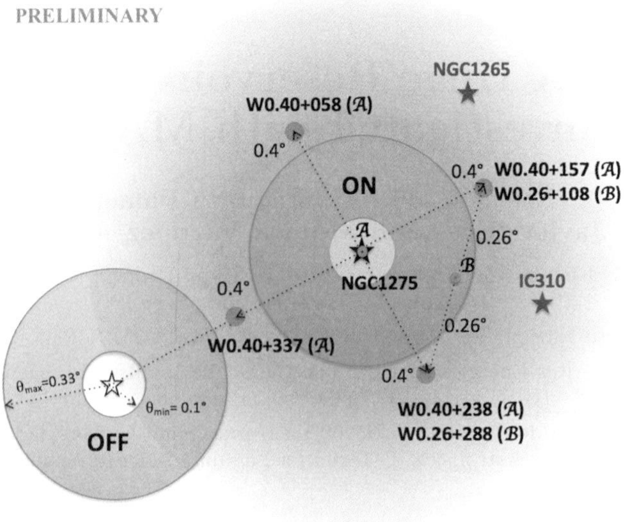

Figure 1. Sketch of the geometry of the region around Perseus observed by MAGIC. Stars marks the galaxies NGC 1275, IC 310 and NGC 1265. The circles mark the signal where the DM signal is searched and the background control region. From MAGIC Coll. (2018).

than 400 h through the years, as well as has allowed for a series of interesting discoveries: the extremely fast and strong activity of the two brightest radiogalaxies (see Fig. 1) hosted in the cluster: NGC 1275 and IC 310, which so far challenged any explanation. Furthermore, relevant limits on the magnetic field and cosmic ray to thermal pressure where obtained by searching for the cosmic ray induced emission toward the center, caused by re-acceleration of electrons or shocks. This communication summarizes the results obtained in search for decaying DM signals and recently published in MAGIC Coll. (2018).

2. MAGIC data reconstruction

The search for decaying DM is performed by selecting a signal region centered around NGC 1275 (see Fig. 1) which encompass a significant fraction of the DM halo (red shadowed area). A central circle of 0.1 deg is excluded due to the presence of the bright NGC 1275. A symmetric background control region is selected at a 0.8 deg from NGC 1275, a value that is related to the specific pointing conditions of MAGIC. Signal leakage into the background region was computed. Periods in which NGC 1275 and IC 310 were active were excluded to control the systematics. A specific Monte Carlo re-weighting was used in order to provide accurate estimation of the instrument response functions (IRFs), especially the effective area and in the energy resolution (Palacio (2018)). The reason is that the response of the instrument is not completely flat over the field of view, mostly due to the parabolic shape of the primary mirror which introduce spherical aberrations which in turns make the performance worsening for region far from the optical axis of the telescope. Furthermore, the center of the DM halo is never observed in the center of the camera, but always at a rotating offset, as shown in Fig 1 (four symmetric red circles). Therefore, Monte Carlo are re-weighted in order to match the exposure of the telescope. Furthermore, the target was observed using 6 different instrument hardware setups, each with its own IRFs. At the end, more than 24 different samples were combined to provide the global analysis.

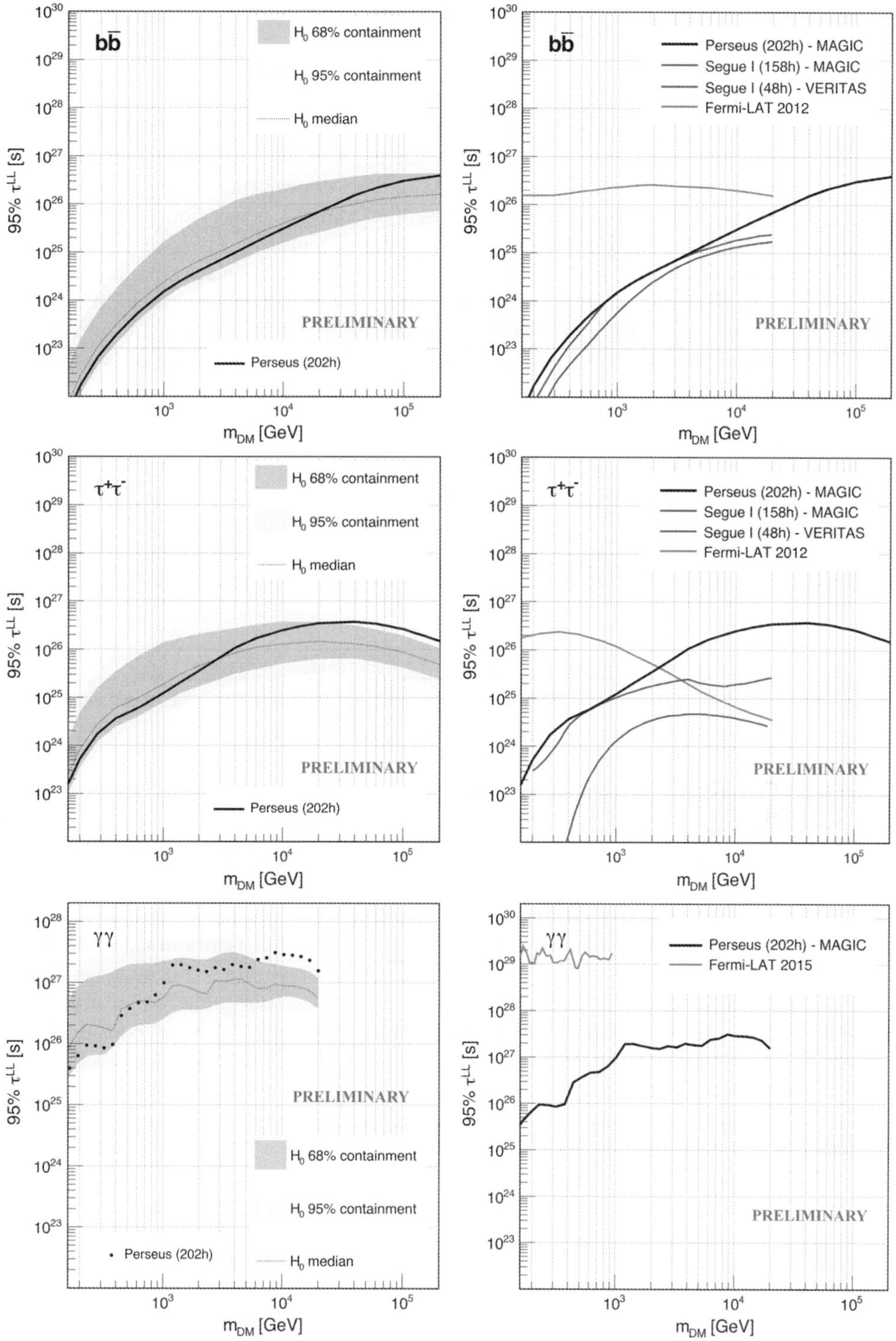

Figure 2. Lower limits on the decay lifetime of a dark matter particle as obtained with 202 hours of selected data in the direction of the Perseus Galaxy Cluster. Continuous decay spectra as pure decay into $b\bar{b}$ and $\tau^+\tau^-$ are searched (topmost plots). The lowest plots indicates limits for a $\gamma\gamma$ decay. The left column plots show the actual limits compared with the distribution of limits for the null hypothesis (median, one and two sigma bands). The right column plot show the limits compared with published results. From MAGIC Coll. (2018).

The expected photon flux from decaying DM from Perseus can be written as:

$$\frac{\mathrm{d}^2\Phi}{\mathrm{d}E\,\mathrm{d}\Omega} = \frac{1}{4\pi}\,\frac{1}{\tau_{\mathrm{DM}}\,m_{\mathrm{DM}}}\,\frac{\mathrm{d}N_\gamma}{\mathrm{d}E}\cdot\int_{\mathrm{l.o.s.}}\mathrm{d}l\,\rho(l,\Omega). \tag{2.1}$$

where m_{DM} is the DM mass, τ_{DM} the DM particle lifetime, $\mathrm{d}N_\gamma/\mathrm{d}E$ is the average decay spectrum per reaction, and the last factor is often called the differential J-factor $\mathrm{d}J_{\mathrm{dec}}/\mathrm{d}\Omega$ and is obtained integrating the DM density ρ over the line-of-sight (l.o.s.) for the decay reaction. The total J-factor enclosed in a given sky region can be obtained integrating over a solid angle $\Delta\Omega$, and the total flux integrating also over energy.

The parameter of interest is the decay lifetime τ_{DM}, therefore a specific likelihood was built in order to infer on this value and consider all other parameters, including the DM mass, photon spectra, instrument effective area and energy resolution as nuisance parameters. The null hypothesis was tested with mock Monte Carlo simulation of backgrounds. The likelihood reads as:

$$\begin{aligned}
&\mathcal{L}\left(1/\tau_{\mathrm{DM}};\boldsymbol{\nu}\,|\,\boldsymbol{\mathcal{D}}\right)\\
&= \prod_{i=1}^{N_{\mathrm{samples}}}\mathcal{K}(\kappa_i|\kappa_{\mathrm{obs},i},\sigma_{\kappa,i})\\
&\quad\times\prod_{j=1}^{N_{\mathrm{bins}}}\left[\frac{\left(g_{ij}(\tau_{\mathrm{DM}})+b_{ij}+f_{ij}\right)^{N_{\mathrm{ON},ij}}}{N_{\mathrm{ON},ij}!}e^{-(g_{ij}(\tau_{\mathrm{DM}})+b_{ij}+f_{ij})}\right.\\
&\quad\times\left.\frac{\left(\kappa_i b_{ij}+g_{ij}^{\mathrm{OFF}}(\tau_{\mathrm{DM}})\right)^{N_{\mathrm{OFF},ij}}}{N_{\mathrm{OFF},ij}!}e^{-\left(\kappa_i b_{ij}+g_{ij}^{\mathrm{OFF}}(\tau_{\mathrm{DM}})\right)}\right],
\end{aligned} \tag{2.2}$$

where $\boldsymbol{\nu}$ collectively refers to the nuisance parameters and $\boldsymbol{\mathcal{D}}$ to the data.

3. MAGIC results and discussion

The main results of the study are shown in Fig. 2 for two pure decay channels: into $b\bar{b}$ and into $\tau^+\tau^-$, which roughly brackets the minimum and maximum number of photons expected from decay. Constrain for line-like emission is also shown. Fig. 2 reports both the distribution of upper limits for the Monte Carlo generated background null hypothesis, as well as the comparison of MAGIC limits with other relevant limits in the same energy range. The limits are of the order of 10^{26} s, the most constraining so far achieved for DM particle masses of the order of few hundreds GeV and above. It is unlikely that observation on different DM targets can provide stronger limits in the close future, considering the large observation time needed. In the future, the Cherenkov Telescope Array will instead be significantly more sensitive to decay DM signals (CTA Consortium (2017)).

References

G. Bertone, D. Hooper and J. Silk, Phys. Rept. **405**, 279 (2005)

M. A. Sanchez-Conde, M. Cannoni, F. Zandanel, M. E. Gomez and F. Prada, JCAP **1112** (2011) 011

J. Palacio, Phd Thesis. Institut de Fisica d'Altes Energies, Universitat Autonoma de Barcelona (2018).

MAGIC Collaboration (Acciari *et al.*), arXiv:1806.11063 [astro-ph.HE]. CTA Consortium (2017)

CTA Consortium, B. S. Acharya *et al.*, arXiv:1709.07997 [astro-ph.IM].

Perseus in Sicily: From Black Hole to Cluster Outskirts
Proceedings IAU Symposium No. 342, 2018
K. Asada, E. de Gouveia Dal Pino, M. Giroletti,
H. Nagai & R. Nemmen, eds.

doi:10.1017/S1743921318005124

The AGN dependence on cluster mass

Elias Koulouridis[1,2], Marina Ricci[3] and the XXL collaboration

[1]IRFU, CEA, Université Paris-Saclay, F-91191 Gif-sur-Yvette, France
[2]Univ. Paris Diderot, AIM, Sorbonne Paris Cité, CEA, CNRS, F-91191 Gif-sur-Yvette, France
email: ilias.koulouridis@cea.fr

[3]Laboratoire Lagrange, Université Cte dAzur, Observatoire de la Cte dAzur, CNRS,
Blvd de l'Observatoire, CS 34229, 06304, Nice cedex 4, France

Abstract. We present the results of a study of the AGN density in a homogeneous and well studied sample of 167 bona-fide X-ray galaxy clusters ($0.1 < z < 0.5$). Our aim is to study the AGN activity in 167 XXL X-ray galaxy clusters as a function of the cluster mass and the location of the AGN in the cluster. We report a significant AGN excess in our low-mass cluster sub-sample between $0.5r_{500}$ and $2r_{500}$. In contrast, the high-mass sub-sample presents no AGN excess. The AGN excess in poor clusters indicates AGN triggering, supporting previous studies that reported enhanced galaxy merging in the cluster outskirts. This effect is probably prevented by high velocity dispersions in high-mass clusters. Comparing also with previous studies of massive or high-redshift clusters, we conclude that the AGN fraction in cluster galaxies anti-correlates strongly with cluster mass.

Keywords. galaxies: active – galaxies: Clusters: general – X-rays: galaxies: clusters – galaxies: interactions – galaxies: evolution – cosmology: large scale structure of Universe

1. Introduction

As the most massive self-gravitating entities of the universe, clusters are ideal laboratories to investigate the impact of dense environments on AGN demographics. However, the effect of the group and cluster environment on the activity of the central supermassive black hole (SMBH) of galaxies is still fairly undetermined. Several studies reported a significant lack of AGNs in rich galaxy clusters with respect to the field (e.g. Haines *et al.* 2012; Ehlert *et al.* 2013, 2014; Koulouridis & Plionis 2010), while spectroscopic studies of X-ray point-like sources in rich galaxy clusters have concluded that low-X-ray-luminosity AGNs are equally present in cluster and field environments (e.g. Martini *et al.* 2007; Haggard *et al.* 2010). Furthermore, most of the X-ray sources found in clusters present weaker optical AGN spectrum than AGNs in the field (Marziani *et al.* 2017), or show no signs of an optical AGN (e.g. Martini *et al.* 2006; Davis *et al.* 2003), while luminous AGNs are rarely found in clusters (e.g. Popesso *et al.* 2006; Caglar & Hudaverdi 2017).

Although AGNs avoid cluster cores (e.g. Ehlert *et al.* 2014; Melnyk *et al.* 2017, XXL paper XXII), there is evidence that X-ray AGNs found in massive clusters are an infalling population located mostly in the outskirts (Haines *et al.* 2012). Furthermore, Ehlert *et al.* (2015) argued that an important part of the cluster AGN population is triggered by galaxy mergers. Indeed, galaxy interaction and merging can lead to AGN triggering (e.g. Kawakatu *et al.* 2006; Koulouridis *et al.* 2006b,a, 2013; Koulouridis 2014; Ellison *et al.* 2011; Hopkins *et al.* 2014), rendering the high-density cluster outskirts a favourable AGN environment. However, in innermost cluster regions the ram pressure stripping from the intra-cluster medium (ICM) is probably able to strip or evaporate the

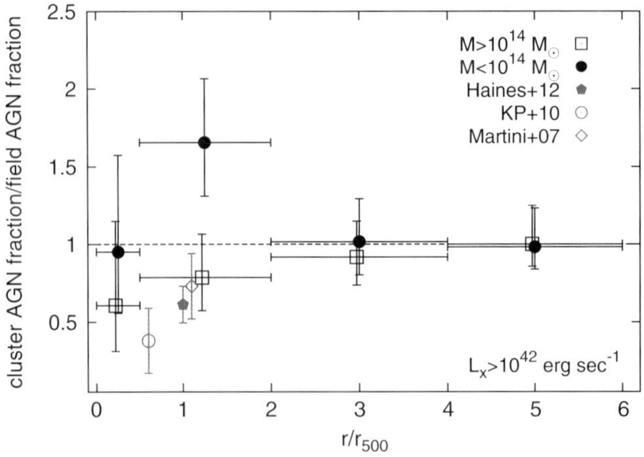

Figure 1. Fraction of XXL bright cluster galaxies hosting an X-ray AGN ($L_{X[0.5-10\,\mathrm{keV}]} > 10^{42}$ erg sec^{-1}), divided by the field fraction. Error bars indicate the 1σ confidence limits for small number of events. For comparison we plot results from the analyses of massive clusters by Martini *et al.* (2007, 8 clusters, 0.06< z <0.31) Koulouridis & Plionis (2010, 16 clusters, 0.07< z <0.28) and Haines *et al.* (2012, 26 clusters, 0.15< z <0.30). A significant AGN excess is found between $0.5r_{500,MT}$ and $2r_{500,MT}$, at the 95% confidence level.

cold gas reservoir of galaxies (e.g. Cowie & Songaila 1977; Poggianti *et al.* 2017) and can strongly affect the fueling of the AGN.

If the AGN activity depends on the gravitational potential, we would expect the AGN presence to anti-correlate with cluster mass. Previous studies of small poor cluster samples reported evidence of this anti-correlation (Arnold *et al.* 2009; Oh *et al.* 2014; Koulouridis *et al.* 2014). In addition, Ehlert *et al.* (2015) argued that the number of AGNs in a sample of 135 clusters scales with $M^{-1.2}$. On the other hand, Bufanda *et al.* (2017) reported no differences between groups and clusters in the DES cluster sample.

In the current work we study the X-ray AGN density in 167 spectroscopically confirmed X-ray selected clusters of the XXL survey (The Ultimate XMM-Newton Survey, Pierre *et al.* 2016). This homogeneous collection of low-mass X-ray clusters provides a far larger sample of such clusters than previous studies. This mass range is largely unexplored but is nevertheless an essential link between optical galaxy groups and massive clusters.

2. The XXL survey and the sample selection

The XXL Survey is the largest *XMM-Newton* project approved to date (>6 Msec) with a median exposure of 10.4 ks and a depth of $\sim 5\times10^{-15}$ erg sec^{-1} cm^{-2} in the (0.5-2 keV) X-ray band (completeness limit for point-like sources). The two fields have extensive multi-wavelength coverage from X-ray to radio. To date some 450 new galaxy clusters have been detected out to redshift $z\sim2$, as well as more than 25000 AGNs.

We selected all spectroscopically confirmed XXL clusters spanning a redshift range of $0.1<z<0.5$. In addition, we limited our sample to these clusters with direct X-ray temperature and luminosity measurements. Our sources cover an estimated mass range of $10^{13}-5\times10^{14}M_{\odot}$, which classifies them as poor clusters (groups) or moderately rich clusters. This is an important feature of our sample for the current study, in order to investigate the role of the cluster mass in triggering and suppressing AGN activity.

The X-ray point source catalogue is described in Chiappetti *et al.* (2018, XXL paper XXVII). Spectroscopic redshifts were obtained with large spectroscopic surveys or from

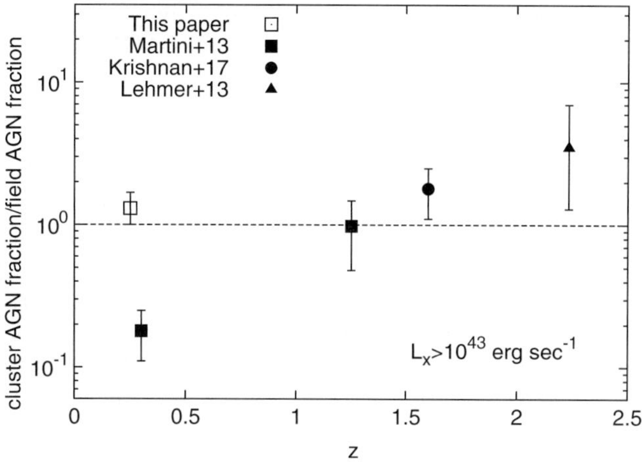

Figure 2. Fraction of XXL bright cluster galaxies hosting an X-ray AGN ($L_{X[0.5-10\,\mathrm{keV}]} > 10^{43}$ erg sec^{-1}) within $2r_{500}$ radius, divided by the field fraction. We overplot the results of similar analyses of cluster samples from Martini *et al.* (2013), or individual high-redshift proto-clusters from Krishnan *et al.* (2017) and Lehmer *et al.* (2013).

a large campaign with the AAOmega spectrograph on the Anglo-Australian Telescope (see Lidman *et al.* 2016, XXL paper XVI). Other smaller-scale spectroscopic observations (e.g. with WHT, Koulouridis *et al.* 2016, XXL paper XII) complement the sample.

3. Results and conclusions

For the presentation of the results we computed the AGN fraction in bright cluster member galaxies, having an absolute magnitude M_i in the i band within $M_i^* - 2 < M_i < M_i^* + 1$, to the respective field AGN fraction. Our results are compared with similar studies of the AGN fraction in bright cluster galaxies selected in various bands. The majority of these studies use an $M^* + 1$ magnitude cut, consistent with the one used in the current study.

Our results show a significant excess of X-ray AGNs in low-mass clusters ($M_{500,MT} < 10^{14} M_{\odot}$) up to a radius of $2r_{500}$ (Overdensity radius with respect to the critical density), at the 95% confidence level. Nevertheless, in the innermost cluster region ($< 0.5r_{500}$) a sharp decrease drives the AGN fraction back to the field level. On the other hand, the high-mass subsample presents a gradually decreasing AGN fraction towards the cluster centres, which reaches down to a factor of three lower than the field level.

We argue that the AGN excess found in the cluster outskirts of low-mass clusters supports an in-falling AGN population scenario. The excess may be attributed to high galaxy densities, which can lead to higher merger rates (Ehlert *et al.* 2015) and finally to AGN triggering. On the other hand, high velocity dispersions in massive clusters may diminish the probability of galaxy interactions, reversing this effect. As an example, Arnold *et al.* (2009) found that 10 groups with velocity dispersion $\sigma < 500$ km sec^{-1} present higher AGN fractions than six clusters above this velocity dispersion limit.

We also argue that the decrease of the AGN fraction in cluster cores, both in low- and high-mass clusters, support the ram pressure stripping scheme produced by the hot ICM. The hotter gas in deep gravitational potentials would be expected to strip the galaxy more effectively than the colder gas in our poor cluster population. Indeed, the density of AGNs within the first $< 0.5r_{500}$ is consistent with the background level, in contrast to the well established lack of AGNs in massive clusters.

Both the effective galaxy merging and gas stripping may affect differently the AGN activity in poor and massive clusters. The "cluster mass – AGN activity" anti-correlation provides evidence of how deeper gravitational potentials prevent AGN triggering in the outskirts and cause more effective ram pressure gas stripping in the cores.

Finally, we compare our results with similar studies in high-redshift cluster samples and proto-clusters. To this end, we computed the fraction of AGNs with $L_{X[0.5-10\,\mathrm{keV}]} > 10^{43}$ erg sec^{-1} within a $2r_{500,MT}$ radius. In Fig. 2 we plot our results along with two samples of low- and high-redshift clusters, a proto-cluster at $z = 1.6$, and a candidate proto-cluster at $z = 2.23$. The 13 clusters between $z = 1$ and 1.5 in Martini *et al.* (2013) have estimated masses $M > 10^{14} M_{\odot}$ and up to a few times $10^{14} M_{\odot}$, consistent with the XXL high mass subsample. Nevertheless, their low-redshift sample at $z \sim 0.3$ is much more massive, being the compilation of the sample of Haines *et al.* (2012) and Martini *et al.* (2009). On the other hand, the mass of the proto-cluster at $z = 1.6$ has an estimated mass of $5.7 \times 10^{13} M_{\odot}$, within the range of the XXL low-mass subsample. Evidently, at redshifts above $z = 1$ the AGN fraction in clusters is not dissimilar to the AGN fraction in our sample, but this probably correlates with cluster mass and not evolution. We note however, that the cluster AGN fraction does evolve, but at the same rate as the field AGN fraction.

References

Arnold, T. J., Martini, P., Mulchaey, J. S., Berti, A., & Jeltema, T. E. 2009, *ApJ*, 707, 1691

Bufanda, E., Hollowood, D., Jeltema, T. E., *et al.* 2017, *MNRAS*, 465, 2531

Caglar, T. & Hudaverdi, M. 2017, *MNRAS*, 471, 4990

Chiappetti, L., Fotopoulou, S., Lidman, C., *et al.* 2018, XXL paper XX, accepted in *A&A*

Cowie, L. L. & Songaila, A. 1977, *Nature*, 266, 501

Davis, D. S., Miller, N. A., & Mushotzky, R. F. 2003, *ApJ*, 597, 202

Ehlert, S., Allen, S. W., Brandt, W. N., *et al.* 2015, *MNRAS*, 446, 2709

Ehlert, S., Allen, S. W., Brandt, W. N., *et al.* 2013, *MNRAS*, 428, 3509

Ehlert, S., von der Linden, A., Allen, S. W., *et al.* 2014, *MNRAS*, 437, 1942

Ellison, S. L., Patton, D. R., Mendel, J. T., & Scudder, J. M. 2011, *MNRAS*, 418, 2043

Haggard, D., Green, P. J., Anderson, S. F., *et al.* 2010, *ApJ*, 723, 1447

Haines, C. P., Pereira, M. J., Sanderson, A. J. R., *et al.* 2012, *ApJ*, 754, 97

Hopkins, P. F., Kocevski, D. D., & Bundy, K. 2014, *MNRAS*, 445, 823

Kawakatu, N., Anabuki, N., Nagao, T., Umemura, M., & Nakagawa, T. 2006, *ApJ*, 637, 104

Koulouridis, E. 2014, *A&A*, 570, A72

Koulouridis, E., Chavushyan, V., Plionis, M., *et al.* 2006a, *ApJ*, 651, 93

Koulouridis, E. & Plionis, M. 2010, ApJ, 714, L181

Koulouridis, E., Plionis, M., Chavushyan, V., *et al.* 2013, *A&A*, 552, A135

Koulouridis, E., Plionis, M., Chavushyan, V., *et al.* 2006b, *ApJ*, 639, 37

Koulouridis, E., Plionis, M., Melnyk, O., *et al.* 2014, *A&A*, 567, A83

Koulouridis, E., Poggianti, B., Altieri, B., *et al.* 2016, *A&A*, 592, A11

Krishnan, C., Hatch, N. A., Almaini, O., *et al.* 2017, *MNRAS*, 470, 2170

Lehmer, B. D., Lucy, A. B., Alexander, D. M., *et al.* 2013, *ApJ*, 765, 87

Lidman, C., Ardila, F., Owers, M., *et al.* 2016, *PASA*, 33, e001

Martini, P., Kelson, D. D., Kim, E., Mulchaey, J. S., & Athey, A. A. 2006, *ApJ*, 644, 116

Martini, P., Miller, E. D., Brodwin, M., *et al.* 2013, *ApJ*, 768, 1

Martini, P., Mulchaey, J. S., & Kelson, D. D. 2007, *ApJ*, 664, 761

Martini, P., Sivakoff, G. R., & Mulchaey, J. S. 2009, *ApJ*, 701, 66

Marziani, P., D'Onofrio, M., Bettoni, D., *et al.* 2017, *A&A*, 599, A83

Melnyk, O., Elyiv, A., Smolcic, V., *et al.* 2017, [arXiv:1712.01872]

Oh, S., Mulchaey, J. S., Woo, J.-H., *et al.* 2014, *ApJ*, 790, 43

Pierre, M., Pacaud, F., Adami, C., *et al.* 2016, *A&A*, 592, A1

Poggianti, B. M., Moretti, A., Gullieuszik, M., *et al.* 2017, *ApJ*, 844, 48

Popesso, P., Biviano, A., Böhringer, H., & Romaniello, M. 2006, *A&A*, 445, 29

Perseus in Sicily: From Black Hole to Cluster Outskirts
Proceedings IAU Symposium No. 342, 2018
K. Asada, E. de Gouveia Dal Pino, M. Giroletti,
H. Nagai & R. Nemmen, eds.

doi:10.1017/S1743921318005379

AGN Jets, Bubbles, and Heat Pumps

Yi-Hao Chen and Sebastian Heinz

Department of Astronomy, University of Wisconsin-Madison,
275 N. Charter Street, Madison, WI 53705, USA
email: ychen@astro.wisc.edu

Abstract. Radio-mode feedback from relativistic jets is one of the prominent heating mechanisms in clusters of galaxies. We present a long-term evolution of high-resolution MHD simulation of jets interacting with an environment modeled to represent the Perseus cluster. We investigate the thermodynamics of the ICM due to the gas motion triggered by the action of the jets and show that low-entropy gas is lifted efficiently in the wake of the inflating radio lobe. We look into the uplift mechanism and estimate the energy budget and the rate of thermal conduction. The redistribution of entropy suggests that heat conduction can play a more significant role in the thermal evolution of the cluster core in the presence of jets, which act effectively as a *heat pump*, thus heating the ICM more efficiently than jets would by themselves in an isentropic cluster.

Keywords. method: numerical, galaxies: clusters: general, cooling flow, jets, conduction

1. Introduction

The core of galaxy clusters often exhibits strong x-ray emission with short cooling time, which should lead to substantial radiative cooling and the formation of a *cooling flow* (Fabian 1994). The lack of such flows in observations stimulates the study of feedback processes that could counteract the cooling. Mechanical energy deposited by the jets could be one of the important mechanisms regulating the thermal evolution of galaxy clusters (McNamara & Nulsen 2012). X-ray observations of galaxy clusters often show radio-filled cavities that are likely inflated by jets from the supermassive black holes in the central galaxy (see e.g. Birzan *et al.* 2004). In some cases, multiple cavities caused by episodic jet activities are observed. The most detailed observations of this kind include the Perseus Cluster (Fabian *et al.* 2011) and M87 (Forman *et al.* 2007). Churazov *et al.* (2000) show that observed cavities are energetically to offset the cooling in most systems. Subsequent detailed observations support the claim (see e.g. Rafferty *et al.* 2006).

However, the processes through which the energy of AGN jets couples to the ICM are still under debate. Possible mechanisms include turbulence (Zhuravleva *et al.* 2014), sound waves (Fabian *et al.* 2017), mixing (Hillel & Soker 2016), shocks (Fabian *et al.* 2006; Li *et al.* 2017), and internal waves (Zhang *et al.* 2018), among others.

Although most of the discussion about the heating focuses on the direct energy coupling between the AGN output and the ICM, a few authors have pointed out that the removal of the cool gas from the core of the cluster could possibly prevent catastrophic cooling. Pope *et al.* (2010) calculate the amount of gas transported by a rising bubble and conclude that the mass transport by the bubble wake could prevent the core from overcooling. We propose that this uplift mechanism along with heat exchange through conduction could provide the necessary mechanism to extract additional energy from the hot atmosphere of the cluster.

Buoyantly rising bubbles are identified and studied by many works. Churazov *et al.* (2001) conduct hydrodynamic simulations of bubbles to model the radio and x-ray

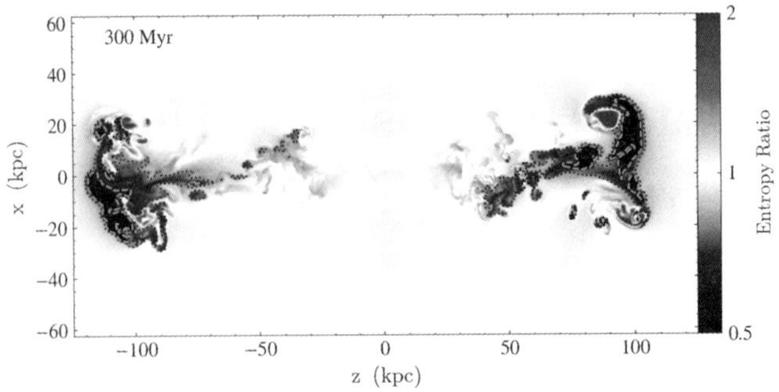

Figure 1. Slice of the simulation showing the entropy of the gas relative to the initial entropy value at the same location. Dotted and dashed contours indicate the ratio of 0.8 and 0.6 respectively.

arms in M87. In a deep *Chandra* observation, Forman *et al.* (2007) detect many filamentary structures that could be associated by independent bubbles in M87. Recently, Gendron-Marsolais *et al.* (2017) report observations of cool x-ray gas rims in NGC 4472 that could be gas lifted by the bubbles. These works all suggest that rising bubbles are prevalent and could bring gas from the core to the outskirt of the galaxy cluster.

In this work, we consider the mechanism that rising bubbles bring the low-entropy gas from the cool core to the hot outskirt of the galaxy cluster where thermal conductions are much more efficient. With this mechanism, the AGN could "heat" the ICM more than its own energy budget would indicate if the energy were directly transferred to heat and that this heating process and is much more gentle than direct heating.

2. The Simulation and Implicatons

Here we summarize the important and relevant information of the simulation and leave the details to a separate paper. We conduct full 5D ideal MHD simulations using `FLASH` (Fryxell *et al.* 2000; Dubey *et al.* 2009) with the unsplit staggered mesh scheme (Lee 2013). In a background environment tuned to match the Perseus Cluster observed by Zhuravleva *et al.* (2015), we set up a nozzle through which energy, momentum, and magnetic fluxes are injected into the grid to model the radio jets. The initial condition is in hydrostatic equilibrium and not magnetized. The jet is active for 10 Myr, during which it outputs at a constant power 10^{45} erg/s. After 10 Myr, the jet ceases and the bubbles continue to rise due to buoyancy. Radiative cooling and thermal conduction are not included in the simulation. We study the long-term impact of this one-time activity of the jets on the thermal state of the galaxy cluster.

2.1. *Low-entropy gas transported by the rising bubble*

First, we investigate the occurrence and the entropy structure of the gas displaced by the AGN. During the active phase of the AGN, only a small amount of gas is displaced by the jet. However, as bubbles rise, they drag a significant amount of gas from the core in the wake of the bubbles. This phenomenon can be seen clearly in the entropy ratio map. In Figure 1, the specific entropy, defined as kT/n^{γ}, of the gas is plotted relative to the entropy profile of the initial condition. Since the entropy is conserved during adiabatic processes, it better indicates the origin of the gas. At the time of this figure, most of the hot gas injected by the jet is mixed with the ICM. Only a small fraction of the hot gas is still visible as a ring surrounded by the lower-entropy gas.

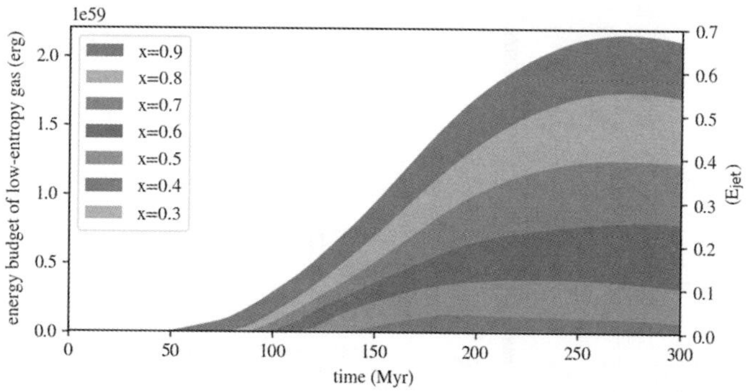

Figure 2. Energy budget of the low-entropy gas. This shows the evolution of the amount of energy needed to bring the low-entropy gas, which is identified by entropy ratio lower than the threshold x, back to the surrounding temperature. The y-axis on the right shows the ratio relative to the total energy injected by the active jet in the first 10 Myr.

2.2. *Energy budget of the low-entropy gas lifted by the rising bubble*

Second, we consider the amount of energy that can be extracted from the hot gas and transferred to the uplifted low-entropy gas by heat conduction. We identify the gas with entropy ratio below a threshold of x in the simulation and estimate the amount of energy needed to bring this gas into thermal equilibrium with its surrounding. This is the amount of energy available for heating low entropy gas through uplift and conduction. Figure 2 shows the evolution of this energy as a function of time with different values of x. The y-axis at the left labels the energy in *erg*, while the y-axis at the right denotes this energy relative to the total energy injected during the active phase of the jet (10 Myr). One can see that the conductive energy budget of the low-entropy gas is comparable to the total energy injected by the jet. It is important to keep in mind that this energy is not from the AGN itself, but from the hot atmosphere at large radii of the clusters. When the gas is brought into thermal contact with the hot reservoir, heat exchange can take place more efficiently. We discuss the rate of the heat exchange next.

2.3. *Heating rates of Spitzer thermal conduction*

Finally, we estimate the rate of thermal conduction between the low-entropy and hot gas. We consider the classic Spitzer conduction in ionized gas (Spitzer 1962, see also Narayan & Medvedev 2001). The Spitzer coefficient can be calculated as

$$\kappa_{\rm Sp} \sim 4 \times 10^{32} \left(\frac{kT}{10\ \rm keV}\right)^{5/2} \left(\frac{n}{10^{-3}\rm cm^{-3}}\right)^{-1} \rm cm^2 s^{-1}. \tag{2.1}$$

The coefficient has a strong dependence on the temperature. This is why thermal conduction is usually considered too slow to offset the cooling in the cool core of the cluster. However, once the cool gas is transported to large radii and placed in close spatial proximity to hot gas, the high temperature and increased temperature gradient both accelerate the thermal conduction rate. We then calculate the heat flux by $q = nkf_{\rm Sp}\kappa_{\rm Sp}\nabla T$, in which $f_{\rm Sp}$ is the fraction relative to the classic Spitzer conduction rate and depends on the orientation of the magnetic fields. Since we do not include magnetic fields in the initial ICM, it is impossible to include the anisotropic conduction in this analysis. The instantaneous heating and cooling rate due to the conduction is the divergence of the heat flux. We then add the thermal conduction heating/cooling rates and x-ray cooling rate in each cell. The summed values are shown in Figure 3 in entropy and radial bins (left panel) and in entropy bins (right panel).

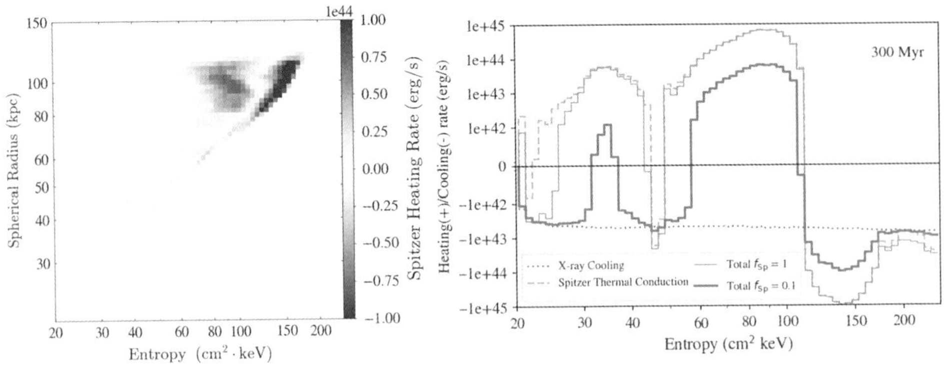

Figure 3. Heating rate histogram at 300 Myr. The total heating and cooling rates are summed in each entropy and radial bins (left panel, $f_{\rm Sp}=1$) and in entropy bins (right panel) with two $f_{\rm Sp}$ values.

At the same radius, heat is transferred from high-entropy to low-entropy gas. Although the details of the conduction need to be scrutinized, we found that at 0.1 Spitzer value ($f_{\rm Sp}=0.1$), the thermal conduction will still be able to bring the low-entropy gas to the temperature of the heat reservoir in a short amount of time. The heating rate is also comparable to the power of the jet (10^{45} erg/s) even at 300 Myr.

3. Discussions and Conclusions

We want to emphasize that this is an exploratory work inspired by the long-term evolution of our simplified numerical experiment. To investigate this heat-pump mechanism rigorously, we will need to re-designed the simulation to include effects like radiative cooling and anisotropy thermal conduction. However, our preliminary analysis indicates that this is a promising process that can provide another avenue in the already hotly debated cooling flow problem.

We consider the idea that the radio AGN could act as a "*heat pump*" by inflating buoyant bubbles that lift the low-entropy gas from the core of the clusters into thermal contact with the hot atmosphere. In this mechanism, the total available energy gained by the cluster core to offset the radiative cooling is not simply limited by the total energy output of the AGN, but also the heat transferred from the hot gas at large radii of the cluster to the uplifted low-entropy gas. Our analysis implies that a 10-Myr active jet could still affect the thermal states of the clusters after 300 Myr. This mechanism has the advantage that bursty AGN activity is transformed into a more gentle and longer-lasting heating process.

References

Birzan L., Rafferty D. A., McNamara B. R., Wise M. W., & Nulsen P. E. J., 2004, *ApJ*, 607, 800

Churazov E., Forman W., Jones C., & Böhringer H., 2000, *A&A*, 356, 788

Churazov E., Bruggen M., Kaiser C. R., Böhringer H., & Forman W., 2001, *ApJ*, 554, 261

Dubey A., Antypas K., Ganapathy M. K., Reid L. B., Riley K., Sheeler D., Siegel A., & Weide K., 2009, *Parallel Computing*, 35, 512

Fabian A. C., 1994, *ARAA*, 32, 277

Fabian A. C., Sanders J. S., Taylor G. B., Allen S. W., Crawford C. S., Johnstone R. M., & Iwasawa K., 2006, *MNRAS*, 366, 417

Fabian A. C., *et al.*, 2011, *MNRAS*, 418, 2154

Fabian A. C., Walker S. A., Russell H. R., Pinto C., Sanders J. S., & Reynolds C. S., 2017, *MNRAS*, 464, L1
Forman W., *et al.*, 2007, *ApJ*, 665, 1057
Fryxell B., *et al.*, 2000, *ApJS*, 131, 273
Gendron-Marsolais M., *et al.*, 2017, *ApJ*, 848, 26
Hillel S., & Soker N., 2016, *MNRAS*, 455, 2139
Lee D., 2013, *Journal of Computational Physics*, 243, 269
Li Y., Ruszkowski M., & Bryan G. L., 2017, *ApJ*, 847, 106
McNamara B. R., & Nulsen P. E. J., 2012, *New Journal of Physics*, 14, 055023
Narayan R., & Medvedev M. V., 2001, *ApJ*, 562, L129
Pope E. C. D., Babul A., Pavlovski G., Bower R. G., & Dotter A., 2010, *MNRAS*, 406, 2023
Rafferty D. A., McNamara B. R., Nulsen P. E. J., & Wise M. W., 2006, *ApJ*, 652, 216
Spitzer L., 1962, *Physics of Fully Ionized Gases*, 2nd ed, Interscience
Zhang C., Churazov E., & Schekochihin A. A., 2018, *MNRAS*, 478, 4785
Zhuravleva I., *et al.*, 2014, *Nature*, 515, 85
Zhuravleva I., *et al.*, 2015, *MNRAS*, 450, 4184

Perseus in Sicily: From Black Hole to Cluster Outskirts
Proceedings IAU Symposium No. 342, 2018
K. Asada, E. de Gouveia Dal Pino, M. Giroletti,
H. Nagai & R. Nemmen, eds.

doi:10.1017/S1743921318005525

Intermediate-Mass Black Hole Feedback in Dwarf Galaxies: a View from Cosmological Simulations

Paramita Barai and Elisabete M. de Gouveia Dal Pino

Instituto de Astronomia, Geofísica e Ciências Atmosféricas - Universidade de São Paulo (IAG-USP), Rua do Matão 1226, São Paulo, 05508-090, Brasil
email: paramita.barai@iag.usp.br

Abstract. Black holes are usually observed to be of stellar-mass or supermassive. By natural extension, there should be a population of Intermediate-Mass Black Holes (IMBHs: with mass between 100 to $10^6 M_\odot$) in the Universe; which has started to been observed. An exciting claim has been made recently by Silk (2017): that early feedback by IMBHs in gas-rich dwarf galaxies at $z = 5 - 8$, can potentially solve multiple dwarf galaxy problems within the Λ-cold-dark-matter cosmology. We are performing Cosmological Hydrodynamical Simulations of $(2Mpc)^3$ volumes, starting from $z = 100$, to test the case for IMBHs in Dwarf Galaxies. Black holes of mass $1000 M_\odot$ are seeded inside halos when they reach a mass of $10^7 M_\odot$. The black holes grow by accretion of gas from their surroundings and by merger with other black holes, and consequently eject feedback energy. We analyze the simulation output in post-processing to study the growth of the first IMBHs, and their impact on star-formation. Our conclusions, based on numerical simulation results, support the phenomenological ideas made by Silk (2017). IMBHs at the centers of dwarf galaxies can be a strong source of feedback to quench star-formation and generate outflows. At the same time, these IMBHs form the missing link between stellar-mass and supermassive BHs.

Keywords. black hole physics, hydrodynamics, methods: numerical, galaxies: active, galaxies: dwarf, galaxies: evolution, galaxies: high-redshift, (cosmology:) large-scale structure of universe

---

1. Introduction

Black holes usually come in two flavours: stellar-mass ($M_{\rm BH} \leqslant 10 M_\odot$), and supermassive ($M_{\rm BH} \geqslant 10^6 M_\odot$). Naturally, there should be a population of Intermediate-Mass Black Holes (IMBHs) of masses between $100 - 10^6 M_\odot$. Analogous to supermassive BHs producing AGN feedback, the IMBHs should also have feedback. In this work we focus on negative BH feedback effects where star-formation is quenched.

AGN feedback mechanism has recently started to been observed in low-mass galaxies. Investigating the presence of AGN in nearby dwarf galaxies using mid-infrared emission, Marleau *et al.* (2017) identified 303 candidates, of which 91% were subsequently confirmed as AGN by other methods. The stellar masses of these galaxies are estimated to be between $10^6 - 10^9 M_\odot$; and the black hole masses in the range $10^3 - 10^6 M_\odot$. Penny *et al.* (2017) presented observational evidence for AGN feedback in a sample of 69 quenched low-mass galaxies ($M_\star < 4 \times 10^9 M_\odot$); including 6 galaxies showing signatures of an active AGN preventing ongoing star-formation.

The concordance ΛCDM cosmological scenario of galaxy formation presents multiple challenges in the dwarf galaxy mass range: e.g. core-cusp, number of DGs. Recently Silk (2017) made an exciting claim that the presence of IMBHs at the centers of essentially all old Dwarf Galaxies (DGs) can potentially solve the problems. Early feedback from these

Table 1. Simulation runs and parameters.

Run name	BH present	Min. Halo Mass for BH Seeding, $M_{\rm HaloMin}[M_\odot]$	Seed BH Mass, $M_{\rm BHseed}[M_\odot]$	BH kinetic feedback kick velocity v_w (km/s)
SN	No	–	–	–
BHs2h1e6	Yes	$h^{-1} \times 10^6$	10^2	2000
BHs2h7e7	Yes	$5h^{-1} \times 10^7$	10^2	2000
BHs3h1e7	Yes	1×10^7	10^3	2000
BHs3h2e7	Yes	2×10^7	10^3	2000
BHs3h3e7	Yes	3×10^7	10^3	2000
BHs3h4e7	Yes	4×10^7	10^3	2000
BHs3h4e7v5	Yes	4×10^7	10^3	5000
BHs3h5e7	Yes	5×10^7	10^3	2000
BHs4h4e7	Yes	4×10^7	10^4	2000

IMBHs output energy and affect the host gas-rich DGs at $z = 5 - 8$. This early feedback can quench star-formation, reduce the number of DGs, and impact the density profile at DG centers. Dashyan *et al.* (2017) studied the same problem analytically, and compared AGN versus SN feedback. They find a critical halo mass below which the central AGN can drive gas out of the host halo. This negative feedback effect of AGN is found to be more efficient than SN in the most massive DGs, where SN is not able to expel the gas.

Here, we investigate the scenario that IMBHs are present at the centers of all dwarf galaxies, by performing cosmological hydrodynamical simulations. Our goals are to (i) test if IMBHs would grow at DG centers, and (ii) quantify the impact on star formation.

2. Numerical Method

We use a modified version of the TreePM (particle mesh) - SPH (smoothed particle hydrodynamics) code GADGET-3 (Springel (2005)). Radiative cooling and heating is incorporated from Wiersma, Schaye & Smith (2009). Eleven element species (H, He, C, Ca, O, N, Ne, Mg, S, Si, Fe) are tracked. Star-formation is implemented following the multiphase effective sub-resolution model by Springel & Hernquist (2003), and chemical enrichment from Tornatore *et al.* (2007).

BHs are collisionless sink particles (of mass $M_{\rm BH}$) in our simulations. A BH (of initial mass $M_{\rm BHseed}$) is seeded at the center of each galaxy more massive than a total mass $M_{\rm HaloMin}$, which does not contain a BH already. We test different values of minimum halo mass and seed BH mass in the range: $M_{\rm HaloMin} = (10^6 - 10^7) M_\odot$, and $M_{\rm BHseed} = (10^2 - 10^3) M_\odot$. The sub-resolution presciptions for BH accretion and *kinetic* feedback are adopted from Barai *et al.* (2014), Barai *et al.* (2016).

2.1. *Simulations*

We perform cosmological hydrodynamical simulations of small-sized boxes to probe dwarf galaxies at high redshifts. The initial condition at $z = 100$ is generated using the MUSIC† software (Hahn & Abel (2011)). The size of the cubic cosmological volume is $(2h^{-1}\ {\rm Mpc})^3$ comoving. Hence the total mass of matter (dark matter + baryons) in the box is $6.86 \times 10^{11} h^{-1} M_\odot$. We use 256^3 dark matter and 256^3 gas particles in the initial condition. The dark matter particle mass is $m_{\rm DM} = 3.44 \times 10^4 h^{-1} M_\odot$, and the gas particle mass is $m_{\rm gas} = 6.43 \times 10^3 h^{-1} M_\odot$. The gravitational softening length is set to $L_{\rm soft} = 0.1 h^{-1}$ kpc comoving. We execute a series of 10 simulations, with characteristics listed in Table 1.

† MUSIC - Multi-scale Initial Conditions for Cosmological Simulations: https://bitbucket.org/ohahn/music

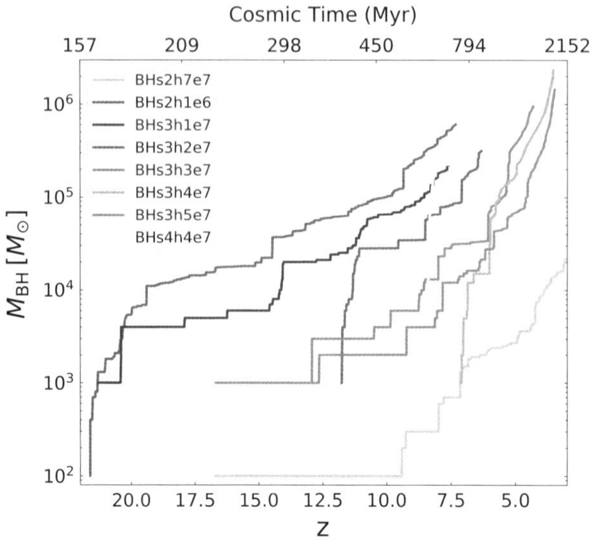

Figure 1. BH mass growth with redshift of the most-massive BH in each run. The different colours discriminate the runs.

3. Results and Discussion

3.1. *Black Hole Accretion and Growth*

We find that first BHs are seeded at different cosmic times depending on the value of minimum halo mass for BH seeding, $M_{\rm HaloMin}$. The seeding epoch varies between $z \sim 22$ to $z \sim 16$ in our simulations, when the first halos reach $M_{\rm halo} = h^{-1} \times 10^6 M_\odot$ to $M_{\rm halo} = 5 \times 10^7 M_\odot$. The redshift evolution of the most-massive BH mass in the BH runs is plotted in Fig. 1. Each BH starts from an initial seed of $M_{\rm BH} = 10^2 M_\odot$ in the runs named *BHs2**, $10^3 M_\odot$ in the runs named *BHs3**, and $10^4 M_\odot$ in the runs named *BHs4**. The subsequent mass growth is due to merger with other BHs (revealed as vertical rises in $M_{\rm BH}$), and gas accretion (visualized as the positive-sloped regions of the $M_{\rm BH}$ versus z curve). The final properties reached depends on the simulation. The most-massive BH, considering all the runs, has grown to $M_{\rm BH} = 2 \times 10^6 M_\odot$ at $z = 5$ in run *BHs3h4e7* (brown curve in Fig. 1).

3.2. *Star Formation*

The Star Formation Rate Density (SFRD in units of $M_\odot yr^{-1} Mpc^{-3}$, counting stars forming in the whole simulation box) versus redshift of the simulation runs is displayed in Fig. 2. The SFRD rises with time in the *SN* run (blue curve in Fig. 2) initially from $z \sim 15$, reaches a peak at $z \sim 4$ (the peak epoch of star-formation activity in the Universe), and decreases subsequently over $z \sim 4 - 2$. The presence of a BH quenches star formation by accreting some gas in, ejecting some gas out of the halo as outflows, and/or heating the gas. The models suppress SF substantially from $z \sim 8$ onwards, when the BHs have grown massive. We find that BHs need to grow to $M_{\rm BH} > 10^5 M_\odot$, in order to suppress star-formation, even in these dwarf galaxies. BH feedback causes a reduction of SFR up to 5 times in the different runs.

The red curve (run *BHs2h1e6*) already quenches SF as early as $z \sim 8$. This is because the BH has already grown to $M_{\rm BH} \sim 5 \times 10^5 M_\odot$ at that epoch, more massive than all the other runs. As another example, the brown (run *BHs3h4e7*) and royal-blue (run

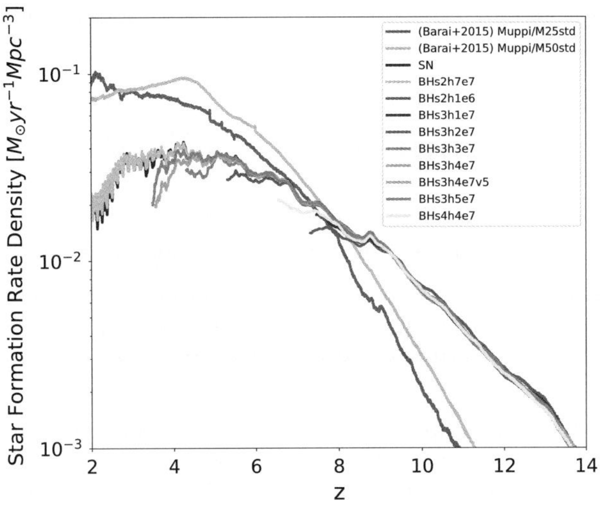

Figure 2. Total star formation rate density in simulation volume as a function of redshift.

BHs3h5e7) curves quench SF from $z \sim 4.5$ to $z \sim 3.5$. This is the epoch when the BH masses in these runs increase from $M_{\rm BH} = 10^5 M_{\odot}$ to $M_{\rm BH} = 10^6 M_{\odot}$.

4. Conclusions

We conclude that (i) IMBHs at DG centers grow from $10^2 - 10^3 M_{\odot}$ to $10^5 - 10^6 M_{\odot}$ by $z \sim 4$ in a cosmological environment. These IMBHs in DGs can become the seeds of supermassive BHs (which grows to $M_{\rm BH} \sim 10^9 M_{\odot}$) in massive galaxies. (ii) Star formation is quenched when the BHs have grown to $M_{\rm BH} > 10^5 M_{\odot}$. We find a positive correlation between the mass growth BHs and the quenching of SF.

References

Barai, P., Viel, M., Murante, G., Gaspari, M. & Borgani, S. 2014, *MNRAS*, 437, 1456

Barai, P., Murante, G., Borgani, S., Gaspari, M., Granato, G. L., Monaco, P. & Ragone-Figueroa, C. 2016, *MNRAS*, 461, 1548

Dashyan, G., Silk, J., Mamon, G. A., Dubois, Y. & Hartwig, T. 2017, eprint arXiv:1710.05900

Hahn, O. & Abel, T. 2011, *MNRAS*, 415, 2101

Marleau, F. R., Clancy, D., Habas, R. & Bianconi, M. 2017, *A&A*, 602, A28

Penny, S. J. *et al.* 2017, submitted to *MNRAS*, eprint arXiv:1710.07568

Silk, J. 2017, *ApJ*, 839, L13

Springel, V. & Hernquist, L. 2003, *MNRAS*, 339, 289

Springel, V. 2005, *MNRAS*, 364, 1105

Tornatore, L., Borgani, S., Dolag, K. & Matteucci, F. 2007, *MNRAS*, 382, 1050

Wiersma, R. P. C., Schaye, J. & Smith, B. D. 2009, *MNRAS*, 393, 99

Perseus in Sicily: From Black Hole to Cluster Outskirts
Proceedings IAU Symposium No. 342, 2018
K. Asada, E. de Gouveia Dal Pino, M. Giroletti,
H. Nagai & R. Nemmen, eds.

doi:10.1017/S1743921318007895

Gamma-ray emission in radio galaxies, from MeV to TeV

Eleonora Torresi[1,2]
On behalf of the Fermi-LAT Collaboration

[1]Dipartimento di Astronomia, Università di Bologna, via Gobetti, 93/2, I-40129, Bologna, Italy
[2]INAF-OAS Bologna, Area della Ricerca CNR, via Gobetti, 101, I-40129, Bologna, Italy
email: eleonora.torresi@inaf.it

Abstract. Thanks to the *Fermi* γ-ray satellite and the current Imaging Atmospheric Cherenkov Telescopes, radio galaxies have arisen as a new class of high- and very-high energy emitters. The favourable orientation of their jets makes radio galaxies extremely relevant in addressing important issues such as: (i) revealing the jet structure complexity; (ii) localising the emitting region(s) of high- and very-high energy radiation; (iii) understanding the physical processes producing these photons. In this review the main results on the γ-ray emission studies of radio galaxies from the MeV to TeV regimes will be presented, and the impact of future Cherenkov Telescope Array observations will be discussed.

Keywords. radiation mechanisms: nonthermal, galaxies: active, galaxies: jets, galaxies: nuclei

1. Introduction

Radio galaxies (RG) are radio-loud Active Galactic Nuclei (AGN) characterised by powerful jets of relativistic plasma whose non-thermal radiation emits from radio frequencies up to TeV energies. RGs have been historically classified by Fanaroff & Riley (1974) on the basis of their radio morphology into FR I and FR II†. Within the Unified Model of AGN (Urry & Padovani 1995), FR Is are considered the parent population of BL Lacs (i.e. their misaligned counterpart) and FR IIs the parent population of Flat Spectrum Radio Quasars (FSRQ). The cause of the FRI/FRII dichotomy is still unknown: external agents, e.g. environment, host galaxy, merging history, etc. (Bicknell 1995) or intrinsic factors, e.g. accretion processes (Ghisellini & Celotti 2001), have been invoked as possible explanations. Jets of RGs are oriented at large inclination angles with respect to the l.o.s. ($\theta >10^{\circ}$), making these sources unique laboratories where to contemporarily study jets and accretion processes and establish a connection between the two. However, the large jet inclination angle geometrically disfavours the detection of RGs above 100 MeV since their non-thermal radiation does not benefit from the strong Doppler amplification typical of blazars. Indeed, the degree of boosting depends on the relativistic Doppler factor δ=1/Γ(1-βcosθ), that relates the intrinsic and observed flux for a source moving at relativistic speed (Urry & Padovani 1995). The flux enhancement is strongly dependent on the viewing angle and decreases very rapidly for θ >8-10°. This trend was confirmed by the detection of only three possible γ-ray counterparts of RGs by the EGRET telescope onboard the CGRO satellite, i.e. NGC 6251, Centaurus A and 3C 111 (Nolan *et al.* 1996,

† A source is considered a FR I if the separation between the points of the peak intensity in the two lobes is smaller than half of the largest size of the source (R<0.5). If R>0.5 the source is a FR II. This morphological criterium corresponds to a separation at a radio luminosity $P_{178\ MHz}=10^{25}$ W Hz^{-1} sr^{-1}. Sources with a radio luminosity below such a threshold are FR Is, above it are FR IIs.

Mukherjee *et al.* 2002, Sguera *et al.* 2005, Hartmann *et al.* 2008). A huge step forward in this field has been made thanks to the *Fermi* satellite, as predicted by several works suggesting these sources as possible targets detectable by the Large Area Telescope (LAT) instrument onboard *Fermi* (Stawarz *et al.* 2003, 2006, Ghisellini *et al.* 2005, Grandi & Palumbo 2007).

In the following, the main recent results on RGs obtained with the *Fermi*-LAT telescope in the MeV-GeV band (Sec. 2) and with the current Imaging Atmospheric Cherenkov Telescopes in the TeV regime (Sec. 3) will be presented. Within this context, the impact of the future Cherenkov Telescope Array (CTA) on the study of RGs at very-high energies will be also discussed.

2. The MeV-GeV band: Fermi-LAT results

The *Fermi* γ-ray space telescope (Atwood *et al.* 2009) was launched in 2008 June 11th and carries onboard two instruments: (i) the LAT, that operates in the nominal energy band 20 MeV-300 GeV† and (ii) the Gamma-ray Burst Monitor (GBM), that covers the energy range 8 keV-40 MeV. Results presented throughout the paper are based on LAT data.

Radio galaxies are γ-ray emitters

In only 15 months of survey the LAT could detect 11 radio objects, 7 FR Is and 4 FR IIs (e.g. 2 radio galaxies and 2 steep spectrum radio quasars‡), generally referred to as Misaligned AGN (MAGN¶; Abdo *et al.* 2010a), confirming that RGs are GeV emitters (Table 1). Most of the sources are faint ($F_{>0.1\ \mathrm{GeV}}$ $\sim 10^{-8}$ ph cm^{-2} s^{-1}) and have steep power-law spectra ($\Gamma > 2.4$), in agreement with the Unified Models of AGN in which radio galaxies are considered a de-boosted version of blazars (Fig. 1 *left panel*). The number of radio galaxies caught by the LAT has continuously increased, passing from 11 to 22 in 4 years of operations (3 LAC; Acero *et al.* 2015), with FR Is preferentially detected with respect to FR IIs (Table 1). The discrepancy in the FR I/FR II detection rate does not seem to be related to the larger distance of FR IIs, generally found at higher redshifts (Grandi & Torresi 2012). If a correlation between the radio core flux and the γ-ray flux (Ackermann *et al.* 2011, Ghirlanda *et al.* 2011, Lico *et al.* 2017) is assumed, implicitly implying that the emission in the two bands is mainly from the jet, several FR IIs are expected to be detected above the LAT sensitivity threshold (see also Kataoka *et al.* 2011). This result suggests that the predominance of FR Is in the GeV sky should be related to other causes, for example intrinsic differences in the jet structure, or in the vicinity of the black hole (see, e.g. de Gouveia Dal Pino et al. 2018, these Proceedings).

Localisation of the γ-ray emitting region

One challenging aspect of the high-energy study of MAGN, and of blazars in general, is the localisation of the γ-ray emitting region. Indeed, localising where high-energy photons are dissipated with respect to the black hole has a strong impact on the physical models invoked to explain the γ-ray radiation. The site of emission is not unique: it can occur at sub-pc scales, within the broad-line region (BLR), or outside it on pc

† With the new Pass8 data the LAT energy range extends up to 1 TeV (Atwood *et al.* 2013). For more information about Pass8 https://fermi.gsfc.nasa.gov/ssc/data/analysis/documentation/Pass8_usage.html

‡ Steep Spectrum Radio Quasars are usually considered as FR II broad line radio galaxies at high redshift.

¶ With the term Misaligned AGN we consider objects characterised by steep radio spectra ($\alpha > 0.5$) and possibly showing symmetrical extension in the radio maps, i.e. radio galaxies and steep spectrum radio quasars.

Table 1. 3LAC MAGN: (i) Source name; (ii) 3FGL name; (iii) right ascension; (iv) declination; (v) redshift; (vi) radio classification; (vii) detected TeV counterpart. 3C 120 has been added to the list since it is a confirmed γ-ray source, and also Tol1326-379 and PKS1718-379 recently associated with 3FGL sources.

Source	3FGL Name	RA (J2000)	Dec (J2000)	z	FR type	TeV source
3C 78	3FGLJ0308.6+0408	03 08 26.22	+04 06 39.3	0.02865	FR I	no
IC310	3FGLJ0316.6+4119	03 16 42.97	+41 19 29.61	0.019	FR I	yes
NGC1275	3FGLJ0319.8+4130	03 19 48.16	+41 30 42.1	0.0175	FR I	yes
ForA(lobes)	3FGLJ0322.5-3721	03 21 37.75	−37 12 49.1	0.00587	FR I	no
4C+39.12	3FGLJ0334.2+3915	03 34 18.41	+39 21 24.4	0.02059	FR I	no
Pictor A	3FGLJ0519.2-4542	05 19 49.72	−45 46 43.85	0.03506	FR II	no
PKS0625-35	3FGLJ0627.0-3529	06 27 06.72	−35 29 15.33	0.05494	FR I	yes
3C 189	3FGLJ0758.7+3747	07 58 28.1	+37 47 11.8	0.04284	FR I	no
3C 221	3FGLJ0934.1+3933	09 35 06.63	+39 42 06.7	–	FR II/SSRQ	no
3C 264	3FGLJ1145.1+1935	11 45 05.0	+19 36 22.74	0.02172	FR I	yes
M87	3FGLJ1230.9+1224	2 30 49.42	+12 23 28.04	0.00428	FR I	yes
CenA(core)	3FGLJ1325.4-4301	13 25 27.61	−43 01 08.8	0.0018	FR I	yes
3C 303	3FGLJ1442.6+5156	14 43 02.76	+52 01 37.29	0.14119	FR II	no
NGC6251	3FGLJ1630.6+8232	16 32 31.96	+82 32 16.39	0.02	FR I	no
3C 111	3FGLJ0418.5+3813c	04 18 21.27	+38 01 35.8	0.0485	FR II	no
CenB	3FGLJ1346.6-6027	13 46 49.04	−60 24 29.35	0.01292	FR I	no
TXS0348+013	3FGLJ0351.1+0128	03 50 57.36	+01 31 04.91	1.12	FR II/SSRQ	no
3C 207	3FGLJ0840.8+1315	08 40 47.58	+13 12 23.56	0.681	FR II/SSRQ	no
PKS1203+04	3FGLJ1205.4+0412	12 06 19.92	+04 06 12.04	0.536	FR II/SSRQ	no
3C 275.1	3FGLJ1244.1+1615	12 43 57.64	+16 22 53.39	–	FR II/SSRQ	no
3C 380	3FGLJ1829.6+4844	18 29 31.78	+48 44 46.16	0.695	FR II/SSRQ	no
3C 120	–	04 33 11.1	+05 21 16	0.033	FR I	no
Tol1326-379	3FGL1330.0-3818	13 29 19.2	−38 14 18	0.028	FR 0	no
PKS1718-649	3FGLJ1728.0-6446	17 23 41.0	−65 00 37	0.014	CSO	no

scales, as attested by several studies performed on blazars (Abdo *et al.* 2010b, Jorstad *et al.* 2010, Agudo *et al.* 2011a, 2011b). MAGN are particularly suitable for this kind of study since they offer the opportunity to observe γ-ray emission processes from sub-pc up to kpc scales: different emission sites imply different seed photons involved in the Inverse Compton process producing the GeV emission‖ (Tavecchio *et al.* 2011). In order to establish where the high-energy radiation is emitted, a multi-wavelength (MW) approach is necessary, as demonstrated by the successful results obtained for the broad-line radio galaxies 3C 111 and 3C 120 ††: in these two sources the radio-to-γ-ray study allowed to establish a clear link between intense γ-ray flares and the expulsion of bright superluminal knots in the radio core. In 3C 111, a MW outburst occurred in late 2008 (Fig. 1 *right panel*) was attributed to the ejection of a new radio blob (Grandi *et al.* 2012): the emitting region was compact ($\leqslant$0.1 pc) and the event was localised in the radio core at ~0.3 pc from the black hole. A similar result was obtained by Tanaka *et al.* (2015) from a six-year MW monitoring of 3C 120. Also in this case the γ-ray flare was related to the ejection of a knot from the radio core at sub-pc distance from the black hole. This finding was further confirmed by Casadio *et al.* (2015) who could better constrain the site of the emitting region within ~0.13 pc by exploiting mm-VLBI data. In the examples above the emitting region is located within the BLR, however, in other cases like M87 different dissipation zones have been found, depending on the event producing the outburst‡‡ (Aharonian *et al.* 2006, Harris *et al.* 2006, Acciari *et al.* 2009,

‖ In the standard leptonic scenario.

†† 3C 111 is a FR II, while 3C 120 although morphologically classified as FR I, is characterised by a powerful accretion disk. Interestingly, these are the only AGN for which a disk/jet connection has been established (Marscher *et al.* 2002, Chatterjee *et al.* 2011).

‡‡ In M87 a long-term MW campaign and very-high energy data have allowed to identify different emitting regions: the TeV flare occurred in 2005 was associated with an X-ray burst of

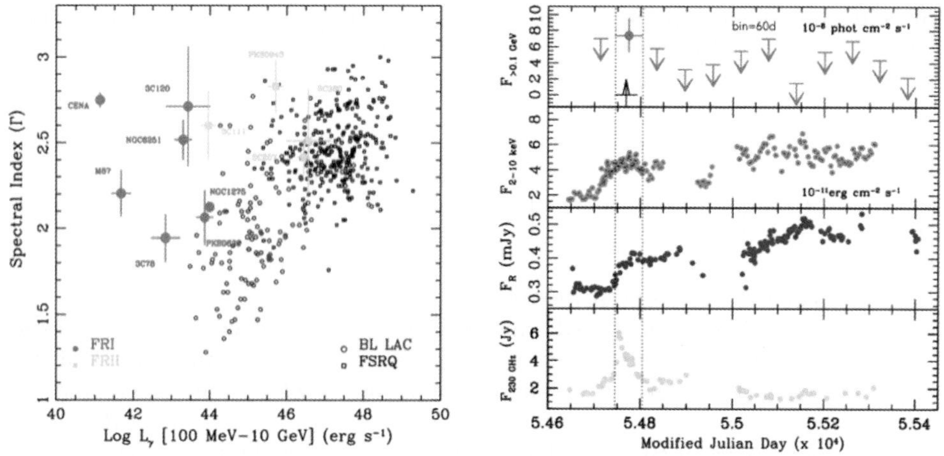

Figure 1. *Left panel:* Power-law photon index (Γ) plotted as a function of the 100 MeV-10 GeV γ-ray luminosity for: BL Lacs (open blue circles), FSRQ (open black squares), FR I (red circles), FR II (green squares). Blazars and MAGN (radio galaxies + SSRQ) occupy different regions of the plot. The two green squares falling in the FSRQ region are the two more distant sources of the sample, i.e. the two SSRQ 3C 207 and 3C 380. *Right panel:* mm (230 GHz; *green*), optical (R; *blue*), X-ray (2.4-10 keV; *magenta*) and γ-ray (0.1-100 GeV; *red*) light curves of 3C 111. mm-to-X-ray data are from Chatterjee *et al.* (2011), γ-ray data are from Grandi *et al.* (2012). The *black dotted lines* limit the flare event in late 2008 (October 4-December 4). The *black arrow* marks the time of ejection of the radio knot. *Red downward arrows* are γ-ray 2σ upper limits.

Abramowski *et al.* 2012). Although the potentiality of such studies is great, MAGN having dedicated long-term MW campaigns are a few. Multi-frequency monitoring is a difficult task that requires a huge effort to collect (quasi-)simultaneous data necessary to find connections among different wavebands. However, it would be very important to have ongoing MW campaigns on MAGN when the CTA observatory will be operative. This would represent a great step forward in the comprehension of the jet structure and evolution.

Extended γ-ray emission

Extended radio lobes represent another possible site of production of high-energy photons (Cheung 2007, Hardcastle *et al.* 2009) as confirmed by the LAT detection of the lobes in the FR I radio galaxy Centaurus A (Abdo *et al.* 2010c). The detection and imaging of extended γ-ray emission is challenging given (i) the PSF of the LAT (the 68% containment PSF is ~0.8 deg at 1 GeV) and (ii) because the emission is isotropic and not boosted by relativistic effects. Indeed, in Centaurus A it was possible to separate the contribution of the two lobes from the core in 10 months of data thanks to the proximity of the source (z=0.001825) and to the total angular extent of its lobes (~10 deg). The flux above 100 MeV of the two lobes accounts for more than half of the total γ-ray flux of the source and the emission is likely produced through Inverse Compton (IC) scattering of the Cosmic Microwave Background Radiation (CMB). Very recently, γ-ray extended emission was also found in the FR I Fornax A (Ackermann *et al.* 2016) in 6.1 years of data. The source is very nearby (z=0.005871) and the radio lobes

the knot HST-1, at 0.85″ from the nucleus. In 2008, the coincidence of another TeV flare with radio flares confined the VHE photon production region within 10 Schwarzschild radii. Finally, the 2010 TeV flare was not accompanied by an enhancement of the radio flux but by an increase in the *Chandra* X-ray flux 3 days after the flare, making hard to unambiguously interpret the data.

span $\sim$50′, but differently from Cen A, the core flux accounts for less than 14% of the total γ-ray flux. However, an IC/CMB-EBL model (Georganopoulos *et al.* 2008, Domínguez *et al.* 2011) alone is not sufficient to reproduce the extended emission because it underestimates the high-energy fluxes. A second component, in addition to the leptonic one, is required and this could be of hadronic origin (McKinley *et al.* 2015). Finally, the *Fermi*-LAT detection of a young Compact Symmetric Object (CSO), i.e. PKS 1718-649 (Migliori *et al.* 2016) further supports the non-thermal γ-ray production in the lobes (for a detailed discussion see Migliori 2018, these Proceedings).

FR I Spectral Energy Distributions

Observations of FR Is at high- (HE) and very-high (VHE) energies† have questioned the pure one-zone SSC model, generally applied to reproduce their spectral energy distributions (SED), because it implies bulk velocities of the jet much lower than BL Lacs ($\Gamma \leqslant 3$). This is at odd with the Unified Scenario in which FR Is are the misaligned counterpart of BL Lacs. The problem is overcome if stratified jets, consisting of different regions moving at different velocities and mutually interacting, are assumed (Georganopoulos *et al.* 2003, Ghisellini *et al.* 2005, Böttcher *et al.* 2010): these models were successfully applied to the SED of M87 (Abdo *et al.* 2009a), NGC 6251 (migliori *et al.* 2011) and NGC 1275 (Abdo *et al.* 2009b), suggesting that in FR Is the presence of slower components could play an important role in amplifying the GeV emission. Interestingly, a limb-brightened jet structure has been observed in M87 (Kovalev *et al.* 2007) and recently in NGC 1275 (Giovannini *et al.* 2018 and Giovannini 2018, these Proceedings). In FR IIs the external layers could be less prominent or even absent and/or the deceleration processes less efficient, explaining their rarity in the GeV sky. Other possible sources of γ-ray photons could be magnetic reconnection events along the jets or in the vicinity of the black hole (Giannios *et al.* 2010, Khiali *et al.* 2015). Hadronic models based on proton-photon interactions have been also developed to provide connections among AGN, ultra-high-energy cosmic rays (UHECR) and neutrinos (Böttcher 2012, Becker & Biermann 2009, Khiali & de Gouveia Dal Pino 2016, Ramirez-Rodriguez et al. 2018, these Proceedings).

Temporal variability

Studying γ-ray variability is a powerful tool to further explore the jet structure issue in FR Is and FR IIs. The two classes show a different temporal behaviour: FR Is are detected, on average, for most of the time by the LAT and their light curves do not show variability structures, as suggested by the 3FGL variability index‡ $TS_{var} < 72.4$ (Grandi *et al.* 2013). On the contrary, FR IIs are observed only when intense and rapid flares occur (Fig. 1 *right panel*). Within this picture an outstanding exception is NGC 1275 which shows a blazar-like behavior and a rapid variability on timescales of days to weeks (Tanada *et al.* 2018).

The FR 0 radio galaxy Tol1326-379

The increased exposure time of the LAT survey combined with recent large-area surveys such as SDSS/NVSS/FIRST (Best & Heckman 2012) allowed to explore for the first time the GeV emission of RGs at low radio fluxes ($\sim$mJy). This is the realm of FR 0 sources (Baldi *et al.* 2015), i.e. compact radio galaxies (radio size $\leqslant$5 kpc) that represent the bulk of the RL AGN population in the local Universe (Baldi *et al.* 2018). Grandi *et al.* (2016)

† HE: >100 MeV; VHE: >100 GeV (see e.g. Aharonian 2012).

‡ The variability index is an indicator of the variability of a source on timescales of months. For the 3FGL catalog, an index >72.4 indicates that the source is variable at a 99% confidence level.

proposed the association of a 3LAC source with the FR 0 Tol1326-379, offering the opportunity to investigate for the first time the behavior of these objects in the high-energy domain. FR 0s are similar to FR Is in the optical and X-ray properties, i.e. they are classified as low-excitation galaxies and their X-ray radiation is of non-thermal origin, likely produced by the jet (Torresi *et al.* 2018). In γ-rays, the flux of Tol1326-379 is similar to FR Is, $F_{(>1\ GeV)} = (3.1 \pm 0.8) \times 10^{-10}$ ph cm^{-2} s^{-1} but the spectrum is steeper, Γ=2.78$\pm$0.14. The SED of Tol1326-379 resembles that of Cen A, known to lie close to the plane of the sky, $\theta \sim 50 - 80°$. Therefore, also for the compact FR 0 source the orientation angle could be large and the process responsible for the γ-ray emission is likely the same as extended FR Is.

3. Radio galaxies in the TeV regime

Up to now, only a few radio galaxies have a very-high energy counterpart (Table 1): all of them are nearby (z<0.06) FR Is. Despite their limited number, observations of radio galaxies in the TeV band have already produced intriguing results. For example, (i) the synergy between VHE data and MW observations have allowed to localise the emitting regions in M87 (Sec. 2); (ii) the fast (ΔT$\sim$4.8 minutes) TeV variability, recently caught by MAGIC in IC 310 (Aleksić *et al.* 2014), has questioned the classical shock-in-jet scenario (Blandford *et al.* 1979) invoking sub-horizon scale structures as a possible explanation for the extreme temporal behavior (Glawion 2018, these Proceedings); (iii) the extrapolation of *Fermi*-LAT data to higher energies has evidenced that a second hard component, whose nature is still debated, is necessary to account for the H.E.S.S. flux (Sahakian *et al.* 2013). It would be important to observe such a structure in other radio galaxies in order to understand if it is (or not) a common characteristic of MAGN. A major problem at the moment, is the difficulty in extrapolating the model from the GeV to the TeV band since data are often not simultaneous. Hopefully, this issue will be overcome with CTA that working as an observatory in synergy with other facilities, will provide simultaneous data in a very wide energy band from a few MeV up to hundreds TeV. Thanks to the improvement in sensitivity by a factor of 5x to 20x, depending on the energy range, with respect to the current Cherenkov telescopes, CTA will give a significant contribution in characterising MAGN at very-high energies (Angioni *et al.* 2017). Fig. 2 is a diagnostic plot showing the detectability of a source by CTA (northern and southern arrays) given a LAT spectral slope and a flux in the 1-100 GeV band: sources with $\Gamma_{\rm LAT} \leqslant 2.1$ can be easily revealed for fluxes down to 10^{-10}ph cm^{-2} s^{-1}. As the slope steepens, larger fluxes are required to overcome the sensitivity threshold of the array. Incidentally, we note that this plot predicted the possible detection of PKS 0625-35 and 3C 264, later revealed by H.E.S.S. and VERITAS, respectively (Dyrda *et al.* 2015, H.E.S.S. Collaboration 2018, Mukherjee 2018).

4. Summary and conclusions

In the last decade our view of RGs above 100 MeV has greatly changed. Indeed, the *Fermi* satellite confirmed RGs as a class of γ-ray emitters, in spite of the unfavourable inclination angle of their jets. Although they are a small number with respect to the blazars detected by *Fermi* (<2% in the 3 LAC), studying RGs at high- and very-high energies is of great importance to understand the jet phenomenon and to reveal the jet structure complexity. In particular, the observation of RGs in the GeV band has evidenced that: (i) γ-ray photons can be produced in different sites and at different scales along the jet and in the radio lobes; (ii) a one-zone SSC model used to reproduce the SED of FR Is is oversimplified: structured jets, magnetic reconnection events or hadronic processes have

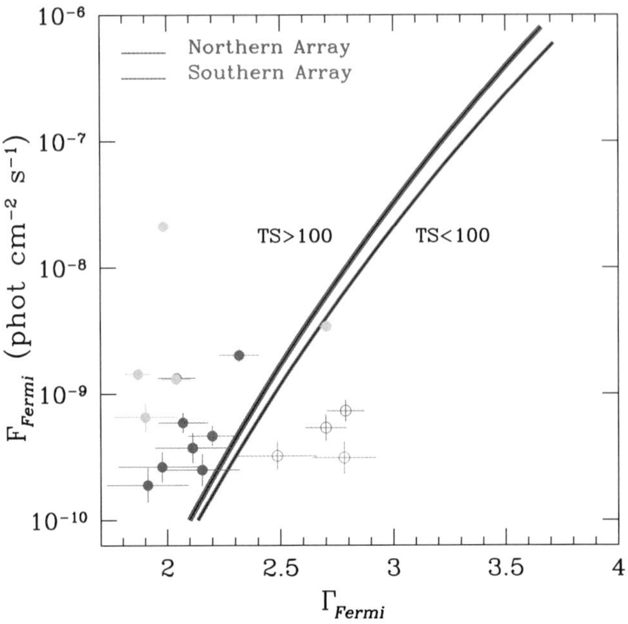

Figure 2. Diagnostic plot for MAGN with known *Fermi*-LAT flux and spectral slope reproduced from Angioni *et al.* (2017). The *blue* and *red lines* define the region where a source can be detected by CTA North and South, respectively, assuming a threshold TS=100. *Green filled circles* are radio galaxies already detected by current IACT telescopes (before the detection of PKS 0625-35 and 3C 264). *Blue filled circles* are TS>100 detections with the northern array and *red filled circles* with the southern array. *Empty circles* represent TS<100 detections.

been invoked to reconcile the data with the Unified Models; (iii) the different detection rate and temporal variability of FR Is and FR IIs suggest possible intrinsic jet differences between these two classes.

RGs in the TeV band are almost unexplored, however the few detections obtained so far with the current IACTs already produced promising results, such as the sub-horizon scale variability in IC 310 or the second hard component found in Cen A. CTA will have a strong impact on our comprehension of RGs as a class by detecting additional sources and better characterising those already known to emit TeV γ-ray photons. This will be very important to shed light on the mechanism producing the VHE radiation, e.g. leptonic or hadronic processes or both. Indeed, RGs are considered good candidates as extragalactic cosmic-ray accelerators, e.g. the relativistic jets and the radio lobes offer an ideal environment where particles can be accelerated, and thus they could be sources of high-energy neutrinos (Fraija *et al.* 2016, Hooper 2016, Murase 2018, Matthews *et al.* 2018).

Acknowledgments

The CTA Consortium is gratefully acknowledged. ET would like to thank the two reviewers of the CTA SAPO for their useful comments. ET thanks P. Grandi, G. Migliori and G. Giovannini for reading the manuscript and for insightful suggestions. The Fermi LAT Collaboration acknowledges generous ongoing support from a number of agencies and institutes that have supported both the development and the operation of the LAT as well as scientific data analysis. These include the National Aeronautics and Space Administration and the Department of Energy in the United States, the Commissariat 'a l'Energie Atomique and the Centre National de la Recherche Scientifique / Institut

National de Physique Nucleaire et de Physique des Particules in France, the Agenzia Spaziale Italiana and the Istituto Nazionale di Fisica Nucleare in Italy, the Ministry of Education, Culture, Sports, Science and Technology (MEXT), High Energy Accelerator Research Organization (KEK) and Japan Aerospace Exploration Agency (JAXA) in Japan, and the K. A. Wallenberg Foundation, the Swedish Research Council and the Swedish National Space Board in Sweden. Additional support for science analysis during the operations phase is gratefully acknowledged from the Istituto Nazionale di Astrofisica in Italy and the Centre National d'Etudes Spatiales in France

References

Abdo, A., A., *et al.* 2009a, *ApJ*, 707, 55
Abdo, A., A., *et al.* 2009b, *ApJ*, 699, 31
Abdo, A., A., Ackermann, M., Ajello, M., *et al.* 2010a, *ApJ*, 720, 912
Abdo, A., A., Ackermann, M., Ajello, M., *et al.* 2010b, *ApJ*, 722, 520
Abdo, A., A., *et al.* 2010c, *Science*, 328, 725
Abramowski, A., Acero, F., Aharonian, F., *et al.* 2012, *ApJ*, 746, 151
Acciari, V. A., *et al.* 2009, *Science* 325, 5939
Acero, F., Ackermann, M., Ajello, M., *et al.* 2015, *ApJS*, 218, 23
Ackermann, M., Ajello, M., Allafort, A., *et al.* 2011, *ApJ*, 741, 30
Ackermann, M., Ajello, M., Baldini, L., *et al.* 2016, *ApJ*, 826, 1
Aharonian, F., *et al.* 2006, *Science*, 314, 1424
Aharonian, F. 2012, *The Twelfth Marcel Grossmann Meeting*, 368A
Agudo, I., *et al.* 2011a, *ApJ*, 726, L13
Agudo, I., *et al.* 2011b, *ApJ*, 735, L10
Aleksić, J., *et al.* 2014, *Science*, 346, 1080
Angioni, R., Grandi, P., Torresi, E., *et al.* 2017, *APh*, 92, 42
Atwood, W. B., Abdo, A. A., Ackermann, M., *et al.* 2009, *ApJ*, 697, 1071
Atwood, W. B., Albert, A., Baldini, L., *et al.* 2013, *arXiv e-prints*, [arXiv:1303.3514]
Baldi, R. D., Capetti, A., & Giovannini, G. 2015, *A&A*, 576, 38
Baldi, R. D., Capetti, A., & Massaro, F. 2018, *A&A*, 609, 1
Becker, J. K., & Biermann, P. L. 2009, *Astroparticle Physcis*, 31, 138
Best, P. N., & Heckman, T. M. 2012, *MNRAS*, 421, 1569
Bicknell, G. V. 1995, *ApJS*, 101, 29
Blandford, R. D., & Königl, A. 1979, *ApJ*, 232, 34
Böttcher, M., & Dermer, C. D. 2010, *ApJ*, 711, 445
Böttcher, M. 2012, *Proceedings of Fermi & Jansky: Our evolving understanding of AGN*, [arXiv:1205.0539]
Casadio, C., Gómez, J. L., Grandi, P., *et al.* 2015, *ApJ*, 808, 162
Chatterjee, R., Marscher, A. P., Jorstad, S. G., *et al.* 2011, *ApJ*, 734, 43
Cheung, C. C. 2007, *AIPC*, 921, 325
Domínguez, A., Primack, J. R., Rosario, D. J., *et al.*. 2011, *MNRAS*, 410, 2556
Dyrda, M., *et al.* 2015, *ICRC*, 34, 801
Fanaroff, B. M., & Riley, J. A. 1974, *MNRAS*, 167, 31
Fraija, N., & Marinelli, A. 2016, *ApJ*, 830, 81
Georganopoulos, M., & Kazanas, D. 2003, *ApJ*, 589, 5
Georganopoulos, M., Sambruna, R. M., Kazanas, D., *et al.* 2008, *ApJL*, 686, L5
Ghirlanda, G., Ghisellini, G., Tavecchio, F., *et al.* 2011, *MNRAS*, 413, 852
Ghisellini, G., & Celotti, A. 2001, *A&A*, 379, 1
Ghisellini, G., Tavecchio, F., & Chiaberge, M. 2005, *A&A*, 432, 401
Giannios, D., *et al.* 2010, *MNRAS*, 402, 1649
Giovannini, G., Savolainen, T., Orienti, M., *et al.* 2018, *NatAs*, 2, 472
Grandi, P., & Palumbo, G. G. C. 2007, *ApJ*, 659, 235
Grandi, P., Torresi, E., & Stanghellini, C. 2012, *ApJL*, 715, L3

Grandi, P. & Torresi, E. 2012, *Proceedings of Fermi & Jansky: Our evolving understanding of AGN*, [arXiv:1205.1686]
Grandi, P., Torresi, E., De Rosa, A., *et al.* 2013, *EPJWC*, 61, 04007
Grandi, P., Capetti, A., & Baldi, R. D. 2016, *MNRAS*, 457, 2
Hardcastle, M. J., *et al.* 2009, *MNRAS*, 393, 1041
Harris, D. E., *et al.* 2006, *ApJ*, 640, 211
Hartmann, R. C., Kadler, M., & Tueller, J. 2008, *ApJ*, 688, 852
H. E. S. S. Collaboration 2018, *MNRAS*, 476, 4187
Hooper, D. 2016, *JCAP*, 09, 002
Jorstad, S. G., *et al.* 2010, *ApJ*, 715, 362
Kataoka, J., *et al.* 2011, *ApJ*, 740, 29
Khiali, B., de Gouveia Dal Pino, E. M., Sol, H. 2015, *arXiv e-print*, [arXiv:1504.07592]
Khiali, B., & de Gouveia Dal Pino, E. M. 2016, *MNRAS*, 455, 838
Kovalev, Y. Y., *et al.* 2007, *ApJ*, 668, 27
Lico, R., Giroletti, M., Orienti, M., *et al.* 2017, *A&A*, 606, A138
Marchesini, D., Celotti, A., & Ferrarese, L. 2004, *MNRAS*, 351, 733
Marscher, A. P., *et al.* 2002, *Nature*, 417, 625
Matthews, J. H., Bell, A. R., Blundell, K. M., *et al.* 2018, *MNRAS*, 479, 76
McKinley, B., *et al.* 2015, *MNRAS*, 446, 3478
Migliori, G., Siemiginowska, A., Sobolewska, M., *et al.* 2016, *ApJ*, 821, 31
Mukherjee, R., *et al.* 2002, *ApJ*, 574, 693
Mukherjee, R. 2018, *The Astronomer's Telegram*, #11436
Murase, K. 2018, *PhysRevD*, 97, 081301
Nolan, P., *et al.* 1996, *ApJ*, 459, 100
Ostorero, L., Moderski, R., Stawarz, Ł., *et al.* 2010, *ApJ*, 715, 1071
Sahakian, N., *et al.* 2013, *ApJ*, 770, 6
Sguera, V., *et al.* 2005, *A&A*, 430, 107
Stawarz, Ł., Sikora, M., & Ostrowski, M. 2003, *ApJ*, 597, 186
Stawarz, Ł., Kneise, T. M., & Kataoka, J. 2006, *ApJ*, 637, 693
Stawarz, Ł., Ostorero, L., Begelman, M. C., *et al.* 2008, *ApJ*, 680, 911
Tanada, K., *et al.* 2018, *ApJ*, 860, 74
Tanaka, Y. T., *et al.* 2015, *ApJL*, 799, 18
Tavecchio, F., Becerra-Gonzalez, J., Ghisellini, G., *et al.* 2011, *A&A*, 534, A86
Torresi, E., Grandi, P., Capetti, A., *et al.* 2018, *MNRAS*, 476, 5535
Urry, C. M., & Padovani, P. 1995, *PASPJ*, 107, 803

Perseus in Sicily: From Black Hole to Cluster Outskirts
Proceedings IAU Symposium No. 342, 2018
K. Asada, E. de Gouveia Dal Pino, M. Giroletti,
H. Nagai & R. Nemmen, eds.

doi:10.1017/S1743921318008165

Search for QPOs in Perseus with *Fermi* LAT

Rodrigo Nemmen[1], Raniere de Menezes[1] and Vasileios Paschalidis[2]
on behalf of *Fermi*-LAT Collaboration

[1]Universidade de São Paulo, Instituto de Astronomia, Geofísica e Ciências Atmosféricas,
Departamento de Astronomia,
São Paulo, SP 05508-090, Brazil
email: rodrigo.nemmen@iag.usp.br

[2]Depts. of Astronomy & Physics, University of Arizona

Abstract. We report the analysis of the gamma-ray variability of NGC 1275–the radio galaxy at the center of the Perseus cluster. NGC 1275 has been observed continuously with the *Fermi* Large Area Telescope over the last nine years. We applied different time-domain analysis methods including Fourier, wavelets and Bayesian methods, in order to search for quasi-periodic oscillations in the gamma-ray emission. We found no evidence for periodicities of astrophysical origin.

Keywords. gamma rays: observations, galaxies: nuclei, galaxies: jets, galaxies: active

1. Introduction

NGC 1275 is the brightest central galaxy in the nearby Perseus galaxy cluster. It is a radio galaxy producing collimated, relativistic jets of particles outflowing from its center at velocities $0.3 - 0.5c$ (Walker, Romney & Benson 1994). These jets are strongly interacting with the intracluster medium, heating it up, offsetting radiative cooling and creating X-ray bright cavities where star formation is quenched (e.g. contributions by Churazov, Zuravleva in these proceedings).

Importantly, the center of NGC 1275 is the *brightest gamma-ray emitter* (e.g. Abdo *et al.* 2009) among nearby active galactic nuclei (AGN). The availability of a high-quality, high-cadence light curve (LC) observed with *Fermi*-LAT for NGC 1275 over the last nine years with uniform sampling make this galaxy not only a remarkable laboratory for high-energy astrophysical processes around black holes in the nearby universe but also an ideal place for mining for interesting time-domain signals due to e.g. potential massive binary black holes.

In this contribution, we will not address the gamma-ray emission due to the galaxy cluster itself; instead, our focus is on the AGN activity. Galaxy clusters are not detected in gamma-rays, an observation that challenges the simplest hadronic models and implies a low energy density in cosmic rays (Ackermann *et al.* 2014).

2. Gamma-ray flares in NGC 1275

The broadband emission of NGC 1275 has been varying dramatically over the last forty years in radio (Dutson *et al.* 2014). More recently, NGC 1275 has been showing increased activity and flaring in the gamma-ray band (Tanada *et al.* 2018), with a broad correspondence between the high-frequency radio data and gamma-rays. NGC 1275 is also detected at TeV energies during epochs of γ-ray flaring (Aleksic *et al.* 2012).

Recently, Tanada *et al.* (2018) presented an analysis of eight years of *Fermi*-LAT (0.1-300 GeV) γ-ray observations for NGC 1275 (cf. Figure 1 in Tanada *et al.*). Besides finding that the gamma-flux has been steadily increasing and is highly variable on short timescales (days to weeks–cf. also contribution of Sahakyan who reports variability on timescales of hours; Baghmanyan *et al.* 2017), Tanada *et al.* also found two distinct periods of flaring activity.

Period A From $\approx$2009 to 2011, strong variability in spectral index, quiescent flux
Period B From $\approx$2013 to 2016, spectral index does not change much, strong variation in flux

Based on these observations, Tanada *et al.* (2018) argue that the source is in two different states in each of the periods above. During *period A*, NGC 1275 would be in a state where there is a fresh production of nonthermal electrons (e.g. internal shocks or magnetic reconnection). This period could be associated with the C3 hotspot observed in radio (cf. contributions by Hodgson, Savolainen in these proceedings). *Period B* can be explained by variations in the Doppler factor of the emitting plasma, possibly due to a small change in the jet orientation.

3. Radio galaxies and QPOs

Quasi-periodic oscillations (QPOs) correspond to strong, narrow peaks in the Fourier power spectrum of a time series indicative of the presence of a deterministic process with a characteristic period. For this reason, QPOs are very useful as indirect probes for separating stochastic versus deterministic processes in accreting black holes in X-ray binaries and active galactic nuclei Remillard & McClintock (2006); Gierlinski *et al.* (2008).

In the case of radio galaxies such as NGC 1275 or blazars, there are different processes associated with the central black hole(s) that could introduce a characteristic period in the jet light curve:

• *Jet-disk instabilities* with QPO period $\tau_{\rm inst}$ due to the unstable magnetospheric interface between the accretion flow and the jet (McKinney, Blandford & Tchekhovskoy 2012).

• *Jet precession* due to the Lense-Thirring precession (period $\tau_{\rm prec}$) caused by an accretion flow misaligned with the black hole spin vector (Liska *et al.* 2018).

• An *accreting massive binary black hole system* where the main period corresponds to the binary orbital period $\tau_{\rm BBH}$, with possible harmonics (e.g. Roedig *et al.* 2012; Gold *et al.* 2014).

We expect roughly

$$\tau_{\rm inst} < \tau_{\rm prec} \lesssim \tau_{\rm BBH}. \tag{3.1}$$

All of these processes can introduce characteristic QPOs in the NGC 1275 gamma LC, thus motivating a detailed study of this time series.

4. Search for gamma-ray QPOs in NGC 1275

We used three methods to search for the presence of QPOs in the gamma LC of NGC 1275: Fourier transform, continuous wavelet transform and a Bayesian QPO search framework. In our experience, the most robust method for QPO search was the Bayesian method as we will describe below. As pointed out before (Vaughan *et al.* 2016 and references therein), one must be very careful in properly modeling the power spectrum (e.g. red noise, power-law) when searching for QPOs.

Figure 1 shows NGC 1275's LC extracted with standard analysis parameters in the energy range 0.1-300 GeV from a circular region of radius 15° (Pass 8). The spectral index was left free to vary. We used the *Fermi* Science Tools version v10r0p5. Whereas

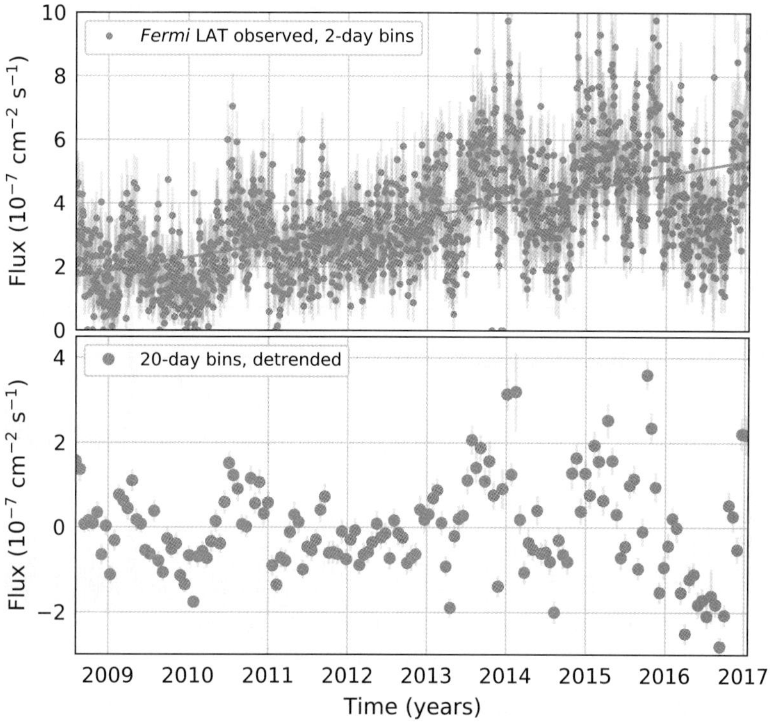

Figure 1. NGC 1275 γ-ray light curve observed with *Fermi*-LAT. Upper panel: LC with bins of 2 days. Lower panel: LC with bins of 20 days after a linear detrending.

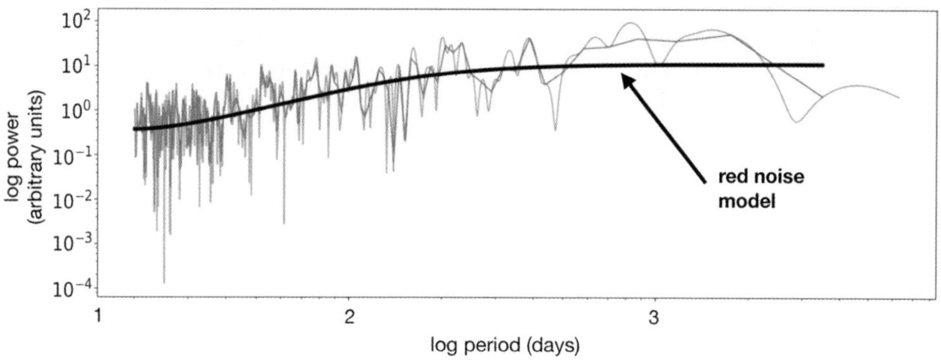

Figure 2. Power spectra computed with the Fourier (blue line) and Lomb-Scargle methods (red line). The black line shows a red noise fit to the LC.

the upper panel clearly demonstrates the progressive increase in the gamma flux over time, an inspection of the lower panel might suggest the potential presence of oscillatory behavior.

Figure 2 shows the LC power spectrum compared with a red noise fit (cf. Torrence & Compo 1998 for details on the red noise model). No QPOs are clearly evident in this figure.

To settle the issue, we used the Bayesian QPO search procedure outlined in Huppenkothen *et al.* (2013). This method consists of comparing the broadband noise model to a more complex model combining both the noise model and a Lorentzian to

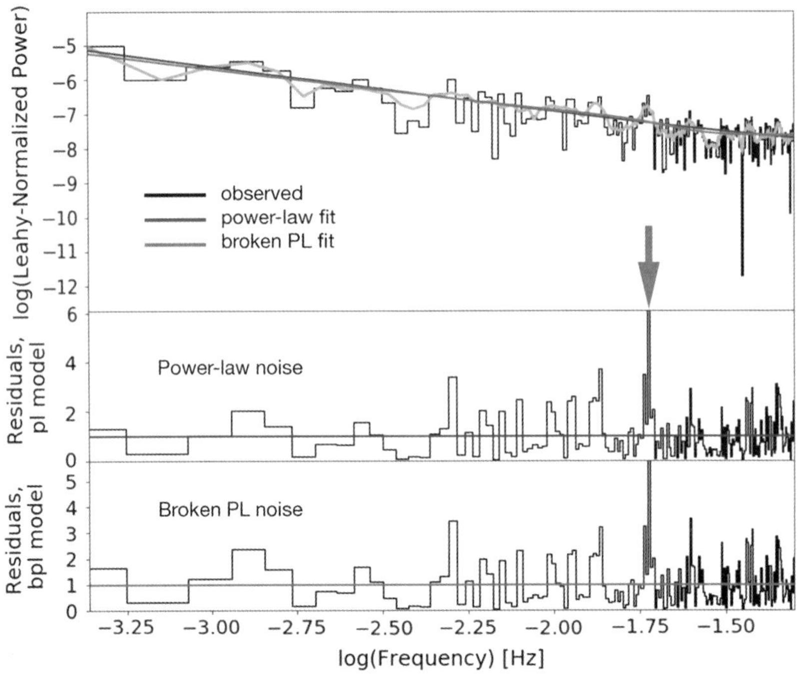

Figure 3. Results of Bayesian QPO search applied to NGC 1275's light curve. The arrow indicates one potential QPO, which is consistent with the rocking period of the *Fermi* satellite.

account for possible QPOs. The noise model is assumed to be either a power-law or broken power-law models, similarly to the noise models assumed in time-domain analysis of X-ray binaries and Seyfert 1 X-ray data. The results from this analysis are shown in Figure 3 demonstrating that the signal is broadly consistent with noise. There is one significant QPO which is indicated with an arrow; however, this feature is compatible with the rocking period of the *Fermi* satellite.

5. Summary

NGC 1275 is a rich lab for investigating high-energy phenomena and black hole physics, particularly in the time domain. It has been brightening in radio and gamma-rays and displayed gamma-ray flares over the last nine years. Processes such as Lense-Thirring precession and the presence of accreting binary black holes could introduce QPOs with periods in the ~months to years range in the light curves of radio galaxies. Motivated by these prospects, we searched for the presence of QPOs in the *Fermi*-LAT light curve of NGC 1275 using a range of methods including a powerful Bayesian QPO search framework. We found no evidence for periodicities of astrophysical origin. The Bayesian QPO search method employed in this work should be a powerful tool for future searches of QPOs in AGNs and blazars.

Acknowledgements

The *Fermi*-LAT Collaboration acknowledges support for LAT development, operation and data analysis from NASA and DOE (United States), CEA/Irfu and IN2P3/CNRS (France), ASI and INFN (Italy), MEXT, KEK, and JAXA (Japan), and the K.A. Wallenberg Foundation, the Swedish Research Council and the National Space Board (Sweden). Science analysis support in the operations phase from INAF (Italy) and

CNES (France) is also gratefully acknowledged. This work performed in part under DOE Contract DE-AC02-76SF00515.

References

A. A. Abdo, *et al.* *ApJ*, 699, 31, 2009
M. Ackermann, *et al.* *ApJ*, 787, 18, 2014
J. Aleksić, *et al.* *A&A*, 539, L2, 2012
V. Baghmanyan, S. Gasparyan, & N. Sahakyan. *ApJ*, 848, 111, October 2017
K. L. Dutson, A. C. Edge, J. A. Hinton, M. T. Hogan, M. A. Gurwell, & W. N. Alston. *MNRAS*, 442, 2048, 2014
M. Gierlinski, M. Middleton, M. Ward, & C. Done. *Nature*, 455, 369, 2008
R. Gold, V. Paschalidis, M. Ruiz, S. L. Shapiro, Z. B. Etienne, & H. P. Pfeiffer. *Phys. Rev. D*, 90, 104030, 2014
D. Huppenkothen, *et al.* *ApJ*, 768, 87, 2013
M. Liska, C. Hesp, A. Tchekhovskoy, A. Ingram, M. van der Klis, & S. Markoff. *MNRAS*, 474, L81, 2018
J. C. McKinney, A. Tchekhovskoy, & R. D. Blandford. *MNRAS*, 423, 3083, 2012
R. A. Remillard & J. E. McClintock. *ARA&A*, 44, 49, 2006
C. Roedig, A. Sesana, M. Dotti, J. Cuadra, P. Amaro-Seoane, & F. Haardt. *A&A*, 545, A127, 2012
K. Tanada, J. Kataoka, M. Arimoto, M. Akita, C. C. Cheung, S. W. Digel, & Y. Fukazawa. *ApJ*, 860, 74, 2018
C. Torrence & G. P. Compo. *Bulletin of the American Meteorological Society*, 79, 61, 1998
S. Vaughan, P. Uttley, A. G. Markowitz, D. Huppenkothen, M. J. Middleton, W. N. Alston, J. D. Scargle, & W. M. Farr. *MNRAS*, 461, 3145, 2016
R. C. Walker, J. D. Romney, & J. M. Benson. *ApJ*, 430, L45, 1994

Perseus in Sicily: From Black Hole to Cluster Outskirts
Proceedings IAU Symposium No. 342, 2018
K. Asada, E. de Gouveia Dal Pino, M. Giroletti,
H. Nagai & R. Nemmen, eds.

doi:10.1017/S1743921318003873

High Energy γ-ray variability of NGC 1275 and 3C 120

Narek Sahakyan

ICRANet-Armenia, Marshall Baghramian Avenue 24a, Yerevan 0019, Armenia
email: narek@icra.it

Abstract. The recent observations in the high energy γ-ray band show that the extragalactic γ-ray sky is dominated by the emission from blazars. However, γ-ray emission from other types of AGNs, e.g., radiogalaxies, also have been detected. These sources were not considered as favored GeV emitters because the nonthermal emission from their jets is less Doppler boosted. Now, the γ-ray emission from more than 25 non-blazar AGNs has been already detected which opened a new window to have an insight into the particle acceleration and emission processes in different components of AGNs. Here, I will present the γ-ray variability of two well-known radiogalaxies, NGC 1275 and 3C 120, which show a rapid flux increase in the γ-ray band.

Keywords. galaxies: active, galaxies: individual (NGC 1275, 3C 120), gamma rays: observations

1. Introduction

The recent observations in the High Energy (HE; $\geqslant$ 100 MeV) γ-ray band show that the extragalactic γ-ray sky is dominated by the emission of Active Galactic Nuclei (AGNs) of different types (Acero *et al.* (2015)). Dominant in these are blazars - an extreme class of AGNs which have jets that are forming a small angle with respect to the line of sight (Urry & Padovani (1995)). Blazars are very bright and luminous sources known to emit electromagnetic radiation in almost all frequencies that are currently being observed, ranging from radio to very high energy γ-ray bands. Their broadband spectrum is mainly dominated by non-thermal emission produced in a relativistic jet pointing toward the observer. Other important class of γ-ray emitting AGNs observed by Fermi Large Area Telescopes (Fermi LAT) are radio galaxies with relativistic jets at systematically larger angles (Sahakyan *et al.* (2018)). Due to larger jet inclination angle as compared with blazars, the jet emission is not significantly Doppler boosted, making it less prevalent over such components as the radiation from mildly relativistic outflows or emission from extended structures. This providing a chance to investigate also the emission from extended non-boosted regions, moderately relativistic plasmas, etc.

Only a very small fraction of the total AGNs detected by Fermi-LAT (2%) are non-blazar AGNs, including fourteen radio galaxies, six SSRQs, one CSSs and five NLSy1s. The radio galaxies detected in the γ-ray band can be sub-divided into two morphological types, FRI and FRII (Fanaroff & Riley (1974)), according to their radio luminosities. FRIs have a typical luminosity of $< 10^{41}$ erg s^{-1} (at 178 MHz) while that of the luminous FRIIs is $> 10^{41}$ erg s^{-1} (at 178 MHz). FRIs have slower, more turbulent and less collimated jets, while FRIIs have powerful collimated jets (often only one is visible) which terminate into well-defined lobes with prominent hot spots. Increasingly aligned versions of FRIIs are SSRQs, which are powerful radio sources with large-scale radio structures and which appear at intermediate angles between FRIIs and FSRQs. SSRQs are sometimes classified as CSS quasars, since most of the radio flux is emitted within galactic scales (< 10 kpc) rather than at hundreds of kilo-parsec scales.

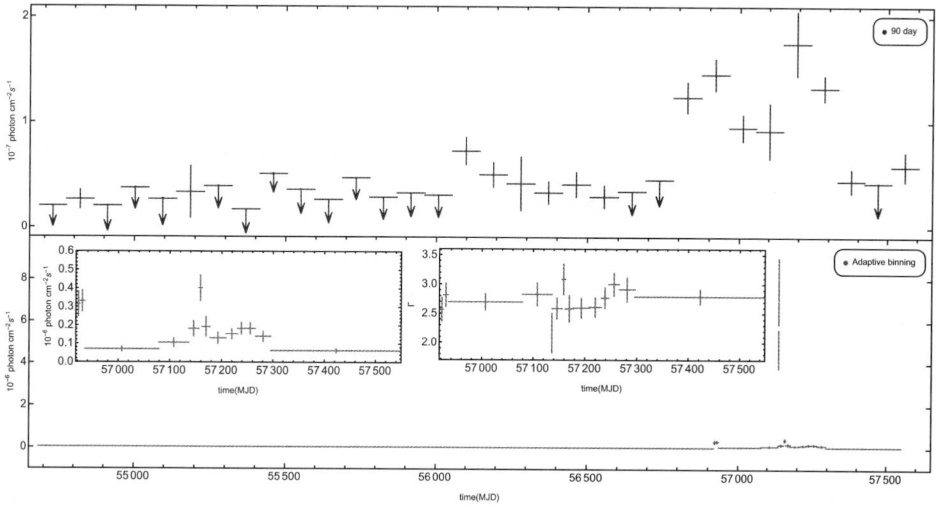

Figure 1. The γ-ray light curve of 3C 120 from August 4, 2008, to August 4, 2016 with 90 days bin (a) and 20 % of uncertainty (b).

The recent studies showed that the γ-ray photon index of non-blazar AGNs changes within $1.84 - 2.86$ (Sahakyan *et al.* (2018)). The hardest photon indexes of $\Gamma = 1.84 \pm 0.17$ and $\Gamma = 1.89 \pm 0.04$ were obtained for TXS 0331+3915 and PKS 0625-35, respectively. The measured faintest flux is of the order of a few times $10^{-9}\,\mathrm{photon\,cm^{-2}s^{-1}}$, while the flux from bright sources exceeds $10^{-7}\,\mathrm{photon\,cm^{-2}s^{-1}}$. In the $\Gamma - F_{\gamma}$ plane, the FRIs have a harder γ-ray photon index with $\Gamma_{\mathrm{m}} \simeq 2.22$ and a flux ranging from $\simeq 1.5 \times 10^{-9} \times \mathrm{photon\,cm^{-2}s^{-1}}$ to $3.6 \times 10^{-8} \times \mathrm{photon\,cm^{-2}s^{-1}}$. The mean photon index of FRIIs/SSRQs shifts to a higher value of $\Gamma_{\mathrm{m}} \simeq 2.58$ (i.e., softer spectra) and the flux to the range of $(0.9 - 4) \times 10^{-8}\,\mathrm{photon\,cm^{-2}s^{-1}}$.

The γ-ray emission from most of the non-blazars AGNs is variable in a long-timescales and only for NGC 1275 and 3C 120 extreme short time scale variabilities are found. Here, I briefly review the main results of the γ-ray observations of NGC 1275 (Baghmanyan *et al.* (2017)) and 3C 120 (Sahakyan *et al.* (2015), Zargaryan *et al.* (2017)), mainly discussing the temporal variation of the γ-ray flux.

2. 3C 120

At the red shift $z = 0.033$, 3C 120 is a nearby Seyfert 1 radio galaxy that is an active and powerful emitter of radiation at all the observed wavebands. With bright continuum and broad optical emission lines 3C 120 is usually classified as a broad line radio galaxy. Hosting a black hole with a mass $5.5 \times 10^{7} M_{\odot}$ (Peterson *et al.* (2004)), well constrained from the reverberation mapping, 3C 120 has a radio morphology more similar to the FR I radio sources. The source has a powerful one-sided radio jet extending from a sub-*pc* up to 100 *kpc* scales (Walker *et al.* (1987)). Observations with the Very Long Baseline Array at frequencies (22, 43, and 86 GHz) reveal a very rich inner jet structure containing several superluminal components with apparent speed up to 4-6 c (e.g., Homan *et al.* (2001)) that can be investigated with better resolution than most other extragalactic superluminal sources because of the relatively low redshift. The jet inclination angle to the line of sight is constrained to be 14° by the measured apparent motion. Recently, using X-ray and radio observations, Marscher *et al.* (2002) found that dips in the X-ray emission are followed by ejections of bright superluminal knots in the radio jet, which clearly establishes an accretion-disk-jet connection.

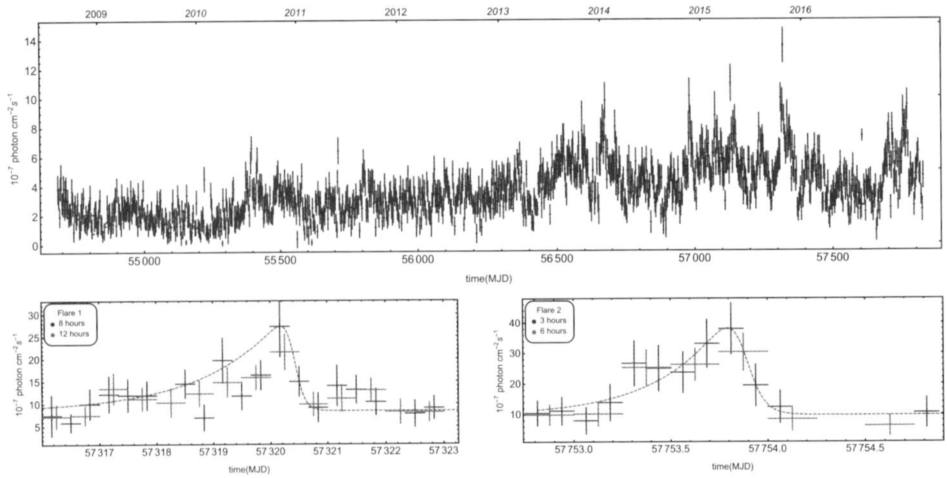

Figure 2. The γ-ray light curve of NGC 1275 from August 4, 2008 to March 5, 2017, with 3-day (blue) binning. *Lower panels:* Sub intervals covering F1 (left) and F2 (right).

The γ-ray light curve of 3C 120 from August 4, 2008, to August 4, 2016 obtained for 90-day binning is shown in the upper panel of Fig. 1 (from Zargaryan *et al.* (2017)). Before $\approx$ MJD 56000 (March 14, 2012), the source is mostly undetectable by Fermi LAT. Then the source flux was high enough to be detected by Fermi LAT, and up to $\approx$ MJD 56800 it remained constant with no significant changes. Starting from $\approx$ MJD 56800, the flux substantially increased up to a few times 10^{-7}photon cm^2 s^{-1} and remained so till $\approx$ MJD 57350. The same light curve 20% adaptively binned intervals is shown in lower panel in Fig. 1. As it is expected, initially it took a long time to reach the necessary 20 % uncertainty. Indeed, the first bin contains the data from the start of the mission to MJD 56919.31 (19 September 2014), amounting to more than 6 years. Afterwards, it took shorter time to reach the required uncertainty. The most dramatic increase in the γ-ray flux was observed on April 24, 2015. First, within 19.0 min the flux reached $(7.46 \pm 1.56) \times 10^{-6}$ photon cm^{-2} s^{-1} with $\Gamma = 2.29 \pm 0.21$ and 11.2σ detection significance. Then for another 3.15 hours it was as high as $(4.71 \pm 0.92) \times 10^{-6}$ photon cm^{-2} s^{-1} with $\Gamma = 1.97 \pm 0.14$ and 12.7σ. Then the flux slowly decreased down to a few times 10^{-7} photon cm^{-2} s^{-1} with the bin size varying within 10 to 35 days. The source was in an active state up to $\approx$ MJD 57300 and then turned again into its quiescent state, in which case the data should be accumulated for $\simeq$254 days.

3. NGC 1275

Due to its proximity (z = 0.0176, $\simeq$ 75.6 Mpc) and brightness, the radio galaxy NGC 1275 has been a target for observations in almost all energy bands. Core-dominated asymmetrical jets at both kpc (Pedlar *et al.* (1990)) and pc (Asada *et al.* (2006)) scales have been detected in the radio band with characteristics more similar to those of FR1 sources. The emission in the X-ray band is mostly dominated by the thermal emission from the cluster, although a nonthermal component in the energy range 0.5-10 keV with a photon index of $\Gamma_X = 1.65$ has been observed (Churazov *et al.* (2003)).

Fig. 2 (upper panel) shows the γ-ray light curve with three-day bin size. Despite the fact that the flux sometimes exceeded the averaged value presented in 3FGL ($\approx 2.26 \times 10^{-7}$ photons cm^{-2} s^{-1}), pronounced flaring activities were detected in October 2015 (hereafter Flare 1 [F1]) and in December 2016/January 2017 (hereafter Flare 2 [F2]). Starting from 22 October 2015 the daily averaged flux of NGC 1275 was above

10^{-6} photons cm^{-2} s^{-1} and remained high for 5 days with a daily averaged maximum of $(1.48 \pm 0.20) \times 10^{-6}$ photons cm^{-2} s^{-1} observed on 24 October 2015. Another substantial increase in the γ-ray flux was observed on December 31, 2016 when the flux increased from about $(4 \sim 5) \times 10^{-7}$ photons cm^{-2} s^{-1} to $(2.21 \pm 0.26) \times 10^{-6}$ photons cm^{-2} s^{-1} within a day with a detection significance of $\sim 21.5\sigma$.

During F1 and F2, the flare time profiles are investigated by fitting them with double exponential functions (Abdo *et al.* (2010)). The corresponding fit is shown in Fig. 2 lower panels (from Baghmanyan *et al.* (2017)). The time profiles show asymmetric structures in both flares, showing a slow rise and a fast decay trend. The time peak of the flares calculated by $t_{\mathrm{p}} = t_0 + t_{\mathrm{r}}\, t_{\mathrm{d}}/(t_{\mathrm{r}} + t_{\mathrm{d}}) ln(t_{\mathrm{d}}/t_{\mathrm{r}})$ is MJD 57320.18 for F1 with the maximum intensity of $(2.39 \pm 0.31) \times 10^{-6}$ photon cm$^{-2}s^{-1}$. The rise time is 32.49 ± 7.20 hours with a sudden drop within 2.22 ± 1.19 hours. The parameters of F2 are better estimated and are characterized with a shorter rise time, when within 8.03 ± 1.24 hours the flux reaches its maximum of $(4.20 \pm 0.48) \times 10^{-6}$ photon cm$^{-2}s^{-1}$ on MJD 57753.79 and drops nearly 4 times in ~ 6 hours. The minimal e-folding time is $t_{\mathrm{d}} = 1.21 \pm 0.22$ hours, using the decay time scale of F2, and it is the most rapid γ-ray variability observed for NGC 1275. We note that even if the rise time of F2 is used, the flux e-folding time of about 8.03 ± 1.24 hours will still be shorter than any previously reported value.

References

Abdo, A. A., Ackermann, M., Ajello, M., *et al.* 2010, *ApJ*, 722, 520
Acero, F., Ackermann, M., Ajello, M., *et al.* 2015, *ApJ*, 218, 23
Asada, K., Kameno, S., Shen, Z.-Q., *et al.* 2006, *PASJ*, 58, 261
Baghmanyan, V., Gasparyan, S., & Sahakyan, N. 2017, *ApJ*, 848, 111
Churazov, E., Forman, W., Jones, C., & Böhringer, H. 2003, *ApJ*, 590, 225
Fanaroff, B. L., & Riley, J. M. 1974, *MNRAS*, 167, 31P
Homan, D. C., Ojha, R., Wardle, J. F. C., *et al.* 2001, *ApJ*, 549, 840
Marscher, A. P., Jorstad, S. G., Gómez, J.-L., *et al.* 2002, *Nature*, 417, 625
Pedlar, A., Ghataure, H. S., Davies, R. D., *et al.* 1990, *MNRAS*, 246, 477
Peterson, B. M., Ferrarese, L., Gilbert, K. M., *et al.* 2004, *ApJ*, 613, 682
Sahakyan, N., Zargaryan, D., & Baghmanyan, V. 2015, *A&A*, 574, A88
Sahakyan, N., Baghmanyan, V., & Zargaryan, D. 2018, *A&A*, 614, A6
Urry, C. M., & Padovani, P. 1995, *PASP*, 107, 803
Walker, R. C., Benson, J. M., & Unwin, S. C. 1987, *ApJ*, 316, 546
Zargaryan, D., Gasparyan, S., Baghmanyan, V., & Sahakyan, N. 2017, *A&A*, 608, A37

Perseus in Sicily: From Black Hole to Cluster Outskirts
Proceedings IAU Symposium No. 342, 2018
K. Asada, E. de Gouveia Dal Pino, M. Giroletti,
H. Nagai & R. Nemmen, eds.

doi:10.1017/S1743921318003903

Young radio sources at high-energies and the γ-ray CSO PKS 1718−649

Giulia Migliori[1,2]

[1]DIFA, Dipartimento di Fisica e Astronomia, Alma Mater Studiorum, Università degli Studi di Bologna, Via Gobetti 93/2, 40129 Bologna, Italy

[2]IRA, INAF Istituto di Radioastronomia, via Gobetti 101, 40129 Bologna, Italy
email: g.migliori@ira.inaf.it

Abstract. Observations at high-energies are important to define the first stages of the evolution of extragalactic radio sources and to characterize the interstellar medium of their host galaxies. In some of the X-ray-observed Compact Symmetric Objects (CSOs, among the youngest and most compact radio galaxies), we measured values of the total hydrogen column densities large enough to slow or prevent the radio source growth. The γ-ray window has the potential to constrain the non-thermal contribution of jets and lobes to the total high-energy emission. However, so far, young radio sources remain elusive in γ-rays, with only a handful of detections (or candidates) reported by *Fermi*. I present our γ-ray study of the CSO PKS 1718−649, and draw comparison with the restarted, γ-ray detected, radio galaxy 3C 84.

Keywords. galaxies: active — galaxies: individual (PKS 1718−649) — radiation mechanisms: non-thermal — gamma-rays: galaxies

1. Young radio sources at high-energies

Compact radio sources with radio structures fully contained within the central regions of their host galaxies (< 1 kpc) are thought to be the progenitors of the large-scale radio galaxies (see a review by Orienti 2016). Among young radio sources (YRSs), Compact Symmetric Objects (CSOs) have sub-kpc-scale structures, symmetric radio morphologies, total radio emission dominated by the mini-lobes, and kinematical ages smaller than a few thousand years.

Radiation at the high-energies (X-ray to γ-rays) traces the most energetic processes taking place in the YRSs: this energy band is important to define the intrinsic properties of the expanding source, such as the accretion process and the power channeled into the radio structures, but also to characterize the environment where the source is embedded.

1.1. *X-rays*

Chandra and *XMM-Newton* observations of samples of CSOs have demonstrated that these sources are X-ray emitters and short observations are enough to obtain a detection (Tengstrand *et al.* 2009, Siemiginowska *et al.* 2016, hereafter S16, and references therein). The CSOs in the X-ray sample of S16 have X-ray luminosities in the range of 10^{41} to 10^{45} erg s^{-1}. X-ray observations allow us to probe the host galaxy environment via measurements of the total hydrogen column densities, N_H. The S16 sample displays moderate or low N_H for the majority of the targets ($N_H \approx 10^{22}$ cm^{-2}), except for a few sources whose X-ray spectra are compatible with being Compton-thick candidates ($N_H > 10^{24}$ cm^{-2}). Investigating the relation between the linear sizes of the radio structure and the density of ambient gas can help us to understand how the source will evolve: a predominance of obscured sources among the most compact and young ones could be indicative of confinement by a dense environment (Sobolewska *et al.* 2019), which could ultimately frustrate the radio source growth. Indeed, X-ray observations of large and complete samples of CSOs are necessary to test this scenario. Interestingly, the positive correlation between X-ray/N_H and radio/N_{HI} absorbing columns reported in Ostorero *et al.* (2017) suggests that the X-ray and radio absorbers are physically connected and could be part of the same, possibly perturbed, hundred-parsec scale structure. A multi-phase gas with complex kinematics, perturbed by the interaction with the YRSs is also supported by the detection of outflowing HI and molecular (CO) gas (see e.g. Morganti *et al.* in these proceedings).

The observed X-ray emission can be produced by thermal Comptonization in the disc-corona of the AGN powering the radio source, by shock-heated ISM by the expanding radio structure or via non-thermal processes within the radio jets and lobes (Tengstrand *et al.* 2009, Heinz *et al.* 1998, Migliori *et al.* 2014, Stawarz *et al.* 2008, Kino *et al.* 2009). In most of the cases, in lack of distinctive spectral features, the spectral and morphological X-ray studies cannot typically discriminate among the different phenomena. However, radiative models predict copious non-thermal, high-energy emission being produced in the jets and lobes of YRSs via inverse Compton (IC) scattering of the synchrotron photons and of the thermal photons radiated by the accretion disc, the torus, and the stellar population of the host galaxy. Depending on the jet and photon fields' parameters, this process could yield luminosities up to 10^{45} erg s^{-1} in the 0.1 – 100 GeV band (Stawarz *et al.* 2008). Given that the accretion disc and corona emission drastically drops in the γ-ray band, a detection of a YRS at these energies is a robust proof of a non-thermal, high-energy component.

1.2. *Gamma-rays*

The launch of the *Fermi* telescope in 2008 has opened the quest for YRSs in the γ-ray window. However, so far searches performed on selected samples have mainly reported upper limits (Migliori *et al.* 2014; D'Ammando *et al.* 2016; Migliori 2016a). A handful of γ-ray sources with uncertain classification show CSO-like features and are currently under investigation (see for example PMN J1603−4904, Müller *et al.* 2015). The notable exceptions are the bona-fide CSO, PKS 1718−649 (here after PKS 1718, Migliori *et al.* 2016b) and the restarted radio source 3C 84 at the center of the Perseus cluster (Abdo *et al.* 2009). In the following, I will focus on PKS 1718, highlighting similarities and differences with 3C 84.

2. PKS 1718−649

PKS 1718 is the closest (z=0.014) confirmed CSO (Tingay *et al.* 1997). It displays a compact ($\sim$2 pc), double-lobed, radio structure and has a kinematically estimated age of about 100 yrs (Giroletti & Polatidis 2009). The third *Fermi* Gamma-ray Large Area Telescope (LAT) catalog (3FGL, Acero *et al.* 2015) reported an unassociated source, 3FGL J1728.0−6446, close to PKS 1718. The analysis of $\sim$7 yrs of LAT data, using the Pass 8 calibration, allowed us to confirm that PKS 1718 lies 0.13° from the best fit position of the γ-ray source, within the 68% confidence radius, and is the most likely counterpart to the unassociated LAT source (Fig. 1, left). In γ-rays, the source yields a $>5\sigma$ detection with a best fit photon spectral index $\Gamma = 2.9 \pm 0.3$ and a 0.1-100 GeV photon flux density $F_{0.1-100\mathrm{GeV}} = (11.5 \pm 0.3) \times 10^{-9}$ ph cm^{-2} s^{-1} (Migliori *et al.* 2016b). Clues on the nature of the γ-ray emission of PKS 1718 come from the comparison of its γ-ray properties with those of blazar sources and radio galaxies reported in the 3FGL. The γ-ray emission of blazars, i.e. AGN with jets oriented close to the line of sight of the observer, is produced

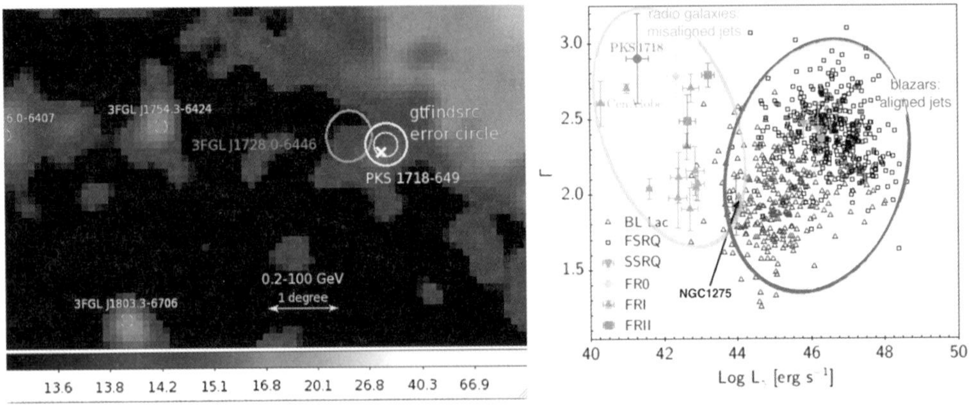

Figure 1. Left: *Fermi*-LAT 0.2-100 GeV count map of the sky. The yellow cross is the radio position of PKS 1718, the magenta ellipse gives the 3FGL position (95% confidence level) of 3FGL J1728.0−6446. The cyan circles are the best fit position (68% and 95% confidence levels) defined with *gtfindsrc*. Right: γ-ray spectral index versus 1-100 GeV luminosity. PKS 1718 is located in the MAGN region of the diagram, while 3C 84 (NGC 1275) is at the border with BL Lac sources (figures adapted from Migliori *et al.* 2016b).

in a compact, relativistically moving jet region close to the AGN. The emission is boosted by relativistic effects and in fact, they represent the vast majority of AGNs in the 3FGL. Conversely, there is a limited number of LAT detected misaligned AGNs (MAGNs), whose jets point away from the observer, such as radio galaxies (see Abdo *et al.* 2010a and the review by Torresi in these proceedings). In MAGNs, the site of production of the γ-ray emission, or the radiative mechanism responsible for it, could be possibly different from blazars. The two classes populate different regions of the radio (1.4 GHz) vs. γ-ray (>1 GeV) luminosity plot and of the Γ vs. L_γ plot. PKS 1718 is located in the region of low γ-ray luminosity and soft photon index which is predominantly occupied by MAGNs (Fig. 1, right). For a comparison, 3C 84 is the brightest radio galaxy detected in the GeV band and displays a harder γ-ray spectrum, which place it in the border region between low-power aligned and misaligned sources.

The γ-ray emission of PKS 1718 appears faint and steady in time, whereas the source is not significantly ($\geq 3\sigma$) detected on yearly timescales, suggesting a low level of variability. The broad-band radio spectrum is shaped by the combination of two effects, synchrotron self-absorption and free-free absorption (Tingay *et al.* 2015). Engulfment of gas clouds of difference size and density by the adiabatically expanding radio lobes (see Bicknell *et al.* 2018, and references therein) can account for the observed moderate radio flux and spectral variability. This scenario is also in agreement with the dynamical model that predicts isotropic high-energy emission from the compact lobes (Stawarz *et al.* 2008). Modeling of PKS 1718's non-thermal broadband spectral energy distribution (SED) can provide us with constraints on the jet's physical parameters (Sobolewska *et al.*, in prep). Estimates of the jet power and characterization of the source environment via multi-band, high-resolution observations (see e.g. Maccagni *et al.* 2014) are crucial to understand the radio source's fate.

3C 84 exhibits a more complex temporal behavior than PKS 1718. A progressive brightening of the γ-ray flux on long timescales is accompanied by short-term (days to week) flaring activity. The source has been also detected at very high energies (>100 GeV, Aleksić *et al.* 2014). Very long baseline interferometry (VLBI) has revealed that the radio outburst, which is ongoing since 2005, coincided with the ejection of a new, possibly hot spot-like component (Nagai *et al.* 2010). The small-scale jet displays an evolving, limb-brightened radio morphology down to few hundreds of gravitational radii from the central engine (Giovannini *et al.* 2018, and references therein), pointing to a stratified jet structure. These features suggest the intriguing possibility of multiple sites of production of the γ-rays (Hodgson *et al.* 2018, Tanada *et al.* 2018).

3. Conclusions

While YRSs remain elusive in γ-rays, the CSO PKS 1718, the restarted radio galaxy 3C 84 and a few other candidates display a range of different γ-ray properties. This fact makes it difficult to identify the most favorable CSOs for a detection in γ-rays. Interestingly, such diversity could reflect the different stages in the evolution of the sources, with 3C 84 being in a more unstable and rapidly evolving phase than PKS 1718. The faint and steady γ-ray emission of PKS 1718 is compatible with an origin in the compact radio lobes. If confirmed, this would be an interesting case of isotropic γ-ray emission such as that associated with the radio lobes of Cen A (Abdo *et al.* 2010b).

Acknowledgments

The author thanks the members of the SOC and LOC for the organization of a great meeting. It is a pleasure to acknowledge the contribution of my collaborators, in particular A. Siemiginowska and M. Sobolewska.

References

Abdo, A. A., *et al.* 2009, *ApJ*, 699, 31
Abdo, A. A., *et al.* 2010a, *ApJ*, 720, 912
Abdo, A. A., *et al.* 2010b, *Science*, 328, 725
Acero, F., *et al.* 2015, *ApJS*, 218, 23
Aleksić, J., *et al.* 2014, *A&A*, 564, A5
Bicknell, G. V., *et al.* 2018, *MNRAS*, 475, 3493
D'Ammando, F., *et al.* 2016, *AN*, 337, 59
Giovannini, G., *et al.* 2018, *Nature Astronomy*, 2, 472
Giroletti, M., & Polatidis, A. 2009, *AN*, 330, 193
Heinz, S., *et al.* 1998, *ApJ*, 501, 126
Hodgson, J. A., *et al.* 2018, *MNRAS*, 475, 368
Kino, M., *et al.* 2009, *MNRAS*, 395, L43
Maccagni, F. M., *et al.* 2014, *A&A*, 571, A67
Migliori, G., *et al.* 2014, *ApJ*, 780, 165
Migliori, G., 2016, *AN*, 337, 52
Migliori, G., *et al.* 2016, *ApJL*, 821, L31
Müller, C., *et al.* 2015, *A&A*, 574, A117
Nagai, H., *et al.* 2010, *PASJ*, 62, L11
Orienti, M. 2016, *AN*, 337, 9
Ostorero, L., *et al.* 2017, *ApJ*, 849, 34
Siemiginowska, A., *et al.* 2016, *ApJ*, 823, 57
Sobolewska, M., *et al.* 2019 *ApJ*, 871, 71
Stawarz, Ł., *et al.* 2008, *ApJ*, 680, 911
Tanada, K., *et al.* 2018, *ApJ*, 860, 74
Tengstrand, O., *et al.* 2009, *A&A*, 501, 89
Tingay, S. J., *et al.* 1997, *AJ*, 113, 2025
Tingay, S. J., *et al.* 2015, *AJ*, 149, 74

Perseus in Sicily: From Black Hole to Cluster Outskirts
Proceedings IAU Symposium No. 342, 2018
K. Asada, E. de Gouveia Dal Pino, M. Giroletti,
H. Nagai & R. Nemmen, eds.

doi:10.1017/S1743921318007974

Exploring the radio and GeV-TeV γ-ray connection in the different blazar sub-classes

R. Lico[1,2], M. Giroletti[2], M. Orienti[2,3], L. Costamante[4], V. Pavlidou[5], F. D'Ammando[2] and F. Tavecchio[6]

[1]Max-Planck-Institut für Radioastronomie, Auf dem Hügel, 69, D-53121 Bonn, Germany.
email: rlico@mpifr-bonn.mpg.de

[2]INAF - Istituto di Radioastronomia, via Gobetti 101, 40129 Bologna, Italy.

[3]DIFA, Università di Bologna, via Gobetti 93/3, 40129 Bologna, Italy.

[4]ASI Unità Ricerca Scientifica, Via del Politecnico snc, I-00133, Roma, Italy.

[5]Department of Physics and Institute for Plasma Physics, University of Crete, 71003 Heraklion, Greece.

[6]INAF Osservatorio Astronomico di Brera, via E. Bianchi 46, I-23807 Merate, Italy.

Abstract. As revealed by the *Fermi*-LAT, blazars represent the dominant population of γ-ray emitters. An essential step for understanding blazar physics and the emission mechanisms is the investigation of a possible connection between the observed low- and high-energy emission. A number of works report on the existence of a significant correlation between radio emission and 0.1-100 GeV γ rays. How does this correlation evolve when very high energy (VHE, $E > 0.1$ TeV) γ rays are considered? The possible radio-VHE emission connection is still elusive mainly because of the lack of a homogeneous VHE sky coverage. In this work we explore the connection between the parsec-scale radio emission and GeV-TeV γ rays by using two unbiased blazar samples extracted from the 1FHL ($E > 10\,\mathrm{GeV}$) and 2FHL ($E > 50\,\mathrm{GeV}$) *Fermi* catalogs. For comparison, we perform the same analysis by using the 3FGL 0.1-300 GeV γ-ray energy flux. Overall, we find out that there is no significant connection between radio and γ-ray emission above 10 GeV for all the blazar sub-classes with the exception of high synchrotron peaked objects. Conversely, when 0.1-300 GeV γ-ray energies are considered, a strong and significant correlation is found for all of the blazar sub-classes. We interpret these results within the context of the blazar spectral energy distribution properties.

Keywords. Galaxies: active - Gamma rays: galaxies - Radio continuum: galaxies – BL Lacertae objects: general – (Galaxies:) quasars: general – Galaxies: statistics

---

1. Introduction

Among active galactic nuclei (AGNs) blazars represent the most extreme objects. The blazar spectral energy distribution (SED) is dominated by two distinctive non-thermal components: a low-frequency component produced by synchrotron emission from relativistic electrons within the jet, peaking from radio to soft X-ray energies, and a high-frequency component, peaking from X-ray to γ-ray bands, commonly though to be produced by inverse Compton (IC) scattering of low energy photons by relativistic electrons in the jet. For the high-energy emission we invoke the synchrotron-self-Compton (e.g. Maraschi *et al.* 1992) or the external Compton (e.g. Ghisellini & Madau 1996) emission mechanisms when the scattered low-energy photons are the same photons produced by the synchrotron or external photons, respectively. According to the position of the synchrotron component peak $\nu_{\mathrm{peak}}^{\mathrm{Syn}}$ in the SED, blazars are classified as low-synchrotron peaked (LSP) or high-synchrotron peaked (HSP) whether $\nu_{\mathrm{peak}}^{\mathrm{Syn}}$ is $< 10^{14}$ Hz or $> 10^{15}$ Hz, respectively. Blazars with 10^{14} Hz $< \nu_{\rm peak}^{\rm Syn} < 10^{15}$ Hz are classified as intermediate synchrotron peaked (ISP). Blazars are also classified into flat spectrum radio quasars (FSRQs) and BL Lac objects (BL Lacs) according to the presence or not of broad emission lines in their optical spectra (e.g. Stickel *et al.* 1991). In general FSRQs are of LSP type, while BL Lacs can be part of all three classes.

After the large area telescope (LAT), launched in 2008 onboard the *Fermi* satellite, clearly showed that blazars dominate the census of the γ-ray sky (e.g. Acero *et al.* 2015), many efforts were made for the investigation of the possible radio and γ-ray emission up to 100 GeV connection (e.g. Kovalev *et al.* 2009; Giroletti *et al.* 2010; Ackermann *et al.* 2011). In particular, it was found that the correlation strength becomes weaker when higher γ-ray energies are considered (e.g. Ackermann *et al.* 2011; Piner & Edwards 2014; Mufakharov *et al.* 2015). However, owing to the lack of a homogeneous coverage of the very high energy (VHE, $E > 0.1$ TeV) sky, due to the limited operational mode of the imaging atmospheric Cherenkov telescopes, the possible connection between radio and VHE emission connection is still elusive.

In Lico *et al.* (2017), we investigate the connection between very long baseline interferometry (VLBI) radio emission and GeV/TeV γ rays by using two large and unbiased samples blazars from the 1FHL ($E > 10$ Gev, Ackermann *et al.* 2013) and 2FHL ($E > 50$ Gev, Ackermann *et al.* 2016) *Fermi* catalogs. In this proceeding we briefly report and summarize some results obtained for the 1FHL blazar sample.

2. Results and discussion

To understand blazar physics and to discern among the various emission models, the investigation of a possible connection between low- and high-energy connection represents a fundamental step. A significant correlation between radio and 0.1-100 GeV γ-ray emission was found in previous studies (e.g. Ghirlanda *et al.* 2010; Ackermann *et al.* 2011). In this work we investigate how this correlation evolves at higher γ-ray energies. For our analysis we use a sample of 237 blazars for which we both consider the 1FHL (10-500 GeV) and 3FGL (0.1-300 GeV, Acero *et al.* 2015) γ-ray energy ranges. At radio frequencies we use 8 GHz VLBI flux densities from the radio fundamental catalog (http://astrogeo.org/rfc/). For more details on the sample selection see Lico *et al.* (2017).

The scatter plots of the VLBI flux density versus 1FHL and 3FGL γ-ray energy flux for the various blazar sub-classes (BL Lac, FSRQ, HSP, ISP, and LSP) are shown in Fig. 1 (top and bottom panels, respectively). The scatter plots for the full sample are reported in Fig. 2 for both the $E > 10$ GeV 1FHL (left frame) and $E > 0.1$ GeV 3FGL (right frame) energy fluxes. Red, green and blue colors represent LSPs, ISPs and HSPs, respectively. Filled and empty symbols represent those sources with known and unknown redshift, and in black color are indicated those blazars without any spectral classification. The results of our analysis, for which we use the statistical method proposed by Pavlidou *et al.* 2012 that takes into account both the various observational biases and the distance effects, are shown in Table 1. A statistically significant correlation ($r > 0.7$ with $p < 10^{-6}$) is found for the full sample and for each single blazar sub-class when 0.1-300 GeV γ rays from the 3FGL catalog are considered. On the other hand, when 10-500 GeV 1FHL γ-ray energy fluxes are considered the correlation breaks for all of the blazar sub-classes except for HSP objects ($r > 0.6$ with $p < 10^{-6}$). Similar results are obtained for the 2FHL ($E > 50$ GeV) blazar sample (see Lico *et al.* 2017).

To explain this correlation behavior we investigate the SED properties for the different blazar sub-classes. A simplified sketch of LSP, ISP and HSP (top, middle and bottom curve, respectively) SEDs is reported in Fig. 3, where the 0.1-300 GeV 3FGL energy range is represented by the green area, and the 1FHL ($E > 10$ GeV) and 2FHL ($E > 50$ GeV) energy thresholds are represented by the vertical dotted and dash-dotted vertical lines,

Table 1. Correlation analysis results. col(1): sub-classes; col(2): source catalog; col(3): number of sources in each subset; col(4): number of redshift bins (each bin contains > 10 objects); col (5): Pearson correlation coefficient (r); col (6): statistical significance (p).

Source type	Catalog	Number of sources	Number of z-bins	r-Pearson	Significance
All sources	1FHL	147	14	−0.05	0.59
	3FGL	147	14	0.71	$< 10^{-6}$
BL Lac	1FHL	100	9	0.12	0.55
	3FGL	100	9	0.70	$< 10^{-6}$
FSRQ	1FHL	44	4	−0.01	0.99
	3FGL	44	4	0.49	$< 10^{-6}$
HSP	1FHL	60	5	0.57	1.0×10^{-6}
	3FGL	60	5	0.77	$< 10^{-6}$
ISP	1FHL	23	2	0.19	0.40
	3FGL	23	2	0.46	2.5×10^{-2}
LSP	1FHL	52	5	0.21	0.12
	3FGL	52	5	0.43	3.0×10^{-6}

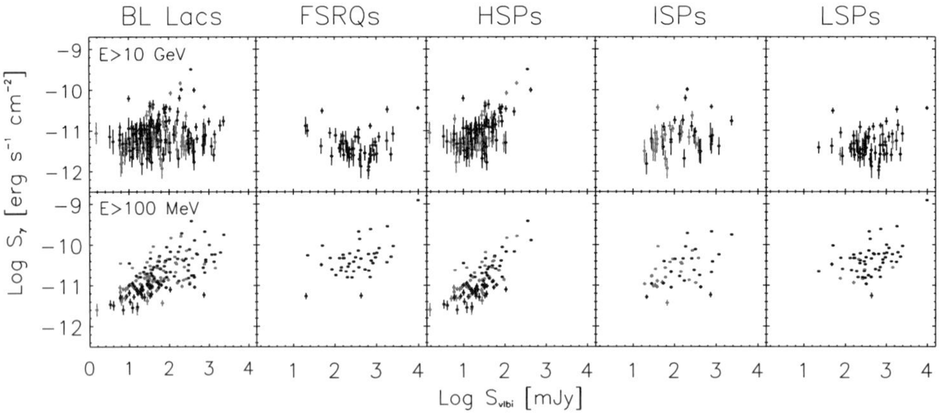

Figure 1. Scatter plot of the VLBI radio emission and 1FHL (upper frames) and 3FGL (lower frames) energy flux the different sub-classes (BL Lacs, FSRQs, HSPs, ISPs, and LSPs). The black and red colors represent sources with or without redshift, respectively.

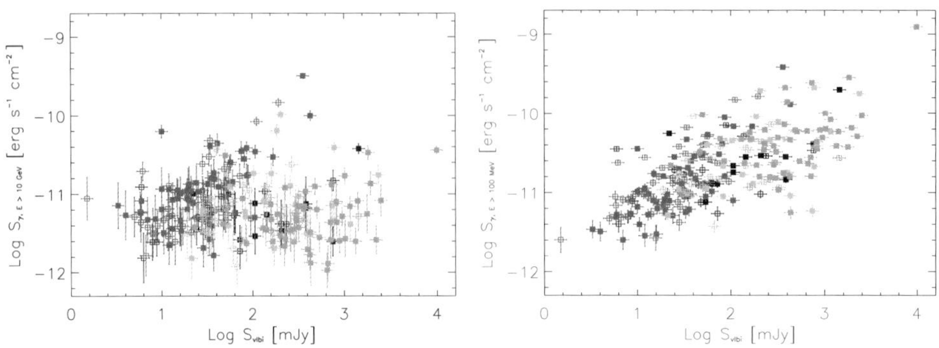

Figure 2. Full sample VLBI radio versus 1FHL (left frame) and 3FGL (right frame) energy flux scatter plots. Red, green and blue colors represent LSPs, ISPs and HSPs, respectively. Filled and empty symbols represent those sources with known and unknown redshift, and in black color are indicated those blazars without any spectral classification.

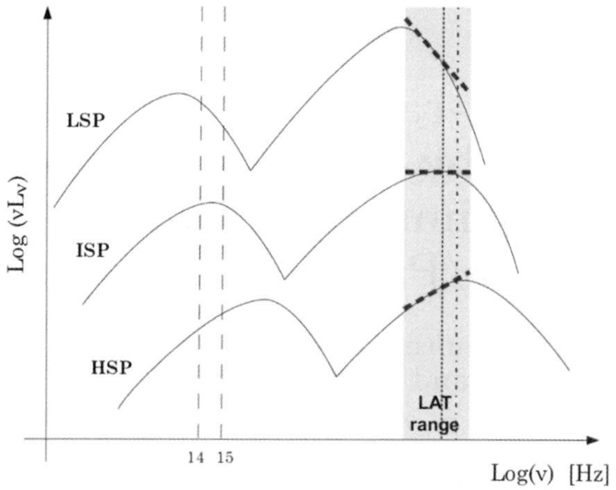

Figure 3. LSP (upper curve), ISP (middle curve) and HSP (lower curve) scchematic SED representation. The 3FGL energy range is represented by the green area, and the black dotted/dash-dotted vertical lines represent the 1FHL/2FHL energy thresholds.

respectively. In the case of LSP blazars both the low- and high-energy SED components peak at lower energies with respect to HSPs, and in general they show softer spectra. By focusing on the high-energy emission SED component, in the case of LSPs within the *Fermi*-LAT energy range (green area in Fig. 3) we are sampling the part of the spectrum where the γ-ray flux is quickly decreasing due to the cooling losses. At $E > 10\,\mathrm{GeV}$, where the radio and γ-ray connection breaks, the steepening is even more pronounced (e.g. Tavecchio & Mazin 2009). On the other hand, in the case of HSP blazars the cooling losses are less severe and their high-energy spectrum peaks at higher energies than LSPs. As a consequence, within the *Fermi*-LAT energy range we are sampling the rising part of the high-energy spectrum, both in the 3FGL and 1-2FHL energy ranges.

This sampling effect, produced by the different SED properties of the different blazar sub-classes, can be related to our correlation results, and can explain why a significant correlation between radio and γ rays at $E > 10\,\mathrm{GeV}$ emission is found only for HSP blazars. Within this simple picture some caveats need to be taken into account (e.g. the variability argument). For the complete correlation analysis, including the 2FHL blazar sample, and the full discussion see Lico *et al.* (2017).

References

Acero, F., Ackermann, M., Ajello, M., *et al.* 2015, *ApJ*, 218, 23
Ackermann, M., Ajello, M., Allafort, A., *et al.* 2011, *ApJ*, 741, 30
Ackermann, M., Ajello, M., Allafort, A., *et al.* 2013, *ApJS*, 209, 34
Ackermann, M., Ajello, M., Atwood, W. B., *et al.* 2016, *ApJS*, 222, 5
Ghirlanda, G., Ghisellini, G., Tavecchio, F., & Foschini, L. 2010, *MNRAS*, 407, 791
Ghisellini, G., & Madau, P. 1996, *MNRAS*, 280, 67
Giroletti, M., Reimer, A., Fuhrmann, L., Pavlidou, V., & Richards, J. L. 2010, arXiv:1001.5123
Kovalev, Y. Y., Aller, H. D., Aller, M. F., *et al.* 2009, *ApJ*, 696, L17
Lico, R., Giroletti, M., Orienti, M., *et al.* 2017, *A&A*, 606, A138
Maraschi, L., Ghisellini, G., & Celotti, A. 1992, *ApJL*, 397, L5
Mufakharov, T., Mingaliev, M., Sotnikova, Y., Naiden, Y., & Erkenov, A. 2015, *MNRAS*, 450, 2658
Pavlidou, V., Richards, J. L., Max-Moerbeck, W., *et al.* 2012, *ApJ*, 751, 149
Piner, B. G., & Edwards, P. G. 2014, *ApJ*, 797, 25
Stickel, M., Padovani, P., Urry, C. M., Fried, J. W., & Kuehr, H. 1991, *ApJ*, 374, 431
Tavecchio, F., & Mazin, D. 2009, *MNRAS*, 392, L40

Perseus in Sicily: From Black Hole to Cluster Outskirts
Proceedings IAU Symposium No. 342, 2018
K. Asada, E. de Gouveia Dal Pino, M. Giroletti,
H. Nagai & R. Nemmen, eds.

doi:10.1017/S1743921318007950

Neutrino and γ-ray Emission from the Core of NGC1275 by Magnetic Reconnection: GRMHD Simulations and Radiative Transfer/Particle Calculations

J. C. Rodríguez-Ramírez, Elisabete M. de Gouveia Dal Pino and R. Alves Batista

Instituto de Astronomia, Geofísica e Ciências Atmosféricas (IAG-USP), Universidade de São Paulo. Cidade Universitaria R. do Matão, 1226 05508-090 São Paulo, SP Brasil
email: juan.rodriguez@iag.usp.br

Abstract. Very high energy (VHE) emission has been detected from the radio galaxy NGC1275, establishing it as a potential cosmic-ray (CR) accelerator and a high energy neutrino source. We here study neutrino and γ-ray emission from the core of NGC1275 simulating the interactions of CRs assumed to be accelerated by magnetic reconnection, with the accreting plasma environment. To do this, we combine (i) numerical general relativistic (GR) magneto-hydrodynamics (MHD), (ii) Monte Carlo GR leptonic radiative transfer and, (iii) Monte Carlo interaction of CRs. A leptonic emission model that reproduces the SED in the [10^3-$10^{10.5}$] eV energy range is used as the background target for photo-pion interactions+electromagnetic cascading. CRs injected with the power-law index =1.3 produce an emission profile that matches the VHE tail of NGC1275. The associated neutrino flux, below the IceCube limits, peaks at ~PeV energies. However, coming from a single source, this neutrino flux may be an over-estimation.

Keywords. galaxies: nuclei, radiation mechanisms: non-thermal, accretion, radiative transfer.

1. Introduction

Very high energy (VHE) emission has been detected from NGC1275 (Aleksić *et al.* 2012), establishing it as potential cosmic-ray (CR) accelerator as well as a source of high energy neutrinos. The non-blazar nature of this radio galaxy suggests that the high energy emission could be produced in the outflows as well as in the accreting core region. The characteristics of potential neutrino emission from this source could be crucial for determining possible regions of CR acceleration.

In this work, we investigate the neutrino and γ-ray emission signatures of CRs accelerated by magnetic reconnection in the core region of NGC1275. This energy release mechanism has been proven to occur in accreting magnetised plasmas (de Gouveia Dal Pino & Lazarian, 2005, Singh, de Gouveia Dal Pino & Kadowaki, 2015, Kadowaki, de Gouveia Dal Pino & Singh 2015, Kadowaki, de Gouveia Dal Pino & Stone 2018a, Kadowaki, de Gouveia Dal Pino & Stone 2018b) and to accelerate particles via first order Fermi reconnection acceleration (Kowal, de Gouveia Dal Pino, & Lazarian, 2011, 2012, del Valle, de Gouveia Dal Pino & Kowal, 2016; see also de Gouveia Dal Pino *et al.* these procs).

We here probe neutrino and γ-ray emission by means of Monte Carlo simulations of the interactions of CRs with the magnetic and photon fields produced by the accreting flaw in the core region of NGC1275. For this aim, we combine three numerical techniques:

(i) Numerical general relativistic (GR) magnetohydrodynamics (MHD) to model the magnetised accreting flow, (ii) Monte Carlo GR leptonic radiation transfer to obtain the environment radiation field, and (iii) Monte Carlo propagation and interaction of CRs assumed to be accelerated by magnetic reconnection, to probe neutrino and hadronic γ-ray emission due to photo-pion processes and electromagnetic cascades.

In the next section we describe the numerical model for the environment in the core of NGC1275 and in Section 3, the CRs propagation stimulations from which we obtain the signatures of neutrino and hadronic γ-ray emission. Finally, in Section 4 we discuss our results and state our conclusions.

2. Background ADAF Model: Numerical GRMHD + Leptonic Radiative Transfer

To study hadronic emission from the core region of NGC1275, we consider a numerical GRMHD, advection dominated accretion flaw (ADAF) model, as the environment where CRs (assumed to be accelerated by magnetic reconnection) and their secondary particles propagate and interact.

The background magnetic field is obtained employing the axi-symmetric version of the publicly available `harm` code for GRMHD around rotating BHs and described by Gammie *et al.* (2003). We use the values of M=3.4 $\times 10^8 M_\odot$ (estimated by Wilman *et al.* 2005) and $a = 0.93$ (arbitrarily chosen) for the mass and dimensionless spin parameter of the central supermassive BH. The accretion of the simulation is triggered by magneto-rotational instability in a hydrostatic equilibrium state, with the initial beta parameter $\beta = 100$ and the specific heat ratio $\gamma = 5/3$. The simulation is performed up to a boundary radius of $R_{max} = 40R_g$ with a computational resolution of 256×256, being $R_g = GM/c^2$ the gravitational radius. We choose the snapshot at the time $t = 2010R_g/c$ as the background environment for the CRs emission simulations.

The background photon field of the chosen snapshot flaw is calculated performing post-processing synchrotron+inverse Compton (IC) radiation transfer assuming the gas in a stationary state. For this calculation we employ the publicly available Monte Carlo `grmonty` code (Dolence *et al.* 2009), using the proton-to-electron temperature ratio $T_p/T_e = 3$. For the mass unit we use the value of $M_u = 3 \times 10^{25}$g that we obtain by fitting the calculated SED, corresponding to a line of sight with $\theta = 20°$ with respect to the direction of the BH spin, to the X-ray component of the observed SED.

The gas density and magnetic field profiles of the chosen snapshot are shown in the left panel of Fig. 1, and its photon field map due to synchrotron + IC radiation are shown in the right panel. The calculated SED of this leptonic emission model is displayed in the left panel of Fig. 2 and reproduces the observed data from X-rays to ~ 10 GeV γ-rays.

3. Emission due to Interactions of CRs with the GRMHD Plasma Environment

To calculate the neutrino and γ-ray emission due to CRs in the accreting plasma profile obtained in the last section, we employ the publicly available `CRPropa3` code (Alves Batista *et al.* 2016). We consider the following interactions: (i) photo-pion process ($p\gamma \rightarrow 2\gamma$ and $p\gamma \rightarrow ... + \nu_s$) that produces γ-rays and neutrinos, (ii) γ-ray absorption due to pair creation ($\gamma\gamma \rightarrow$ e±) that generates secondary electron/positron pairs, and (iii) IC scattering of background photons by the secondary electron/positron pairs. Neutrinos produced due to photo-pion processes will escape the accreting torus travelling virtually unimpeded .

To emulate the release of CRs due to a magnetic reconnection event, we run simulations injecting power-law distributions of CRs with the indices κ =1.3, 1.5 and 1.8, within the energy range [100 TeV − 50 PeV]. We consider the injection of protons within a sphere

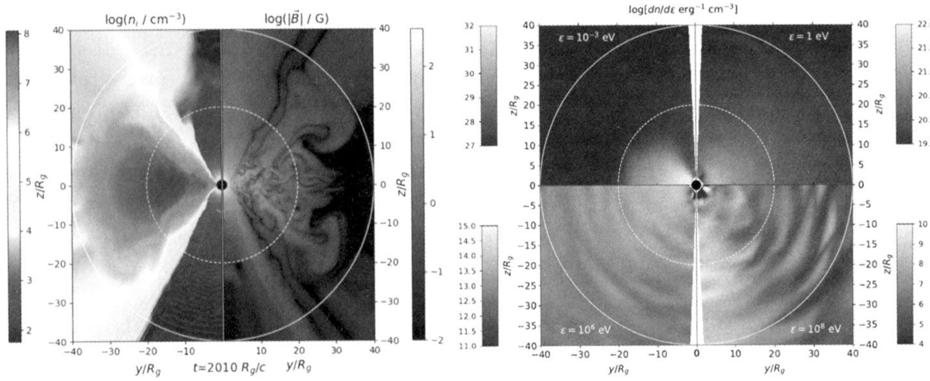

Figure 1. Left: Numerical GRMHD ADAF model for the core region of NGC1275. The figure shows the gas number density and magnetic field profiles for the snapshot, obtained with the axi-symmetric `harm` code. Right: Photon field map of the snapshot flaw in the left panel, calculated with synchrotron+IC radiative transfer, employing the `grmonty` code. The inner white circle represents the sphere inside which we inject the initial CRs accelerated by magnetic reconnection. The outer circle represents the spherical boundary where we detect neutrinos and γ-rays that result from the CR interactions plus electromagnetic cascades (see Section 3).

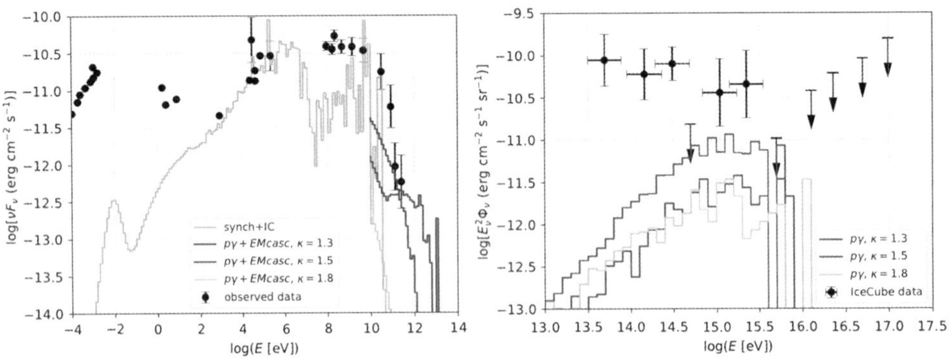

Figure 2. Left: SED for NGC1275, with leptonic synchrotron+inverse IC (cyan histogram, employing the `grmonty` code) and with photo-pion processes plus electromagnetic cascading (blue, purple and orange histograms, employing the `CRPropa3` code). The different histograms of the hadronic models correspond to the power-law index of the injected CRs. The black points are observed data, from *MOJAVE* (Lister, Aller, Aller *et al.* 2009), *HST FOS* (Johnstone & Fabian 1995), *Swift*-BAT (Ajello *et al.* 2009), *Fermi*-LAT (Abdo *et al.* 2009, Ackermann *et al.* 2012) and *MAGIC* (Aleksić et al. 2012a,b). Right: Neutrino fluxes associated to the hadronic γ-rays emission models. The IceCube data (Aartsen et al. 2014) is shown for comparison.

of $20R_g$ (see Fig. 1). Particles stop being tracked when they complete a maximum trajectory length of $150R_g$, or attain a minimum energy of 10 GeV, or cross the detection sphere at $40R_g$. We calculate the neutrino and γ-rays fluxes that would be detected at the Earth as $E_{\nu_s}^2 \Phi_{\nu_s} = \frac{w}{4\pi} \frac{\epsilon N_{\epsilon,\nu_s}}{\Delta \ln \epsilon} (4\pi R_0^2 \Delta t)^{-1}$ and $\nu F_\nu = w \frac{\epsilon N_{\epsilon,\gamma}}{\Delta \ln \epsilon} (4\pi R_0^2 \Delta t)^{-1}$, being $w = N_p / N_p^{sim}$ the ratio of physical to simulated CRs, $N_{\epsilon,\nu}$ and $N_{\epsilon,\gamma}$ the number of neutrinos and γ-rays arriving at the detection sphere within the ϵ energy bin, Δt the detection time interval, and $R_0 = 75$ Mpc the distance from NGC1275 to us.

The number of physical CRs, N_p, is expected to be a very small fraction of the number of thermal ions N_{th} in the background CRs injection region. Thus, for the normalisation factor w we use the upper limit condition $w \leq f_{CR} N_{th}/N_p^{sim}$ with $f_{CR} = 10^{-5}$†. The number of thermal ions inside the injection sphere is estimated as $N_{th} \sim (1/2)(4\pi/3)R_{inj}^3 \langle n_i \rangle$ (the 1/2 factor considers that most of the background ions lie in half of volume of the injection sphere, see the gas density map in Fig. 1). Using $\langle n_i \rangle = 5 \times 10^7$, and $N_p^{sim} = 10^3$, we obtain the condition $w \leq 5 \times 10^{44}$.

In Fig. 2 we show the γ-rays and the associated neutrino fluxes that result from the `CRPropa3` simulations and using $w = 5 \times 10^{44}$. The hadronic γ-ray fluxes models are the curves in the left panel that begin at 10^{10} eV and correspond to different power-law indices of CR injection. These hadronic emission fluxes tend to reproduce better the high energy tail of the SED for smaller power-law index. The models of neutrino fluxes shown in the right panel correspond to the γ-ray curves of the same colour. The three models produce neutrinos peaking at $\sim$ PeV energies, producing fluxes that are compatible with IceCube data (Aartsen et al. 2014).

4. Summary and Conclusions

We have investigated neutrino and γ-ray emission signatures produced in the core region of NGC1275, adopting a numerical GRMHD ADAF model for the BH accreting plasma, together with Monte Carlo simulations of leptonic and hadronic emissions. The essential assumption for the hadronic component is the acceleration of CRs in a Fermi process by turbulent magnetic reconnection events (refs. in Section 1 and de Gouveia Dal Pino et al., these procs.), that we have emulated with the injection of power-law energy distributions of CRs.

We have presented a model of pure leptonic synchrotron+IC emission that reproduces the observed SED in the 10^3-$10^{10.5}$ eV energy range (see Fig. 2, left panel) which we have used as the target photon field for simulations of hadronic interactions. For the hadronic simulations we have considered only photon-pion processes, from which we have obtained neutrino fluxes and their associated emission due to γ-ray cascades (see Figs. 1 and 2).

The very high energy emission tail of NGC1275 is better reproduced with the hadronic models with smaller power-law index of CR injection, (which are compatible with the predictions of reconnection acceleration). The associated neutrino fluxes are consistent with the upper limits from IceCube (see Fig. 2, right panel) and from the models presented here, we conclude that the core of NGC1275 would mostly contribute to neutrino detections at $\sim$PeV energies. However, the neutrino fluxes presented in Fig. 2 may be over-estimations, since they come from a single source.

Possibly, other background plasma configurations as well as the inclusion of other hadronic interactions (e.g. *pp* and Bethe-Heitler processes) will produce better emission models. We will explore these possibilities in a forthcoming paper.

Acknowledgments

We acknowledge support from the Brazilian agencies FAPESP (2013/10559-5 grant) and CNPq (306598/2009-4 grant). The simulations presented in this lecture have made use of the computing facilities of the GAPAE group (IAG-USP) and the Laboratory of Astroinformatics IAG/USP, NAT/Unicsul (FAPESP grant 2009/54006-4). RAB is supported by the FAPESP grant 2017/12828-4 and JCRR by the FAPESP grant 2017/12188-5.

† We note that this factor of CRs is somewhat overestimated with regard to the value one obtains when considering the total magnetic reconnection acceleration power available in the simulated region (see e.g., Singh, de Gouveia Dal Pino & Kadowaki, 2015). Nevertheless, for stronger background magnetic fields, we expect to obtain more realistic results, which will be explored in forthcoming work.

References

Aartsen, M., Ackermann, M., Adams, J. *et al.* 2016, *PhRvL* 113, 101101
Abdo, A. *et al.* 2009, *ApJ* 699, 31
Ackermann, M. *et al.* 2012, *ApJS* 203, 4
Ajello, M. *et al.* 2009, *ApJ* 690, 367
Aleksic, J. *et al.* 2012, *A&A*, 539L, 4
Aleksic, J. *et al.* 2012, *A&A*, 541, 99
Alves Batista, R., Dundovic, A., Erdmann, M., Kampert, K., Kuempel, D., Mller, G., Sigl, G., van Vliet, A., Walz, D. & Winchen, T. 2016, *JCAP* 05, 38
de Gouveia Dal Pino, E. M. *et al.* *These Proceedings*
de Gouveia Dal Pino, E. M. & Lazarian, A. 2005, *A&A*, 441, 845
de Gouveia Dal Pino, E. M., Piovezan, P. P. & Kadowaki, L. H. S. 2010a *A&A*, 518, A5
Dolence, J., Gammie, C., Mocibrodzka & M, Leung, P-K 2009 *ApJS*, 184, 387D
Gammie, C., McKinney, J., Toth, G. 2003, *ApJ*, 598, 444
Johnstone, R. M. & Fabian, A. C. 1995, *MNRAS*, 273, 625
Kadowaki, L. H. S., de Gouveia Dal Pino, E. M. & Singh C. B. 2015, *ApJ*, 802,113
Kadowaki, L. H. S., de Gouveia Dal Pino, E. M. & Stone J. 2018a, arXiv:1803.08557
Kadowaki, L. H. S., de Gouveia Dal Pino, E. M. & Stone J. 2018b, *In prep.*
Kowal, G., de Gouveia Dal Pino, E. M. & Lazarian, A. 2011, *ApJ*, 735, 102
Kowal, G., de Gouveia Dal Pino & E. M., Lazarian, A. 2012, *PhRvL*, 108, 24
Lister, M. L., Aller, H. D., Aller, M. F. *et al.* 2009, *AJ*, 137, 3718
Wilman, R., Edge, A., & Johnstone, R. 2005, *MNRAS*, 359, 755

Perseus in Sicily: From Black Hole to Cluster Outskirts
Proceedings IAU Symposium No. 342, 2018
K. Asada, E. de Gouveia Dal Pino, M. Giroletti,
H. Nagai & R. Nemmen, eds.

doi:10.1017/S1743921318007846

Magnetic Fields in the Relativistic Jets of Active Galactic Nuclei

Denise C. Gabuzda

Physics Department, University College Cork, Cork, Ireland
email: d.gabuzda@ucc.ie

Abstract. An abundance of information about the magnetic (**B**) fields of relativistic AGN jets has been obtained through radio polarization observations made on a wide range of scales, from the parsec scales probed by Very Long Baseline Interferometry to the kiloparsec scales probed by instruments such as the the Jansky Very Large Array and e-MERLIN. The observed radio emission is synchrotron radiation, for which the linear polarization angles in optically thin regions is orthogonal to the local synchrotron **B** fields. The characteristic **B** field structures observed for these jets on parsec scales are described. A key question is whether **B** field structures observed in particular AGN jets come about primarily due to "global" effects such as the intrinsic **B** field of the jet itself, which is expected to be helical, or local phenomena such as shocks, shear and bending of the jets. Observational criteria that can be used to try to distinguish between various possible origins for observed B field structures are described. There is now considerable evidence that the relativistic jets of AGNs do indeed carry helical **B** fields, with the observed polarization also sometimes appreciably influenced by local effects. Patterns seen in the helical **B** field components, indicated for example by Faraday rotation observations, provide unique information about the processes involved in the generation of these helical **B** fields. The collected observations on parsec and kiloparsec scales provide weighty evidence that an important role is played by the action of a cosmic "battery" that acts to generate an inward current along the jet axis and an outward current in a more extended region surrounding the jet.

Keywords.

1. AGN Jets and Helical/Toroidal B fields

Core-dominated AGN almost universally display one-sided core-jet structures on VLBI scales, consistent with a picture in which the jets ejected from the active nucleus are highly relativistic and we are detecting only the Doppler-boosted approaching jet (e.g. Lister & Homan 2005). Apparent superluminal motions are also extremely common, again consistent with relativistic jets oriented at relatively small angles to the line of sight (Kellermann *et al.* 2004).

It has long been believed theoretically that the mechanism launching the jets from the vicinity of the central engine is electromagnetic in nature. The main mechanisms considered are those proposed by Blandford & Znajek (1977), where the jets are launched from the immediate vicinity of the central black hole, and by Blandford & Payne (1982), where the jets are launched from the surrounding accretion disk. Numerical simulations have demonstrated the generation of helical magnetic (**B**) fields associated with the jets, due to the "winding up" of an initial longitudinal field component (e.g., Barniol Duran *et al.* 2017). This leads to the clear expectation on theoretical grounds that AGN jets should carry helical **B** fields, which should have certain observational consequences. These fields may effectively be manifest as toroidal fields when the azimuthal component of the field is very dominant. Observational evidence for the presence of helical jet **B** fields provides direct support for this widely accepted theoretical picture.

The highly relativistic jets of AGN are also typically extremely narrow, obviously well collimated structures, with the jets sometimes remaining collimated all the way out to kiloparsec scales. One natural way for this to come about is self-collimation by the azimuthal component of the helical **B** field that is generated when the jets are launched.

A key physical phenomenon in relation to the observational detection of the presence of helical jet **B** fields is Faraday rotation: a rotation of the plane of polarization of an electromagnetic wave as it passes through a region with free charges and magnetic field (i.e., a magnetized plasma). This rotation of the linear polarization angle comes about due to the different indices of refraction of the right circularly polarized (RCP) and left circularly polarized (LCP) components of the polarized wave, leading to different speeds of propagation through the magnetized medium. This difference in the speeds of the RCP and LCP components induces a delay between them, manifest as a rotation in the plane of linear polarization. The amount of rotation depends on the strength of the ambient magnetic field $\mathbf{B}_{amb}$, the number density of free electrons in the Faraday-rotating plasma n_e, the electron charge e and mass m, and the wavelength of the radiation λ:

$$\chi = \chi_o + RM\lambda^2 \qquad RM = \frac{e^3}{8\pi^2\epsilon_o m_e^2 c^3}\int n_e \mathbf{B}_{amb}\cdot d\mathbf{l} \tag{1.1}$$

where χ is the observed polarization angle, χ_o the intrinsic emitted polarization angle, ϵ_o the permittivity of free space, and c the speed of light; the integral is carried out over the line of sight from the source to the observer.

The magnitude of the Faraday rotation measure RM depends on both n_e and the line-of-sight component of $\mathbf{B}_{amb}$, while the sign of the Faraday rotation is determined purely by the direction of the line-of-sight component of $\mathbf{B}_{amb}$ (toward or away from the observer). The action of Faraday rotation can be identified through the λ^2 dependence of the observed polarization angle χ.

As was pointed out by Blandford (1993), if a jet has a helical **B** field, this should give rise to a Faraday-rotation gradient across the jet, due to the systematically changing line-of-sight component of the **B** field across the jet. Thus, the detection of transverse Faraday-rotation gradients across AGN jets can provide a powerful diagnostic for the presence of an azimuthal field component that may be associated with a helical jet **B** field. Furthermore, the presence of such an azimuthal field component straightforwardly implies the presence of a current in the jet, whose direction is given by the usual right-hand rule from university physics. Figure 1 shows an example of an AGN jet with a clear, statistically significant transverse RM gradient across its jet, with the direction of the associated azimuthal **B** field component shown together with the inferred direction of the jet current. Statistically significant transverse RM gradients, interpreted as reflecting the presence of an azimuthal **B**-field component, have now been detected across about 50 AGN jets on parsec scales (Gabuzda *et al.* 2018) and roughly a dozen AGN jets on larger scales of tens to thousands of parsecs (Christodoulou *et al.* 2016, Knuettel *et al.* 2017). This provides direct evidence for toroidal or helical **B** fields in AGN jets on all scales accessible to direct imaging in the radio.

The presence of helical jet **B** fields will also give rise to certain characteristic polarization (**B**-field) structures in AGN jets, as is discussed in the following section. Other recent observations that have been interpreted as manifestations of helical jet **B** fields include an increase (rather than the expected decrease) in degree of polarization with increasing wavelength (so-called "inverse depolarization", Homan 2011), variability of jet ridge lines (Cohen *et al.* 2015), variability of the Faraday rotation sign (Lico *et al.* 2017), double polarization angle rotations (Cohen 2017), and circular polarization (Gabuzda 2015). This suggests that helical **B** fields carried by AGN jets may provide a framework for understanding a wide range of their properties and the behaviours they display.

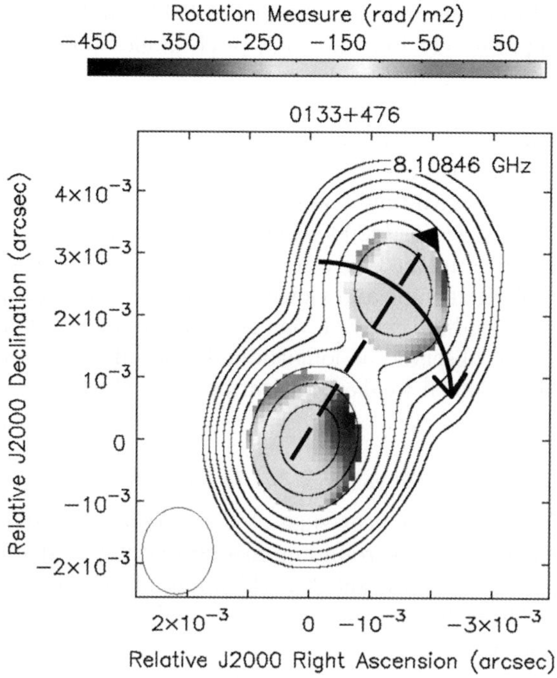

Figure 1. Relationship between an observed transverse RM gradient, the direction of the associated azimuthal **B** field component (solid curved arrow), and the inferred direction of the current (dashed arrow). RM image adapted from Gabuzda (2018).

2. B Field Structures of AGN Jets: Global or Local Origin?

It has been noted since the earliest VLBI polarization images were analyzed that the observed polarization in the jets of AGN on parsec scales tends to lie either parallel to or perpendicular to the local jet direction, with the inferred direction of the associated jet **B** field being orthogonal to the direction of the polarization. Commonly observed polarization patterns correspond to (a) extended regions of longitudinal **B** field, (b) regions of orthogonal **B** field associated with bright, compact regions, (c) extended regions of orthgonal **B** field, (d) spine-sheath transverse **B**-field structure, (e) longitudinal **B** field offset toward one side of the jet, and (f) longitudinal **B** field around a bend in the jet; here, "longitudinal" and "orthogonal" are relative to the direction of the jet (see the schematic of these polarization configurations shown by Gabuzda (2015)). This should come about naturally if the jets locally exhibit approximately cylindrical symmetry: the **B** field can then always be separated into longitudinal and orthogonal components projected onto the sky (e.g. Lyutikov *et al.* 2005).

Thus, the jets are expected theoretically to carry helical **B** fields, and direct observational evidence for such fields has been found through the detection of statistically significant transverse Faraday-rotation gradients across AGN jets on scales from parsecs to kiloparsecs. This makes it is natural to consider which of these commonly observed **B**-field patterns could be associated with these helical **B** fields.

2.1. *Structures associated with helical fields*

Helical jet **B** fields could give rise to many of the characteristic polarization patterns referred to above.

• Extended regions of longitudinal or othogonal **B** field could be associated with helical jet **B** fields with comparatively low and high pitch angles (i.e., comparatively "loosely wound" or "tightly wound" helical fields), respectively.
• Spine–sheath transverse polarization structure implying orthogonal **B** fields near the center of the jet and longitudinal **B** fields at one or both edges of the jet could be associated with a helical field, with the azimuthal component of the field dominant near the center of the jet and the longitudinal component dominant near the jet edges, projected onto the sky (e.g., Pushkarev *et al.* 2005, Lyutikov *et al.* 2005). In this case, this should also give rise to an increase in the degree of polarization toward the jet edges.
• Some combinations of pitch angle and viewing angle give rise to projected helical **B**-field configurations with longitudinal field on one side of the jet and transverse field on the other, which may also be observed as longitudinal field offset toward one side of the jet if the transverse field has much weaker polariation (e.g. Murphy *et al.* 2013).

2.2. *Structures due to local effects*

The main local effects that have been suggested as possible origins of observed polarization features are shear with the surrounding medium acting to stretch out the magnetic field along the flow, and transverse shocks, which amplify the **B** field component in the plane of compression. Other factors that are increasingly believed to play significant roles include turbulence and magnetic reconnection.

Shear will enhance the longitudinal field component, and has been proposed as an explanation for regions of longitudinal **B** field in AGN jets, especially at the jet edges. Note, however, that such regions could also come about due to the presence of helical jet **B** fields. Shocks enhance the local orthogonal field component; compact features with orthogonal **B** field are good candidates for localized regions of transverse shocks, although extended regions of extended orthogonal field are probably more plausibly explained by the presence of a relatively high-pitch-angle helical field.

Turbulence is almost certainly present in AGN jets to some degree, and could be induced by mechanisms such as Kelvin–Helmholtz instability or current-driven (kink) instability in helical or toroidal jet **B** fields (e.g. Singh *et al.* 2016, Striani *et al.* 2016, de Gouveia Dal Pino et al. these proceedings). Turbulence could, in turn, lead to magnetic reconnection. It is likely that both of these effects play important roles, particularly in relation to variability, particle acceleration and very high energy emission (e.g., de Gouveia Dal Pino et al. these proceedings). Magnetic reconnection suggests the presence of magnetic field reversals, but it is not entirely clear how these can be detected in polarization images. The general impact of these effects should be to partially destroy the overall polarization patterns due to coherent helical or toroidal fields. It may be that turbulence acts primarily to reduce the degree of polarizaton, leaving the overall polarization patterns more or less intact, whereas magnetic reconnection would likely bring about appreciably changes in the initial field configuration. Regions that display variability and unusual **B**-field structures may be good candidates for regions of magnetic reconnection.

Another less discussed mechanism that can also potentially play a role is enhancement of the local longitudinal **B** field at the outer edges of bends in the jet, sometimes called "curvature induced polarization". This could either amplify an existing longitudinal field component, for example, associated with a helical jet **B** field, or induce a net longitudinal component in an initially tangled field, analogous to the ability of a shock to induce a net orthogonal component in an initially tangled field.

It is clear that helical jet **B** fields can potentially provide a natural explanation for essentially all of the observed characteristic **B**-field patterns observed in AGN jets.

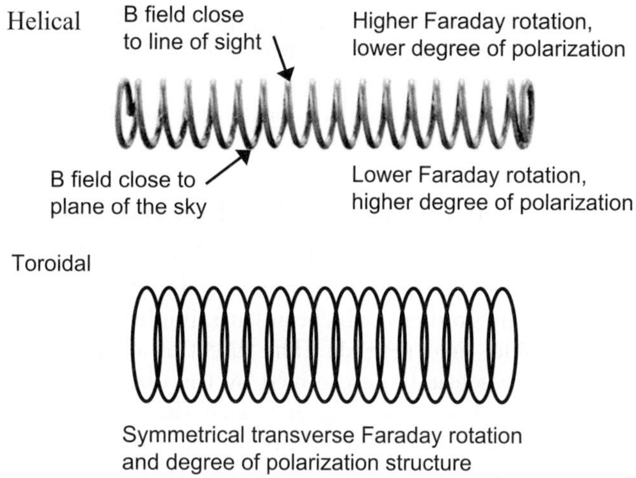

Figure 2. Schematic illustrating differences in polarization asymmetry for jets carrying helical versus toroidal **B** fields.

However, it is also the case that the origin of a given **B**-field structure may not be obvious from the observed polarization-angle distribution alone, and additional information is usually required as a basis for sound interpretation, such as the distribution of the degree of polarization, information about Faraday rotation and Faraday-rotation gradients occurring in the vicinity of the jet etc. Of course, in general, we might expect situations where there is a contribution from both the intrinsic helical **B** field of the jet itself and local phenomena such as shocks and shear.

2.3. *Helical versus toroidal fields*

An important question is whether the azimuthal field component revealed by a transverse RM gradient is associated with a helical or a toroidal **B** field. In general, it is only asymmetry of the intensity and polarization profiles across the jet that can distinguish observationally between a helical field (with an ordered longitudinal component) and a toroidal field (without a net longitudinal component, because this component is either absent or disordered). The key point here is that the synchrotron intensity and degree of polarization are determined by the component of the jet **B** field in the plane of the sky (e.g., a **B** field directed precisely toward the observer would give no linear polarization).

Figure 2 shows a schematic visually illustrating why a helical field generally gives rise to asymmetric intensity and linear polarization profiles across thet jet, while a toroidal field gives rise to symmetric profiles. This figure shows helical and toroidal fields viewed side-on. The toroidal field should give rise to symmetric intensity and polarizaton profiles, independent of viewing angle. However, due to the presence of a longitudinal component in the helical field, the overall field usually lies predominantly along the line of sight along one edge of the jet, where it gives rise to lower degrees of polarization, and predominantly in the plane of the sky at the other edge of the jet, where it gives rise to higher degrees of polarization. Only for particular viewing angles and helical pitch angles will a helical **B** field give rise to symmetrical intensity and polarization structures (see, e.g., Murphy *et al.* 2013).

In addition, the fact that Faraday rotation is proportional to the line-of-sight component of the ambient magnetic field provides additional potential tests of consistency of observations with the presence of helical **B** fields in jets displaying transverse Faraday RM gradients. The picture depicted in Fig. 2 predicts that the side of such a jet with the higher degree of polarization (dominant **B**-field component in the plane of the sky) should display a lower amount of Faraday rotation, whereas the side of the jet with the lower degree of polarization should display a higher amount of Faraday rotation. Indeed, some cases where this pattern is observed have been noted (Murphy *et al.* 2013, Gabuzda *et al.* 2018).

3. Overall Patterns in the B Fields/Axial Currents of AGN Jets

As was noted in the first section of this article, the direction of an observed transverse Faraday RM gradient across an AGN jet implies a direction for the azimuthal field component giving rise to the gradient, which, in turn, implies the direction for a current associated with this toroidal field (Fig. 1). The direction of the azimuthal component of the jet's helical field is essentially determined by the direction of the rotation of the central black hole and accretion disk, together with the direction of the longitudinal component of the initial field that is "wound up" by the rotation.

In this simplest picture, we would expect to see a particular direction for the observed transverse RM gradient all along the jet. However, in fact, reversals in the directions of observed transverse RM gradients, either with distance along the jet or with time, have been observed for a number of AGN jets (Gabuzda *et al.* 2018 and references therein). A recently discovered example of such an RM-gradient reversal is shown in Fig. 3. This can be understood in a picture with a nested helical-field structure, with the directions of the azimuthal field components being opposite in the inner and outer regions of helical field (Mahmud *et al.* 2013). In this case, a change in the direction of the net observed RM gradient could be due to a change in which region of helical field (inner or outer) makes the dominant contribution to the overall Faraday rotation.

Such a nested helical-field configuration is predicted by the "cosmic battery" model described by Contopoulos *et al.* (2009) and Christodoulou *et al.* (2016). Further, this model predicts that the orientations of the azimuthal components of the inner and outer regions of helical field are not random: the inner region of helical field should have a counterclockwise azimuthal component projected on the plane of the sky, associated with an inward current along the jet axis, while the outer helical field should have a clockwise azimuthal component projected onto the sky, associated with a more extended region of outward current (essentially present between the two regions of helical field).

Sufficient statistics to test this picture have recently become available. Gabuzda *et al.* (2018) have demonstrated that the collected data on transverse RM gradients detected across about 50 parsec-scale AGN jets indicate a statistically significant predominance of *inward* currents (probability of the predominance being spurious $\simeq 0.40\%$). In contrast, the results of Christodoulou *et al.* (2016), supplemented by the more recent results of Knuettel *et al.* (2017), indicate a highly statistically significant predominance of *outward* currents on decaparsec–kiloparsec scales (probability of the predominance being spurious $\simeq 0.05\%$). These observations are consistent with a picture in which the jets have a nested helical-field configuration, with current loops oriented roughly orthogonal to the accretion disk, with the current flowing inward along the jet axis, away from the axis in the accretion disk, outward in an extended region around the jet, and back toward the jet axis in a region near the termination of the jet. This is broadly similar to the set of currents and **B** fields in a co-axial cable, but on a grand scale.

4. Summary

AGN jets are expected theoretically to carry helical **B** fields, due to the joint action of the rotation of the central black hole and accretion disk and the jet outflow. In fact, there is now substantial observational evidence that many or all jets do carry helical

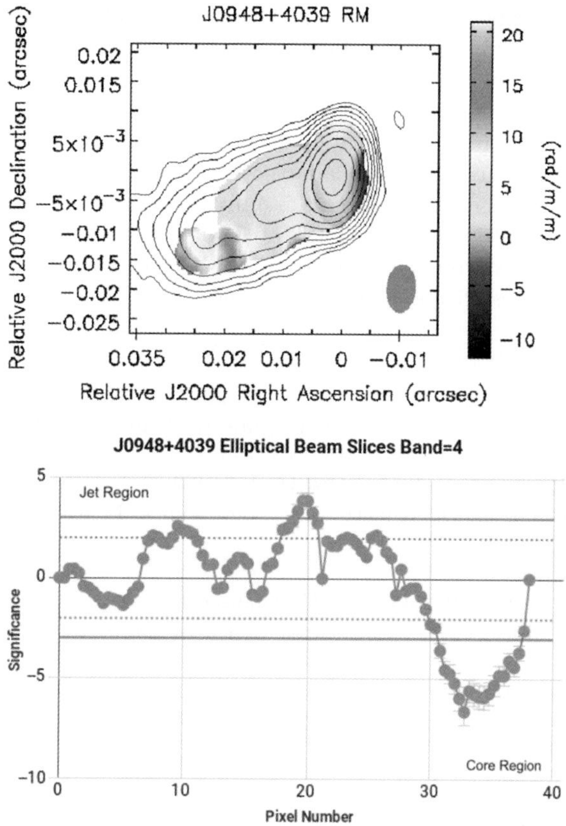

Figure 3. Transverse RM gradients across the VLBI jet of J0948+4039. The top panel shows the Faraday rotation map made using VLBA data obtained at 1.8+1.4+2.2+5.0 GHz, with maximum RM uncertainties of 3 rad/m^2. Oppositely directed transverse RM gradients can be seen in the core region and jet. The lower panel shows the significances of a series of RM slices taken perpendicular to the jet. The gradients in the core region have significances exceeding 5σ, and there are fairly extended regions in the jet with transverse RM gradients with significances of $2-4\sigma$. The dotted green and solid red horizontal lines show significances of 2σ and 3σ, respectively.

B fields, most importantly, the detection of statistically significant transverse Faraday RM gradients across some 50 AGN jets on parsec scales, as well as about a dozen AGN jets on larger scales of tens to thousands of parsec. This realization is now leading to a variety of interpretations of observed properties and behaviour exhibited by AGN jets in the framework of models with helical jet **B** fields.

Furthermore, many of the characteristic polarization/magnetic-field structures observed in AGN jets on parsec scales can be understood as manifestations of a helical jet **B** field. At the same time, some are also consistent with the action of local agents, such as shocks, shear, and jet bending. It can be difficult to unambiguously identify the origin of observed polarization structures in particular jets in practice without incorporating additional information about the distribution of the degree of polarizaton, the morphology of the jet, the distribution of Faraday rotation in the vicinity of the jet, and the possible presence of transverse RM gradients across the jet. In many cases, it is likely that the observed polarization structure is produced by both the intrinsic helical **B** field of the jet and the action of various local factors.

The direction of a transverse RM gradient across an AGN jet implies a direction for the azimuthal **B**-field component producing it, which, in turn, implies the direction for the net current flowing inside the region occupied by this toroidal field. The collected data on transverse RM gradients observed on parsec to kiloparsec scales demonstrate a statistically significant predominance of inward currents along the jet axis on parsec scales, and of outward currents on decaparsec–kiloparsec scales, presumably in a more extended region surrounding the jet. This is fully consistent with the "cosmic battery" model described by Contopoulos *et al.* (2009) and Christodoulou *et al.* (2016). In this picture, the systems of currents and fields in AGN jets and their accretion disks are essentially similar to those for giant co-axial cables.

References

Barniol Duran, R., Tchehovskoy, A. & Giannios, D. 2017, *MNRAS*, 469, 4957

Blandford, R. D. 1993, in *Astrophysical Jets* (Cambridge University Press), p. 26

Blandford, R. D. & Payne, D. G. 1982, *MNRAS*, 199, 883

Blandford, R. D. & Znajek, R L. 1977, *MNRAS*, 179, 433

Christodoulou, D., Gabuzda, D., Knuettel, S., Contopoulos, I., Kazanas, D. & Coughlan, C. 2016, *A&A*, 591, A61

Cohen, M. H. 2017, *Galaxies*, 5, 12

Cohen, M. H., Meier, D. L., ARshakian, T. G., Clausen-Brown, E., Homan, D. C., Hovatta, T., Kovalev, Y. Y., Lister, M. L., Pushkarev, A. B., Richards, J. L. & Savolainen, T. 2015, *ApJ*, 803, 3

Contopoulos, I., Christodoulou, D., Kazanas, D. & Gabuzda, D. C. 2009, *ApJ*, 702, L148

Gabuzda, D. C. 2015, in *The Formation and Disruption of Black Hole Jets, Astrophysics and Space Science Library*, 414, p. 117

Gabuzda, D. C. 2018, *Galaxies*, 6, 9

Gabuzda, D. C., Nagle, M. & Roche, N. 2018, *A&A*, 612, A67

Homan, D. C. 2011, *ApJ*, 757, L24

Kellermann, K. I., Lister, M. L., Homan, D. C., Vermeulen R. C., Cohen, M. H., Ros, E., Kadler, M., Zensus, J. A. & Kovalev, Y. Y. 2004, *ApJ*, 609, 539

Knuettel, S., Gabuzda, D. C. & O'Sullivan, S. P. 2017, *Galaxies*, 5, 61

Lico, R., Gomez, J. L., Asada, K. & Fuentes, A. 2017, *Galaxies*, 5, 57

Lister, M. L. & Homan, D. C. 2005, *AJ*, 130, 1389

Lyutikov, M., Pariev , V. I. & Gabuzda, D. C. 2005, *MNRAS*, 360, 869

Mahmud, M., Coughlan, C.P., Murphy, E., Gabuzda, D. C. & Hallahan, D. R. 2013, *MNRAS*, 431, 695

Murphy, E., Cawthorne, T. V. & Gabuzda, D. C. 2013, *MNRAS*, 430, 1504

Pushkarev, A. B., Gabuzda, D. C., Vetukhnovskaya, Yu. N. & Yakimov, V. E. 2005, *MNRAS*, 356, 859

Singh, C. B., Mizuno, Y. & de Gouveia Dal Pino, E. M. 2016, *ApJ*, 824, 48

Striani, E., Mignone, A., Vaidya, B., Bodo, G. & Ferrari, A. 2016, *MNRAS*, 462, 2970

Perseus in Sicily: From Black Hole to Cluster Outskirts
Proceedings IAU Symposium No. 342, 2018
K. Asada, E. de Gouveia Dal Pino, M. Giroletti,
H. Nagai & R. Nemmen, eds.

doi:10.1017/S1743921318006087

The correlation between the total jet power and the Poynting flux at the jet base

Elena E. Nokhrina

Laboratory of Fundamental and Applied Research of Relativistic Objects of the Universe,
Moscow Institute of Physics and Technology,
Dolgoprudnyy, Institutsky per. 9, 141700, Russia
email: nokhrina@phystech.edu

Abstract. The magneto hydrodynamic models of relativistic jets from active galactic nuclei predict the jet power transported by the Poynting flux at the jet base, setting the correlation between the jet power and the total magnetic flux. For highly collimated jets taking the transversal structure into account allows to rewrite this correlation through the observed jet properties such as spectral flux and core shift. Applying this method we find that, for the sample of 48 sources, their jet power distribution is well peaked at the theoretically predicted level.

Keywords. MHD, BL Lacertae objects: general, quasars: general, galaxies: jets

1. Introduction

It is established that the electromagnetic processes are governing in the relativistic jet launching from a super massive black hole (BH) in a center of active galactic nuclei. The ordered magnetic field $B_{\rm g}$ frozen into accreting disk matter is brought to the black hole by the accretion, with the total magnetic flux $\Psi_0 = \pi B_{\rm g} r_{\rm g}^2$ threading the BH ergosphere of radius of the order of Schwarzschild radius $r_{\rm g} = 2GM/c^2$. The spinning BH with a spin parameter $a_* \in [0;\,1)$ and corresponding angular velocity of field lines $\Omega_{\rm F} = ca_*/2r_{\rm g}(1+\sqrt{1-a_*^2})$ creates in the lab frame the electric magnetic field of the order of $\Omega_{\rm F} r_{\rm g} B_{\rm g}/c$. This electromagnetic field configuration drives the Poynting flux from a black hole (Beskin (2010)).

The energy density flux $E(\Psi)$, conserved on magnetic surfaces designated by the contained magnetic flux Ψ, can be written for the central part of a flow, connected with a black hole, as:

$$E(\Psi) = \frac{\Omega_0 \Psi}{4\pi^2} + E_{\rm part}, \qquad (1.1)$$

where the constant angular velocity $\Omega_{\rm F} = \Omega_0$, and $E_{\rm part}$ is an initial particle energy flux. The first term in the Eq. 1.1 represents the Poynting flux at the jet base, while the second term — the particle kinetic energy flux. The ratio of two terms is equal (Tchekhovskoy *et al.* (2009); Nokhrina *et al.* (2015)) to the Michel's magnetization parameter $\sigma_{\rm M}$, which value may be safely taken as of the order of 10, as supported by analytical results by (Nokhrina *et al.* (2015)) and by the kinematics of bright features in jets on pc scales (Lister *et al.* (2009)). If the observed Lorentz factors represent the jet bulk motion, than the highest Γ corresponds to a half of the initial magnetization $\sigma_{\rm M}$. These arguments allow us to state that the first term in Eq. 1.1 with about 10% accuracy represents the total energy density flux in a jet. The total jet power in this case is approximately equal to the integral of the first term in 1.1:

$$W_\Psi = \frac{1}{c}\int_0^{\Psi_0} E(\Psi)d\Psi = \frac{c}{8}\left(\frac{\Psi_0}{\pi R_{\rm L}}\right)^2. \tag{1.2}$$

The result above means that within well-established model of jet launching, there must be a correlation between the total jet power and the total magnetic flux in a jet. Both values are estimated through the observation only implicitly.

2. Magnetic flux estimate

The magnetic flux contained in a jet is defined as

$$\Psi_0 = \int_S \mathbf{B}_{\rm P}\cdot \mathbf{dS}. \tag{2.1}$$

The analytical modeling (see, e.g., Lyubarsky (2009); Beskin (2010)) of relativistic MHD outflows provide that the ratio of the poloidal magnetic field that determines the magnetic flux to the toroidal magnetic field is

$$|\mathbf{B}_{\rm P}|/|\mathbf{B}_\varphi| = R_{\rm L}/r. \tag{2.2}$$

Thus, the toroidal magnetic field dominates on the scales greater than the light cylinder with typically unresolved dimension of $R_{\rm L} = 4\times 10^{-3}\left(M_{\rm BH}/10^9\,M_\odot\right)$ pc for the black hole spin parameter $a_* = 0.1$. Any measurements of a magnetic field provide the dominating toroidal magnetic field that does not determine explicitly the magnetic flux. The MHD modeling provides an instrument to correlate the toroidal and poloidal components of a magnetic field in a jet, as has been done by (Zamaninasab *et al.* (2014); Zdziarski *et al.* (2015); Nokhrina (2017B)). The jet magnetic field transversal structure comprises the central core with the characteristic scale of $R_{\rm L}$ with dominating constant poloidal magnetic field of a magnitude B_0 and the Poynting-carrying outer jet with the correlation given by Eq. 2.2 and approximate poloidal field profile $B_{\rm P}\propto r^{-2}$ (Nokhrina *et al.* (2015); Bromberg & Tchekhovskoy (2016)). It is important that the amplitudes of both poloidal and toroidal magnetic fields are of the same order B_0. Taking these estimate we obtain the jet magnetic flux

$$\Psi_0 = \pi B_0 R_{\rm L}^2\left(1 + 2\ln\frac{R_{\rm jet}}{R_{\rm L}}\right). \tag{2.3}$$

The field amplitude B_0 introduced into the expression 2.3 does not represent the magnetic field measured in a jet. The magnetic field in a jet may be estimated through core shift measurements (Lobanov (1998)) or the brightness temperature measurements (Zdziarski *et al.* (2015); Nokhrina (2017A)). Both methods provide the amplitude of a uniform model magnetic field in a bright feature.

The spectral flux from a jet with prescribed transversal structure may be readily estimated for the blazars (Nokhrina (2017A)), providing an instrument to obtain the magnetic field amplitude B_0 measuring the brightness temperature. A jet with uniform magnetic field $B_{\rm uni}$ has the same brightness temperature if the magnetic field amplitudes relate as

$$\frac{B_0}{B_{\rm uni}} = 0.86\frac{R_{\rm jet}}{R_{\rm L}}. \tag{2.4}$$

The value $B_{\rm uni}$ may be a magnetic field measured either using core shift method or brightness temperature measurements. Thus, the expression for the magnetic flux through the observables is

$$\frac{\Psi_0}{\pi R_{\rm L}} = 0.86 B_{\rm uni} R_{\rm jet}\left(1 + 2\ln\frac{R_{\rm jet}}{R_{\rm L}}\right). \tag{2.5}$$

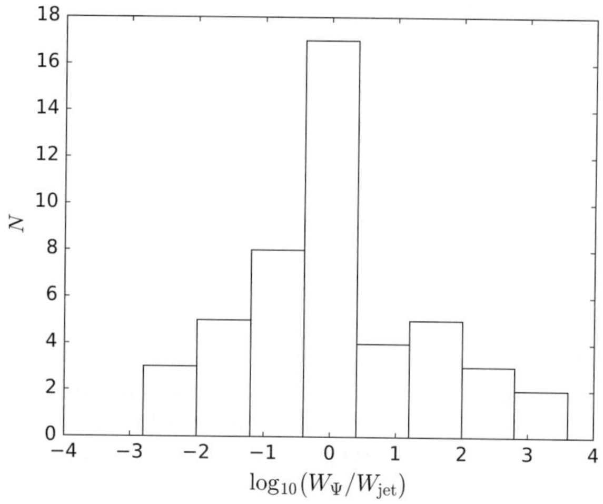

Figure 1. The distribution for the ratio $W_\Psi/W_{\rm jet}$ in log-scale for the sample of 48 blazars.

Here we write it so as to have in the l.h.s. the expression entering in the 1.2. The r.h.s. contains the values that may be estimated through the observables ($B_{\rm uni}$ and $R_{\rm jet}$). Here the light cylinder radius $R_{\rm L}$ is unknown, but the r.h.s. depends on it logarithmically weakly, so we may safely substitute for it the rough estimate $R_{\rm L}=40r_{\rm g}$ for the black hole spin parameter $a=0.1$.

3. Power estimate

In order to check whether the sources obey the correlation

$$W_\Psi = \frac{c}{8}\left[0.86 B_{\rm uni} R_{\rm jet}\left(1+2\ln\frac{R_{\rm jet}}{R_{\rm L}}\right)\right]^2 \qquad (3.1)$$

we use the averaged over lifetime jet power as a proxy for the jet power. The obtained by Cavagnolo *et al.* (2010) correlation of the average jet power needed to fill the cavities in a surrounding gas against the jet spectral flux at frequencies 200 – 400 MHz

$$\left(\frac{W_{\rm jet}}{10^{43}\,{\rm erg/s}}\right) = 3.5\left(\frac{W_{200-400}}{10^{40}\,{\rm erg/s}}\right)^{0.64} \qquad (3.2)$$

allows to use the spectral flux data accumulated by the CATS (Verkhodanov *et al.* (1997)) database to obtain $W_{\rm jet}$.

We use the sample of 48 sources describe in Nokhrina (2017B). We calculate (see details in Nokhrina (2017B)) the uniform magnetic field magnitude $B_{\rm uni}$ and corresponding value for W_Ψ. We have also calculated the average jet power $W_{\rm jet}$ using the data accumulated by the CATS database. The histogram of ratio of these two values is presented in Fig. 1. The contrast with the result in Nokhrina (2017B) is more accurate formula for the jet power estimate using the total magnetic flux: the reasonable energy density flux integral $E(\Psi)$ provides the denominator 8, while the estimate used in Nokhrina (2017B) has the denominator 1.

4. Results and discussions

We observe that the total distribution is well peaked around the ratio $W_\Psi/W_{\rm jet}=1$, with 17 sources falling in the central bin $\log_{10}(W_\Psi/W_{\rm jet})\in[-0.4,\,0.4]$, and 29 sources falling in the range $\log_{10}(W_\Psi/W_{\rm jet})\in[-1.2,\,1.2]$. Thus, we conclude, that for about 60% of the sources under consideration the total jet power can be explained purely by the power carried by the Poynting flux at the jet base, and there is no sufficient input from a possible outflow from a disk. The scatter of the distribution may be attributed to several factors. The uncertainties in the observed parameters estimation may lead to the observed scatter. The second factor may be our using the averaged jet power and the correlation (Cavagnolo *et al.* (2010)). We will address the correlation between W_Ψ and a momentary jet power that can be obtained for the number of sources (Ghisellini *et al.* (2014); Pjanka *et al.* (2017)) in the future work. The symmetry of the distribution might point to the absence of sufficient input of a jet originating from the disk into total jet power, or to correlation between these two sources of jet energy supply.

Acknowledgments

This study has been supported in part by the Russian Science Foundation: grant 16-12-10051. This research has made use of data from the MOJAVE database that is maintained by the MOJAVE team (Lister *et al.* (2009)), and data accumulated by the CATS data base (Verkhodanov *et al.* (1997)).

References

Beskin, V. S. 2010, *Physics Uspekhi*, 53, 1199
Bromberg, O. & Tchekhovskoy, A. 2016, *MNRAS*, 456, 1739
Cavagnolo, K. W., McNamara, B. R., Nulsen, P. E. J., Carilli, C. L., Jones, C., & Bîrzan, L. 2010, *ApJ*, 720, 1066
Ghisellini, G., Tavecchio, F., Maraschi, L., Celotti, A., & Sbarrato T. 2014, *Nature*, 515, 376
Lister, M. L., Aller, H. D., Aller, M. F., Cohen, M. H.,Homan, D. C., Kadler, M., Kellermann, K. I., Kovalev, Y. Y., Ros, E., Savolainen, T., Zensus, J. A., & Vermeulen, R. C. 2009, *Astronomical Journal*, 137, 3718
Lobanov, A. 1998, *A&A*, 330, 79
Lyubarsky, Y. 2009, *ApJ*, 698, 1570
Nokhrina, E. E., Beskin, V. S., Kovalev, Y. Y., & Zheltoukhov, A.A. 2015, *MNRAS*, 447, 2726
Nokhrina, E. E. 2017A, *MNRAS*, 4687, 2372
Nokhrina, E. E. 2017B, *Frontiers in Astronomy and Space Sciences*, 4, 63
Pjanka, P., Zdziarski, A. A., & Sikora, M. 2017, *MNRAS*, 465, 3506
Tchekhovskoy, A., McKinney, J. C., & Narayan, R. 2009, *ApJ*, 699, 1789
Verkhodanov, O. V., Trushkin, S. A., Andernach, H., & Chernenkov, V. N. 1997, in: G. Hunt & H. E. Payne (eds.) *Astronomical Data Analysis Software and Systems VI* (San Francisco: ASP) 125, 322
Zamaninasab, M., Clausen-Brown, E., Savolainen, T., & Tchekhovskoy, A. 2014, *Nature*, 510, 126
Zdziarski, A., Sikora, M., Pjanka, P., & Tchekhovskoy, A. 2015, *MNRAS*, 451, 927

Perseus in Sicily: From Black Hole to Cluster Outskirts
Proceedings IAU Symposium No. 342, 2018
K. Asada, E. de Gouveia Dal Pino, M. Giroletti,
H. Nagai & R. Nemmen, eds.

doi:10.1017/S1743921318007664

Why only a small fraction of quasars are radio loud?

Xinwu Cao[1,2,3]

[1]Shanghai Astronomical Observatory, Chinese Academy of Sciences, Shanghai, China
[2]Key Laboratory of Radio Astronomy, Chinese Academy of Sciences, 210008 Nanjing, China
[3]University of Chinese Academy of Sciences, Beijing 100049, China

Abstract. It is still a mystery why only a small fraction of quasars contain relativistic jets. A strong magnetic field is a necessary ingredient for jet formation. Gas falls from the Bondi radius $R_{\rm B}$ nearly freely to the circularization radius $R_{\rm c}$, and a thin accretion disk is formed within $R_{\rm c}$. We suggest that the external weak magnetic field threading interstellar medium is substantially enhanced in this region, and the magnetic field at $R_{\rm c}$ can be sufficiently strong to drive outflows from the disk if the angular velocity of the gas is low at $R_{\rm B}$. In this case, the magnetic field is efficiently dragged in the disk, because most angular momentum of the disk is removed by the outflows that lead to a significantly high radial velocity. The strong magnetic field formed in this way may accelerate jets in the region near the black hole, either by the Blandford-Payne or/and Blandford-Znajek mechanisms. If the angular velocity of the circumnuclear gas is low, the field advection in the thin disk is inefficient, and it will appear as a radio-quiet (RQ) quasar.

Keywords. accretion, accretion disks, magnetic fields, galaxies: active, galaxies: jets

1. Introduction

Only a small fraction (about one tenth) of quasars exhibit relativistic jets, while their appearance is quite similar to the radio-quiet (RQ) counterparts in almost all wavebands except radio wavebands. The Blandford-Znajek (BZ) and Blandford-Payne (BP) mechanisms are the most favored models of jet formation in quasars (Blandford & Znajek 1977; Blandford & Payne 1982). Although we do not know which mechanism is dominant in the jet formation of quasars, a strong large-scale magnetic field near the black hole is necessary for the jet formation in radio-loud (RL) quasars either with the BP or BZ mechanisms. Strong magnetic field is probably a crucial factor causing the radio dichotomy in quasars (Sikora & Begelman 2013), though the black hole spin may also play an important role in the BZ scenario.

The external weak large-scale poloidal field (e.g., the field threading the interstellar medium) is suggested to be dragged inward by the accretion plasma, which is balanced by the magnetic diffusion in the disk for a steady field case. It was found that the field can hardly be dragged inwards by a thin disk ($H/R \ll 1$) because of its small radial velocity (Lubow *et al.* 1994). In order to solve this problem, Cao & Spruit (2013) suggested that the most angular momentum of the gas in the thin disk can be removed by the magnetically driven outflows, and the radial velocity of the disk is significantly increased. In this case, the external field can be advected efficiently in the thin disk with magnetic outflows.

2. Model

The gas falls almost freely toward the black hole, if the angular momentum of the gas is significantly lower than the Keplerian value at the Bondi radius. The angular momentum of the gas is roughly conserved until it approaches the circularization radius (e.g., Melia *et al.* 2001). The external weak magnetic field is dragged in by the gas in the region between the Bondi radius and the circularization radius, due to the field flux freezing, and the field threading the gas is substantially enhanced at the circularization radius. An optically thick, geometrically thin accretion disk is formed with the circularization radius, if the gas is supplied at an appropriate rate.

The radial velocity of the disk is significantly increased due to the presence of the outflows, if the angular momentum of the disk is removed predominantly by the magnetically driven outflows (see Cao & Spruit 2013). The field in such a disk with outflows is therefore efficiently advected toward the black hole. The strong magnetic field formed in this way may accelerate relativistic jets in the inner region of the disk near the black hole, either by the BP or/and BZ mechanisms. The object containing such an accretion disk with magnetic outflows may appear as an RL quasar, otherwise, it may be an RQ quasar.

The radial velocity of an accretion disk with magnetic outflows is assumed to be $1+f_{\rm m}$ times of that of a conventional turbulent disk. There is a minimal field strength required to maintain such a disk-outflow system, and we can therefore derive a necessary condition for RL quasars as

$$\tilde{\Omega}_{\rm B} < 0.336\left(1+\frac{1}{f_{\rm m}}\right)^{1/3} \times \left(\frac{T_{\rm B}}{\rm keV}\right)^{-5/6} \left(\frac{B_{\rm ext}}{\rm mGauss}\right)^{2/3} \left(\frac{M_{\rm bh}}{10^8 M_\odot}\right)^{1/3} \dot{m}^{-1/3}, \qquad (2.1)$$

where $T_{\rm B}$ is the temperature of the gas at the Bondi radius, and $\dot{m}$ is the mass accretion rate in unit of Eddington value (see Cao 2016 for the detailed calculations).

The local structure of a disk accreting at a mass rate $\dot{M}$ with magnetically driven outflows, is the same as a standard thin disk accreting at $\dot{M}/(1+f_{\rm m})$. In the region beyond a critical radius $R_{\rm T}$, the disk may become a clumpy disk due to the gravitational instability. The accretion of the clumpy disk may be driven by the collisions of the clumpies. The large scale magnetic field threading the clumps may not be able to survive through repeatedly collisions. A clumpy disk region would be an obstacle for accumulation of external magnetic field. To avoid such a clumpy region in the disk, one requires $R_{\rm c} < R_{\rm T}$, i.e.,

$$\tilde{\Omega}_{\rm B} \leq \left(\frac{R_{\rm T}}{R_{\rm B}}\right)^{1/2} = 0.0017 r_{\rm T}^{1/2}\left(\frac{kT_{\rm B}}{1\ {\rm keV}}\right)^{1/2}, \qquad (2.2)$$

where $r_{\rm T} = R_{\rm T}/R_{\rm S}$, and $R_{\rm S} = 2GM_{\rm bh}/c^2$. This is the second condition for efficient field advection in an accretion disk with magnetic outflows Cao 2016).

3. Results and discussion

The external field can be efficiently advected in an accretion disk with magnetic outflows, only if the gas has angular velocity lower than both of these two critical values at the Bondi radius. In Figure 1, we plot the results varying with the black hole mass for different gas temperatures and external field strengths. It is found that the angular velocity $\Omega_{\rm B}$ is always required to be much lower than the Keplerian velocity, in order to form a strong field in the inner region of the disk. The results vary little with the mass accretion rate. The model parameter $f_{\rm m} = 5$ is adopted in all the calculations. It guarantees the field to be significantly inclined to the disk surface, which is required for launching outflows from the disk.

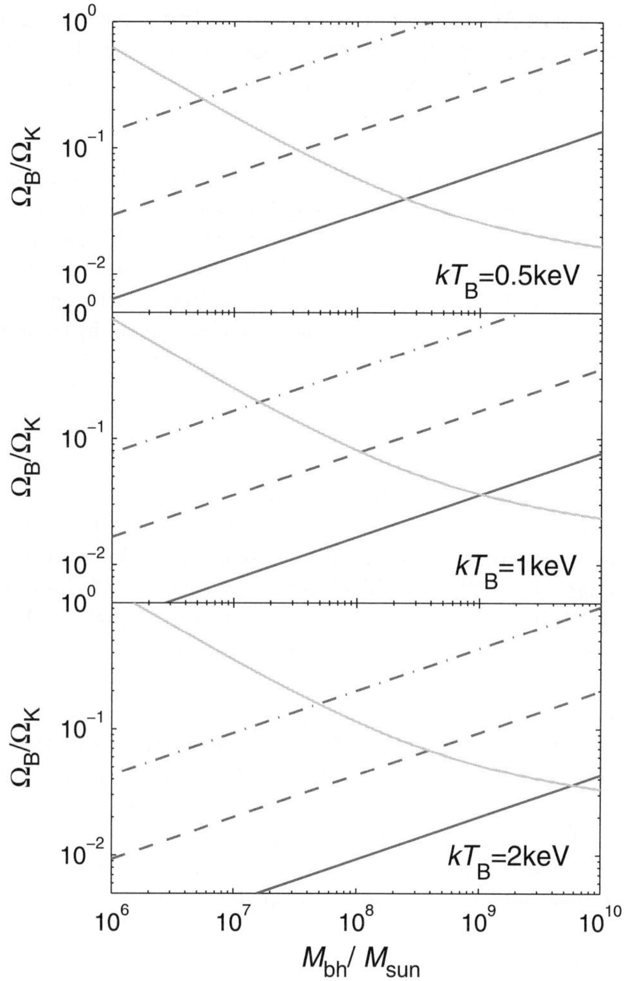

Figure 1. The critical angular velocities of the gas at the Bondi radius as functions of the black hole mass. The external field can be efficiently dragged in by an accretion disk with magnetic outflows only if the angular velocity of the gas is lower than both of the two critical values. All the calculations are carried out with $\dot{m} = 1$. The red lines correspond to the first condition (see Equation 2.1), and the green lines to the second condition for efficient field advection in the accretion disk with magnetic outflows (see Equation 2.2). The viscosity parameter $\alpha = 1$ is adopted in the calculations. The different types of the red lines indicate the results with different values of the external field strength at the Bondi radius (solid: 0.01 mGauss, dashed: 0.1 mGauss, and dash-dotted: 1 mGauss).

Typical magnetic field strengths of galaxy cluster atmospheres are at the order of $\sim \mu$Gauss, and the field strengths could be stronger for the ISM in galaxies. In the central region of our galaxy, the field strength of the gas can be as high as $\sim$mGauss. A weaker magnetic field requires the gas with a lower angular momentum at the Bondi radius for RL quasars, because the circularization radius decreases with decreasing angular momentum of the gas, which increases the amplification of the external magnetic field in the region between the Bondi radius and the circularization radius. Using *Chandra* X-ray observations of nine nearby radio galaxies, Allen *et al.* (2006) measured the temperatures of the gas at the Bondi radius of these galaxies, which are in the range of $\sim 0.5 - 1.3$ keV. We find that our main results are insensitive to the gas temperature (see Figure 1).

The radio properties of AGNs are found to be linked to their host galaxies. It is found that the core galaxies invariably host a RL nucleus (Capetti & Balmaverde 2005). The core galaxies are slowly rotating and have boxy isophotes, while the power-law galaxies rotate rapidly and are disky. This supports the model suggested in this paper that the source accreting the gas with a low angular velocity may preferentially appear as an RL quasar. The mass accretion rate of an RL quasar is much higher than a conventional accretion disk with the same luminosity in an RQ quasar, because most of the gravitational energy released is carried away by the outflows (e.g., Li 2014). This implies that the mass growth of the black holes in RL AGNs is much faster than that in RQ AGNs with the same luminosity. This is consistent with the fact that the massive black holes in RL quasars are systematically a few times heavier than those in their RQ counterparts (Laor 2000).

4. Summary

The circumnuclear gas with a low angular velocity falls nearly freely from the Bondi radius to form an accretion disk within the circularization radius. The external magnetic field threading the gas is strongly amplified in this region due to the field flux freezing. For the gas with an angular velocity lower than a critical value, the field in the disk is strong enough to drive magnetic outflows, which carries away most of the angular momentum of the disk. This strongly increases the radial velocity of the disk, and therefore the field can be efficiently dragged inwards by the disk. Relativistic jets may be driven by such a large scale magnetic field through either the BP or BZ mechanism. In this case, the object may appear as an RL quasar.

If the angular velocity of the circumnuclei gas is larger than a critical value, the circularization radius (i.e., the outer radius of the disk) becomes larger, and the field cannot be amplified to a strong field in the disk to accelerate outflows. Thus, a conventional turbulence driven thin disk is formed within the circularization radius, and the advection of the field in the disk is rather inefficient. In this case, no relativistic jet is formed in the inner region of the disk, which corresponds to an RL quasar.

Acknowledgements

This work is supported by the NSFC (grant 11773050), and the CAS grant (QYZDJ-SSW-SYS023).

References

Allen, S. W., Dunn, R. J. H., Fabian, A. C., Taylor, G. B., & Reynolds, C. S. 2006, *MNRAS*, 372, 21
Blandford, R. D., & Payne, D. G. 1982, *MNRAS*, 199, 883
Blandford, R. D., & Znajek, R. L. 1977, *MNRAS*, 179, 433
Bu, D.-F., & Yuan, F. 2014, *MNRAS*, 442, 917
Cao, X. 2016, *ApJ*, 833, 30
Cao, X., & Spruit, H. C. 2013, *ApJ*, 765, 149
Capetti, A., & Balmaverde, B. 2005, *A&A*, 440, 73
Laor, A. 2000, *ApJ*, 543, L111
Li, S.-L. 2014, *ApJ*, 788, 71
Lubow, S. H., Papaloizou, J. C. B., & Pringle, J. E. 1994, *MNRAS*, 267, 235
Melia, F., Liu, S., & Coker, R. 2001, *ApJ*, 553, 146
Sikora, M., & Begelman, M. C. 2013, *ApJL*, 764, L24

Perseus in Sicily: From Black Hole to Cluster Outskirts
Proceedings IAU Symposium No. 342, 2018
K. Asada, E. de Gouveia Dal Pino, M. Giroletti,
H. Nagai & R. Nemmen, eds.

doi:10.1017/S1743921318003848

Ratio of kinetic-to-bolometric luminosity at the "cold" disk accretion onto black holes

Sergey Bogovalov

National Research Nuclear University (MEPHI), Kashirskoje shosse, 31, Moscow, Russia
email: svbogovalov@mephi.ru

Abstract. In galactic nuclei (AGN), the kinetic energy flux of the jet may exceed the bolometric luminosity of the disk a few orders of magnitude. At the "cold" accretion the radiation from the disk is suppressed because the wind from the disk carries out almost all the angular momentum and the gravitational energy of the accreted material. We calculate an unavoidable radiation from such a disk and the ratio of the kinetic-to-bolometric luminosity from a super massive black hole in framework of the paradigm of the optically thick α-disk of Shakura & Sunyaev. The results confirm that the gravitational energy of the accreted material can be the only source of energy in AGNs.

Keywords. accretion disks, magnetohydrodynamics, galaxies: jets.

1. Introduction

The observations of recent years show that the power of kinetic energy of the jets in some AGN exceeds the bolometric luminosity of the disk (Körding *et al.* 2008; Ma *et al.* 2008; López-Corredoira & Perucho 2012; Daly 2016; Daly *et al.* 2016; Fernandes *et al.* 2011; Punsly 2011; Ghisellini *et al.* 2014). In this case the question about the source of the energy of the jets becomes the central one for understanding of the processes at the accretion onto SMBH (McNamara *et al.* 2011). Presently, the rotational energy of a black hole is considered as the most likely source of energy which is transformed in to the energy of jets due to the so-called mechanism of Blandford & Znajek (1977). This mechanism can provide the energy flux in the jet $\approx 3\dot{M}c^2$, where $\dot{M}$ is the accretion rate (Tchekhovskoy *et al.* 2011; Barkov & Khangulyan 2012).

It was firstly pointed out by Pelletier & Pudritz (1992) and explored by the Grenoble group (Ferreira & Pelletier 1995) that the magnetized wind from the disk can carry out a significant fraction of the angular momentum of the accreted matter rather than the viscous stresses. We call this regime of accretion as "cold" accretion because the disk appears radiationally inefficient. Starting with our first work (Bogovalov & Kelner 2010) we stressed that the "cold" accretion can be applied to the solution of the problem of high ratio of the kinetic luminosity of the jets. In this work we estimate the expected ratio of the kinetic-to-bolometric luminosity to verify can the model of the cold accretion be consistent with the current observations.

2. Physics of the "cold" accretion

The equation of conservation of the angular momentum in framework of the α-disks paradigm Shakura & Sunyaev (1973) has the following form:

$$\dot{M}\frac{\partial r V_k}{\partial r} + \frac{\partial}{\partial r}4\pi r^2 t_{r\varphi} h + r^2 < B_\varphi B_z > |_{wind} = 0, \tag{2.1}$$

where $t_{r\varphi}$ is the tangential stress, V_k is the Keplerian velocity of rotation, $\dot{M}$ is the accretion rate, r is the cylindrical radius and h is the semi-height of the disk. B_φ and B_z are toroidal and z-components of the magnetic field at the base of the wind. The brackets <> mean averaging over time.

The tangential stress in the disk $t_{r\varphi}$ is defined by the turbulent motion and by the magnetic field. According to Shakura & Sunyaev (1973)

$$t_{r\varphi} = -\alpha \rho v_s^2. \tag{2.2}$$

where v_s is the sound velocity, ρ is the matter density and α is the parameter connecting the pressure and tangential stress in the disk.

As it was pointed out by Pelletier & Pudritz (1992), the momentum loss due to the wind will dominate the losses caused by the viscous stresses provided that

$$4\pi r t_{r\varphi} h \ll r^2 < B_\varphi B_z > |_{wind}. \tag{2.3}$$

This is the condition for the "cold" accretion. In the opposite case we have the standard Shakura & Sunyaev (1973) version of the disk accretion. The heat production equals

$$Q = t_{r\varphi} h r \frac{\partial \Omega}{\partial r}. \tag{2.4}$$

The angular momentum carried out by the wind at the "cold" accretion is

$$r^2 < B_\varphi B_z > |_{wind} = 0.5 \dot{M} V_k. \tag{2.5}$$

To connect this torque at the base of the wind with $t_{r\varphi}$ we introduce a parameter

$$\theta = \frac{4\pi t_{r\varphi}}{< B_\varphi B_z > |_{wind}}. \tag{2.6}$$

Below we assume that $\theta \geq 1$ but should satisfy to the condition of "cold" accretion following from eq.(2.3)

$$\frac{\theta h}{r} \ll 1. \tag{2.7}$$

For the geometrically thin disks with $h/r \ll 1$ the "cold" disk accretion can take place even for $\theta > 1$.

3. Thermal radiation from the disk

We consider the case when the gas pressure dominates the radiation pressure. According to Shakura & Sunyaev (1973) the heat conductivity of the disk is defined by the transport of radiation and $\varepsilon = \frac{3}{4}\frac{Q\sigma u_0}{c}$, where $\sigma = 0.4$ cm^2/g is the Thomson opacity and $u_0 = 2\rho h$, $\varepsilon = bT^4$. Then the rate of heating of the disk equals

$$Q = \frac{3\theta \dot{M} V_k v_s}{16\pi r^2}. \tag{3.1}$$

We used here that $h = v_s/\Omega$. The sound velocity is defined as $v_s^2 = kT/m_p$, where m_p is the proton mass. Let us express $\dot{M} = \dot{m}\dot{M}_{crit}$, the radius r in $r = (3r_g)x$ and the mass M in the solar masses $M = mM_\odot$. When the Thomson scattering dominates free-free absorption we obtain

$$Q = 0.75 \cdot 10^{23} (\theta \dot{m})^{5/4} m^{-9/8} x^{-47/16} \alpha^{-1/8}, \quad \mathrm{erg/s/cm^2}. \tag{3.2}$$

$$L_{bol} = 0.84 \cdot 10^{36} (\theta \dot{m})^{5/4} m^{7/8} \alpha^{-1/8}, \quad \mathrm{erg/s}. \tag{3.3}$$

$$T = 2.5 \cdot 10^7 (\theta \dot{m})^{1/2} \alpha^{-1/4} x^{-7/8} m^{-1/4}, \quad \mathrm{K} \tag{3.4}$$

$$\rho = 0.6 (\theta \dot{m})^{1/2} m^{-3/4} x^{-13/8} \alpha^{-3/4}, \quad \mathrm{g/cm^3}. \tag{3.5}$$

The kinetic luminosity of the jets equals to the total energy release at accretion $L_{kin} = 1.4 \cdot 10^{38} m\dot{m}$, erg/s. Then, the ratio of the kinetic luminosity over the bolometric luminosity equals

$$L_{kin}/L_{bol} = 168(m\alpha)^{1/8}\dot{m}^{-1/4}\theta^{-5/4}. \tag{3.6}$$

The aspect ratio of the disk, true optical depth $\tau^* = \sqrt{\sigma \cdot \sigma_{ff}} \cdot u_0$ and surface density are

$$h/r = 3.7 \cdot 10^{-3}(\dot{m}\theta)^{1/4}x^{1/16}(\alpha m)^{-1/8}. \tag{3.7}$$

$$\tau^* = 51(\theta\dot{m})^{1/8}m^{3/16}x^{5/32}\alpha^{-13/16}. \tag{3.8}$$

$$T_s = 6 \cdot 10^6(\theta\dot{m})^{5/16}m^{-9/32}x^{-47/64}\alpha^{-1/32} \text{ K}. \tag{3.9}$$

where $\sigma_{ff} = 0.11 \cdot T^{-7/2}n$, cm^2/g is the free-free opacity of the disk.

At the condition $4.6 \cdot 10^{-3}(\alpha m)^{1/10}x^{23/20}(\theta\dot{m})^{-1} > 1$ the free-free absorption dominates the Thomson scattering opacity and parameters of the disk take a form

$$T = 10^7(\theta\dot{m})^{6/17}x^{-12/17}(\alpha \cdot m)^{-4/17} \text{ K}. \tag{3.10}$$

$$L_{bol} = 0.6 \cdot 10^{36}(\theta\dot{m})^{20/17}m^{15/17}\alpha^{-2/17} \text{ erg/s}. \tag{3.11}$$

$$\tau = 93(\theta\dot{m})^{4/17}m^{3/17}x^{1/34}\alpha^{-14/17}, \tag{3.12}$$

$$\rho = 1.2(\theta\dot{m})^{11/17}(\alpha m)^{-13/17} \cdot x^{-61/34} \text{ g/cm}^3, \tag{3.13}$$

$$h/r = 2.5 \cdot 10^{-3}x^{5/34}(\theta\dot{m})^{3/17}(\alpha m)^{-2/17}. \tag{3.14}$$

$$T_s = 5.5 \cdot 10^6(\theta\dot{m})^{5/17}m^{-19/68}x^{-97/136}\alpha^{-1/34} \text{ K}. \tag{3.15}$$

$$L_{kin}/L_{bol} = 228(\alpha m)^{2/17}\dot{m}^{-3/17}\theta^{-20/17} \tag{3.16}$$

4. Comparison with the fundamental plane of the black holes

The fundamental plane (FP) encapsulates the relationship between the compact radio luminosity, X-ray luminosity, and the black hole mass and provides a good description of the data over a very large range of black hole mass. In the work Daly *et al.* (2016), the position of objects of different masses in the coordinates L_{kin}/L_{bol} and L_{bol}/L_{Edd} has been collected in one FP. If this is true, the FP can be used to extract information about the dependence of θ on $\dot{m}$ and m.

Obviously, at the constant θ the theoretical predictions are not consistent with observations. θ must depend on $\dot{m}$ and m. The most natural option is to assume that θ depends on $\dot{m}$ as a power law

$$\theta = D\dot{m}^{\gamma}. \tag{4.1}$$

Our calculations are in agreement with the FP if in the Thomson and free-free regimes

$$\theta_T = 5 \cdot 10^3\dot{m}^{3/4}(\alpha m)^{1/10} \tag{4.2}$$

$$\theta_{ff} = 11.4 \cdot 10^3\dot{m}^{0.86}(\alpha m)^{1/10}. \tag{4.3}$$

For details see (Bogovalov 2018).

5. Discussion

The cold accretion onto SMBH transforms practically all the gravitational energy into the kinetic energy of jets. On this reason the disk appears radiationally inefficient. This strongly contrasts with radiationally inefficient ADAF models Esin *et al.* (1997). In the last case AGN appears very inefficient machine which transforms almost all the gravitational energy into mass and rotational energy of the SMBH. ADAF model needs an additional source of energy which is believed can be the rotational energy of SMBH.

We show that the "cold" accretion is consistent with the observations presented by FP provided that θ depends on $\dot{m}$ and m in accordance with equations (4.3). θ basically reflects the ratio of the magnetic field pressure inside and at the surface of the disk. The estimated values of θ agree with the recent calculations of the disk structure (Salvesen *et al.* 2016). The obtained ratio of kinetic-to-bolometric luminosity is close but exceeds the value obtained in (Li 2014).

At small accretion rates $\dot{m} < 10^{-2}$, the estimated value of θ is located in the region well below the line were the Shakura-Sunyaev model is valid. The magnetic pressure inside the disk appears less than the magnetic pressure estimated in the model Shakura & Sunyaev (1973). Therefore, it is quite reasonable to assume that at relatively low rates of accretion, $\dot{m} < 10^{-2}$, the accretion occurs predominantly in the regime of the "cold" accretion. At higher values of $\dot{m} > 0.1$ the accretion occurs in the regime of Shakura & Sunyaev. The transition between the regimes takes place at the value of $\dot{m}$ between 0.01 and 0.1 which well agree with location of the transition from very bright to very dim disks around SMBH with powerful outflow deduced in Churazov *et al.* (2005).

Acknowledgement

The work was supported by Russian science foundation, project N 16-12-10443.

References

Barkov, M. V. & Khangulyan, D. V., 2012. *MNRAS*, vol. 421, 1351
Blandford, R. D. & Znajek, R. L., 1977. *MNRAS*, vol. 179, 433
Bogovalov, S. V., 2018. ArXiv e-prints
Bogovalov, S. V. & Kelner, S. R., 2010. International Journal of Modern Physics D, vol. 19, 339
Churazov, E., Sazonov, S., Sunyaev, R. *et al.*, 2005. *MNRAS*, vol. 363, L91
Daly, R. A., 2016. *MNRAS*, vol. 458, L24
Daly, R. A., Stout, D. A., & Mysliwiec, J. N., 2016. ArXiv e-prints
Esin, A. A., McClintock, J. E., & Narayan, R., 1997. *ApJ*, vol. 489, 865
Fernandes, C. A. C., Jarvis, M. J., Rawlings, S. *et al.*, 2011. *MNRAS*, vol. 411, 1909
Ferreira, J. & Pelletier, G., 1995. *A&A*, vol. 295, 807
Ghisellini, G., Tavecchio, F., Maraschi, L. *et al.*, 2014. *Nature*, vol. 515, 376
Körding, E. G., Jester, S., & Fender, R., 2008. *MNRAS*, vol. 383, 277
Li, S.-L., 2014. *ApJ*, vol. 788, 71
López-Corredoira, M. & Perucho, M., 2012. *A&A*, vol. 544, A56
Ma, M.-L., Cao, X.-W., Jiang, D.-R. *et al.*, 2008. Chinese J. Astron. Astrophys., vol. 8, 39
McNamara, B. R., Rohanizadegan, M., & Nulsen, P. E. J., 2011. *ApJ*, vol. 727, 39
Pelletier, G. & Pudritz, R. E., 1992. *ApJ*, vol. 394, 117
Punsly, B., 2011. *ApJ*, vol. 728, L17
Salvesen, G., Simon, J. B., Armitage, P. J. *et al.*, 2016. *MNRAS*, vol. 457, 857
Shakura, N. I. & Sunyaev, R. A., 1973. *A&A*, vol. 24, 337
Tchekhovskoy, A., Narayan, R., & McKinney, J. C., 2011. *MNRAS*, vol. 418, L79

Perseus in Sicily: From Black Hole to Cluster Outskirts
Proceedings IAU Symposium No. 342, 2018
K. Asada, E. de Gouveia Dal Pino, M. Giroletti,
H. Nagai & R. Nemmen, eds.

doi:10.1017/S1743921318008050

Emission modelling of hydrodynamic AGN jet simulations

Izak P. van der Westhuizen, Brian van Soelen and Petrus J. Meintjes

Department of Physics, University of the Free State,
PO Box 339, Bloemfontein 9300, South Africa,
email: vanderwesthuizenip@ufs.ac.za

Abstract. Radio-loud Active Galactic Nuclei (AGN) produce relativistic jets that can be modelled with relativistic hydrodynamic (RHD) simulations. In this study we present two such simulations of jets, used to investigate the parameters required to reproduce structures consistent with both FR I and FRII jets. In the first simulation a Lorentz factor of 10 and supersonic flow of Mach 30 were chosen, while for the second simulation a Lorentz factor of 1.0014 with a supersonic flow of Mach 4 was used. Over similar distances scales the first case shows a well collimated beam with a strong shock at the interface between the jet and ambient medium while the second case shows a less stable beam and a larger cocoon. To determine whether the simulated physical structures are consistent with the observed FR I/II jets, the synchrotron emission has been calculated to produce radio maps at a single frequency of 1.5 GHz.

Keywords. Hydrodynamic, Active Galactic Nuclei, Relativistic jets

1. Introduction

Radio-loud AGN are associated with jet-like structures which originate from the nucleus and can cover hundreds of kiloparsec in distance. These jet structures consist of collimated outflows of plasma with relativistic velocities often on the order of $\Gamma = 10$ (Urry & Padovani 1995). These jets show a complex morphology, which is dependent of both the external environment as well as intrinsic properties such as the kinetic luminousity. In general they are divided into two classes defined as FR I and FR II type jets. FR I type radio jets show the brightest emission close to the nucleus, with the jet brightness generally decreasing with distance. FR II type radio jets show collimated beams which terminate in radio lobes. The brightest components in these jets are observed as hot spots at the termination point of the jets (Fanaroff & Riley 1974; see also Torresi 2018, these Proceedings).

With the advance in computer technology numeric hydrodynamic simulations have become a powerful tool that can be used to investigate the dynamics of these jets. This is especially useful in the search for parameters which can distinguish the two different morphological classes from one another. One of the first comprehensive studies regarding the effects of injection parameters on the structure of kiloparsec relativistic jets was conducted by Martí *et al.* (1997). Several studies since have investigated various features in AGN jets such as the effects of jet composition (Scheck *et al.* 2002) and the formation of superluminal components (Aloy *et al.* 2003). Recent studies such as Rossi *et al.* (2008) and Massaglia *et al.* (2016) investigated the necessary parameters required to form FR I type jets. Both studies suggested that the deceleration of the jet beams is required to obtain physical structures consistent with FR I jets. Massaglia *et al.* (2016) estimated that the division in the FR I/FR II dichotomy occurs at a kinetic injection luminousity of 10^{43} erg.s^{-1}.

Table 1. Jet injection parameters.

Case	Lorentz factor	Density ratio	Mach number
HL	10	10^{-5}	30
LL	1.0014	10^{-4}	4

To fully determine the class of an AGN jet one must not only compare the physical structure of the jet, but also the resulting synchrotron emission. In this paper we present an estimate for the synchrotron emission of two relativistic hydrodynamic (RHD) simulations; a high Lorentz factor injection case and a low Lorentz factor injection case. The jet injection parameters were chosen to compare both the physical structure as well as the resulting emission of the two simulations.

2. Simulation setup

Two simulations were set up for this study, one to represent a decelerated beam with a low injection rate (case LL) and another with a high injection rate (case HL). The simulations were set up and evolved with time using the PLUTO code (Mignone *et al.* 2007). For the simulations a Cartesian grid was used containing $512 \times 512 \times 1024$ cells. A nozzle, with a radius of 8 cells, was created on the lower z boundary to inject jet material into the domain. The jet radius spans 100 pc in physical units, therefore the simulation domain covers an area of $6.4 \times 6.4 \times 12.8$ kpc of the outer edge of the jet.

A spherically stratified background medium was assigned to the grid with a density gradient given by,

$$\rho(r) = \frac{\rho_{am}}{1 - (\frac{r}{40})^2}, \tag{2.1}$$

where ρ_{am} is the ambient density at the injection point and r is the distance from the injection point in beam radii.

The injection parameters of the two models are listed in Table 1. The parameters for case LL were chosen based on those shown in Massaglia *et al.* (2016) to produce FR I type structure, while similar parameters with a high Lorentz factor were chosen for case HL. In both cases a pressure matched model was used. Both simulations were evolved using parabolic interpolation and characteristic tracing time stepping.

3. Synchrotron emission estimates

To compare the physical structures of the numerical simulations to the observed Fanaroff classes, synthetic intensity maps were created of the synchrotron emission at 1500 MHz. It was assumed that the synchrotron radiation is predominantly emitted by non-thermal electrons in a single power-law distribution given by,

$$n(\gamma) = n_0 \gamma^{-p}, \tag{3.2}$$

in which $n(\gamma)$ is the number density of non thermal electrons, γ is the Lorentz factor, $p = 2.1$ is the photon index and n_0 is a normalization factor. The non-thermal spectrum is normalized to the internal energy and number density of the fluid as was done in previous studies such as Gomez *et al.* (1995).

Based on the power-law spectrum emission and absorption coefficients were calculated for each cell in the simulation using the δ-approximation (see e.g. Böttcher 2012) and integrated along a chosen line of sight. Relativistic effects and light travel time were taken into account during integration, however, cosmological corrections were neglected.

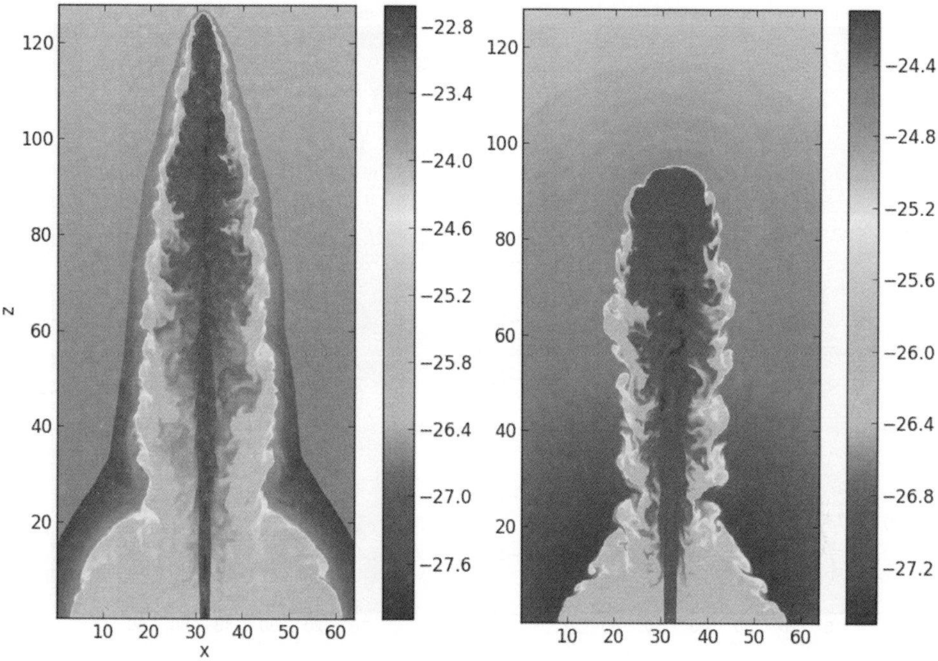

Figure 1. Density slices through the xz-plane of the simulations for Case HL (left) at a time of 9.5×10^4 yr and Case LL (right) at a time of 2.4×10^6 yr. Logarithmic scales are shown for the density plots in units of g.cm^{-3}.

4. Results

The LL simulation was evolved for a time of 2.4×10^6 years. At this stage the propagation of the jet into the ambient medium became negligible, with only the cocoon increasing in size as more matter is injected. A cross-section of the density is shown in Figure 1 and is consistent with the structure obtained by Massaglia *et al.* (2016). In this simulation the jet beam becomes unstable after 40 beam radii, the bulk velocity of the jet fluid is decelerated and the jet breaks down before it can reach the interface with the ambient medium.

The HL model was evolved for a simulation time of 9×10^4 years. At this stage the head of the jet exceeded the upper z boundary and the evolution could no longer be simulated. The high Lorentz factor model shows a well collimated beam which terminates in the interface between the jet and ambient medium, the working surface. This leads to the formation of a shock front at which the kinetic energy of the fluid is converted into internal energy. The cocoon of this model is narrow when compared to AGN radio lobes. It is expected that the jet will propagate over a larger distance into the ambient medium forming a more extended cocoon as more material is injected. A strong bow shock is also observed enveloping the jets structure.

Intensity maps were calculated for the two models at the same time step and are shown in Figure 2. Despite the difference in physical structure both of the simulations show intensity maps more consistent with FR II type radio jets. The LL model shows a region of uniform increased intensity at the jet head while the HL model shows the presence of a more filamentary structure with brighter emission.

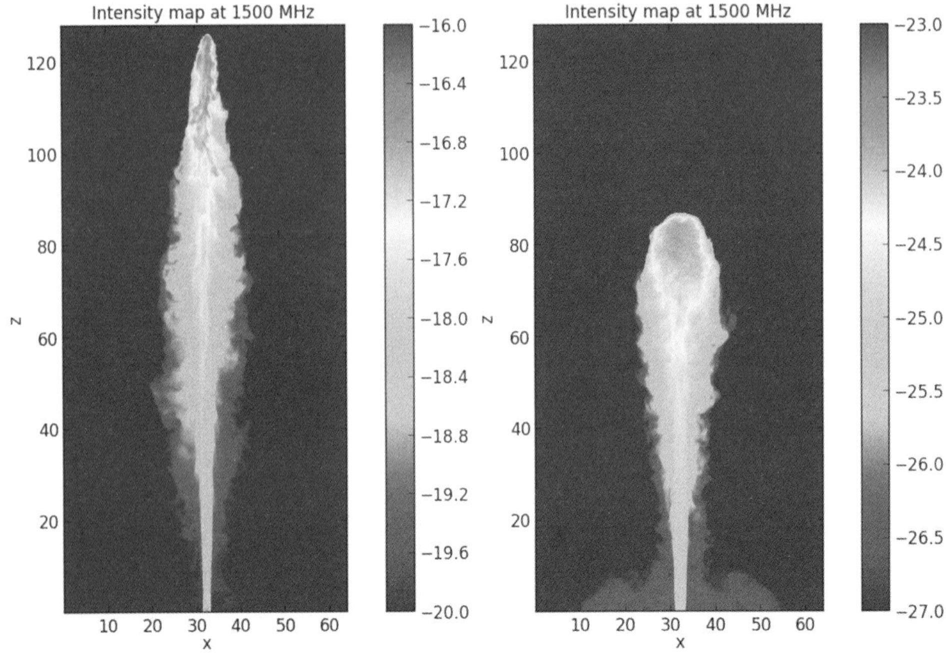

Figure 2. Intensity maps plotted for a viewing angle of 90° relative to the injection direction for Case HL (left) at a time of 9.5×10^4 yr and Case LL (right) at a time of 2.4×10^6 yr. The intensity is show in a logarithmic scale of arbitrary units at a frequency of 1500 MHz.

5. Conclusion

In this paper two RHD simulations of AGN jets are shown. The simulations were set up to study the difference in structure between high and low Lorentz factor jets. The High Lorentz factor (HL) simulation showed physical structures which have been associated with that of FR II type jets and propagated past the domain of the simulation. In the Low Lorentz factor (LL) simulation the propagation of the jet slowed down before it reached the edge of the domain. The beam of the jet also destabilized before it reached the head of the jet. This structure is more consistent with that of FR I type jets.

Intensity maps were calculated for both simulations and showed an increase in emission with distance from the injection nozzle. This is in contrast with the physical structure which was obtained. Additional investigation is ongoing to determine the cause of this effect. The results may suggest that the effects of a decelerated beam alone is not enough to reproduce the FR I type emission structures. It is, however, possible that effects such as radiative cooling may change the observed structure over time. Improved emission modelling is being implemented in the form of Monte-Carlo methods to obtain a self consistent spectrum that will allow us to better investigate the emission structures which form in these simulations.

References

Aloy, M. A., Martí, J. M., Gómez, J. L., Agudo, I., Müller, E., & Ibáñez, J. M., 2003, *ApJl*, 585, 109

Böttcher M., Harris D. E. & Krawczynski H., 2012, *Relativistic Jets from Active Galactic Nuclei*, Wiley-VCH Verlag GmbH & Co. KGaA, Weinheim, Germany, Chapter 3, pp 39-80

Fanaroff B. L. & Riley J. M. 1974, *MNRAS*, 167, 31

Gomez J. L., Marti J. M., Marscher A. P., Ibáñez J. M. & Marcaide J. M., 1995, *ApJ*, 449, 19–21

Martí, J. M., Müller, E., Font, J. A., Ibáñez, J. M. & Marquina 1997, *ApJ*, 479, 151

Massaglia S., Bodo G., Rossi P., Capetti S., & Mignone A., 2016, *A&A*, 596, 10

Mignone A, Bodo G, Massaglia S, Matsakos T, Tesileanu O, Zanni C & Ferrari A 2007, *ApJ*, 170, 228–42

Rossi P, Mignone A, Bodo G, Massaglia S & Ferrari A, 2008 *A&A* 488, 795–806

Scheck, L., Aloy, M. A., Martí, J. M., Gómez, J. L., & Müller, E. 2002, *MNRAS*, 331, 615

Torresi E., 2018, *Proc. IAU Symposium No. 342*, (S342)

Urry, C. M. & Padovani, P. 1995, *PASP*, 107, 803

Perseus in Sicily: From Black Hole to Cluster Outskirts
Proceedings IAU Symposium No. 342, 2018
K. Asada, E. de Gouveia Dal Pino, M. Giroletti,
H. Nagai & R. Nemmen, eds.

doi:10.1017/S1743921318007755

The wind production from black hole hot accretion flow

De-Fu Bu

Key Laboratory for Research in Galaxies and Cosmology, Shanghai Astronomical Observatory,
Chinese Academy of Sciences,
80 Nandan Road, Shanghai 200030, China
email: dfbu@shao.ac.cn

Abstract. Observations of low luminosity active galactic nuclei (LLAGNs) and the hard state of black hole X-ray binaries (BHBs) show that the wind exists. Black hole in LLAGNs and hard state of BHBs accretes gas in hot accretion mode. In this paper, we first use magnetohydrodynamic (MHD) simulations of hot accretion flow around a black hole to study the origin of the wind. We find that the wind is driven by the combination of gradients of gas and magnetic pressure and centrifugal forces. Second, we use simulations with focus on the region around Bondi radius to study whether the wind can be generated outside Bondi radius. In the simulation studying hot accretion flow around Bondi radius, in addition to the black hole gravity, we also take into account the gravity of nuclei stars. We find that the wind can not be generated outside Bondi radius. The absence of the wind is due to the change of gravity potential.

Keywords. accretion, accretion discs, black hole physics, hydrodynamics

1. Introduction

Hot accretion flow is originally proposed and studied analytically by vertically integrated one-dimensional method (e.g., Narayan & Yi 1994). Later on, intensive numerical simulations have been performed to study hot accretion flows. One of the most important findings is that the mass inflow rate is not a constant of radius, it decreases inward (e.g., Stone, Pringle & Begelman 1999).

What is the reason for the inward decrease of mass inflow rate? Two models have been proposed. In the adiabatic inflow-outflow solution (ADIOS; Blandford & Begelman 1999), it is assumed that the wind is present at any radii. The wind takes away mass, therefore, the mass accretion rate decreases inwards. The other model is the convection-dominated accretion flow (CDAF; Narayan *et al.* 2000; Quataert & Gruzinov 2000). This model is based on an assumption that hot accretion flow is convectively unstable. In this model, it is believed that with accretion, more and more gas is locked in convective eddies which are doing convective motions. Gas is locked in convective eddies and can not fall onto the black hole. Therefore, the mass inflow rate decreases inwards.

In Yuan *et al.* (2012), we performed both hydrodynamic (HD) and MHD simulations to study which model is correct. We first analyze the convective stability of an MHD flow. We find that the accretion flow is convectively stable. We also compare various properties of both inflow and outflow. If the CDAF model is correct, the inflow and outflow rates are due to convective turbulent fluctuations. We should expect that the properties of inflow and outflow should roughly be the same. However, we find that the properties of inflow and outflow (including velocities, temperature, Bernoulli parameter) are significantly different. Based on these results, we conclude that the inward decrease of mass inflow rate is because of the wind.

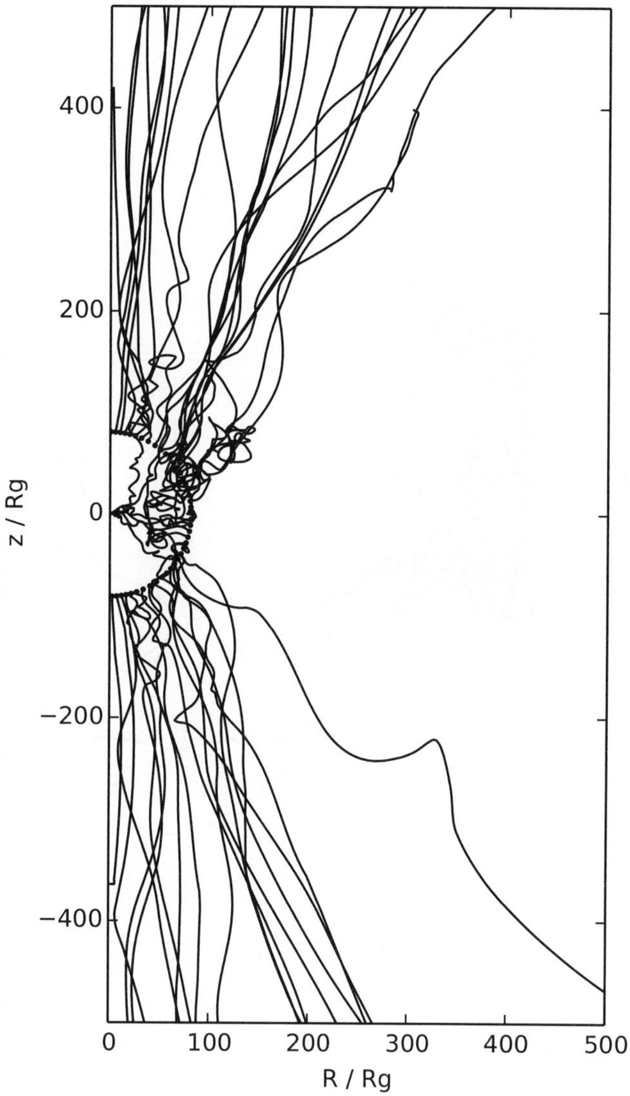

Figure 1. Trajectories of "test particles". The wind is present in the coronal region. The inflow is around the equatorial plane.

2. The wind properties study by trajectory method

In Yuan *et al.* (2012), the conclusion that the wind is present is based on indirect argument. Therefore, it is necessary to show the existence of the wind in a more direct way.

In Yuan *et al.* (2015), we use the"trajectory" method to study the wind. Trajectory is related to the Lagrangian description of fluid. Trajectory is obtained by following the motion of fluid elements at consecutive times. It is different from the streamline. Streamline is associated with the Euler description of fluid, obtained by connecting the velocity vectors of adjacent fluid elements at a given time. Trajectory is only equivalent to streamline for strictly steady motion, which is not the case for accretion flow since it is always turbulent.

Fig. 1 shows the trajectories of some "test particles" originated from 80 Rs (Rs is Schwarzschild radius). In this figure, the black hole is located at $R = z = 0$. It is clear

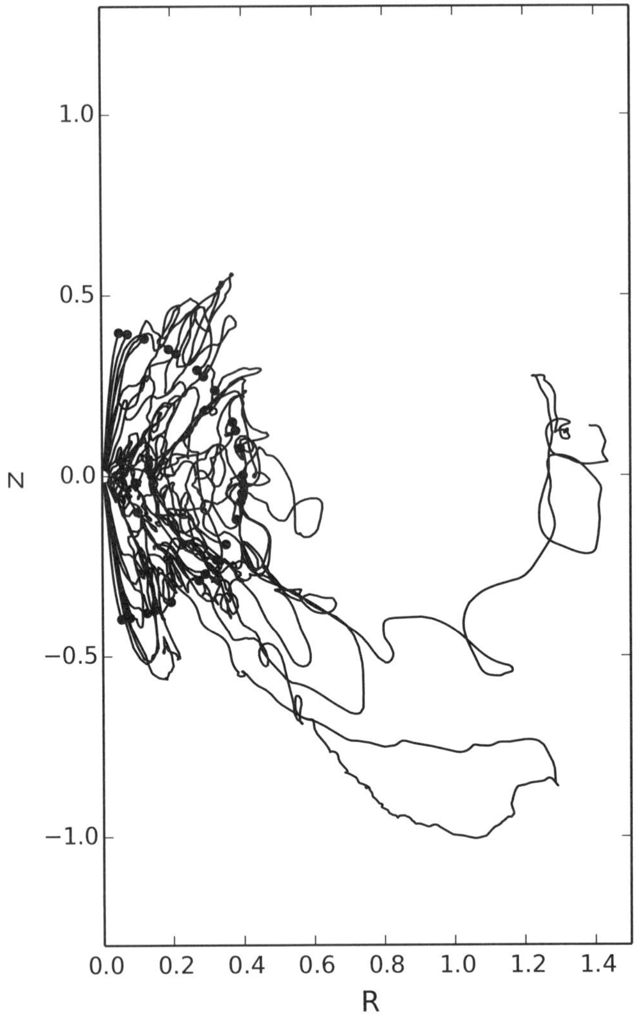

Figure 2. Trajectories of "test particles" for simulations around the Bondi radius. Almost all of the "test particles" are doing turbulent motions. There is no wind.

that the inflow is around the equatorial plane. The wind is present at the coronal region. The wind is driven by the combination of gradients of gas pressure and magnetic pressure and centrifugal forces. The mass flux of the wind can be described by a power law function of radius

$$\dot{M}_{\rm wind}(r) = \dot{M}_{\rm BH}\left(\frac{r}{20Rs}\right) \tag{2.1}$$

3. Can the wind be generated beyond Bondi radius?

The above mentioned works only focus on the region close to the black hole. From Equation (2.1), we know that most of the wind comes from the region of large radius. Then a question is that how large the value of r can be in Equation (2.1). To answer it, we have to study the accretion flow far away from the black hole. In Bu *et al.* (2016a; 2016b), we studied the accretion from around the Bondi radius. Around the Bondi radius, the gravitational potential of nuclear star cluster is important and has been taken into account. Therefore, around the Bondi radius, in our simulations, we have both the black hole and the nuclear star cluster gravities.

The dispersion velocity of the nuclear stars is roughly a constant with radius (e.g., Kormendy & Ho 2013). So, the potential of the star cluster is

$$\psi_{\rm star}(r) = \sigma^2 \ln(r) + C \tag{3.2}$$

σ is the dispersion of velocity. For a 10^8 solar mass black hole, $\sigma \sim 200{\rm kms}^{-1}$. C is a constant. The radial profile of nuclear star potential is much flatter than that of the black hole. Therefore, the total potential (nuclear star's + black hole) is much flatter than that of black hole.

We find that in this case, the mass inflow rate also decreases inward. We use trajectory method to study whether the wind exists. Fig. 2 shows the result. We find that all of the "test particles" are doing turbulent motions. There is no wind. We speculate that the absence of the wind is due to the change of the radial profile of the gravitational potential.

The wind can not be generated "locally" outside the Bondi radius. However, we note that the wind generated inside the Bondi radius can move to the region outside the Bondi radius and interact with the interstellar medium.

4. Implications

There are also some analytical works that study the wind from accretion disks (e.g. Li & Cao 2009; Xie *et al.* 2008; Bu *et al.* 2009; Bu & Mosallanezhad 2018; Abbassi *et al.* 2010; Chen *et al.* 2012; Li & Begelman 2014; Gu 2015; Cao 2016). The topic of the wind is interesting because it is now widely believed that AGN feedback plays an important role in the evolution of their host galaxies (e.g., Ciotti & Ostriker 2001, 2007; Gan *et al.* 2014; Yuan *et al.* 2018). The wind can push away the gas outside an AGN. This can reduce the feeding rate of the central black hole. Also, the star formation rate may be changed.

References

Abbassi, S., Ghanbari, J., & Ghasemnezhad, M. 2010, *MNRAS*, 409, 1113
Bu, D., Yuan, F., & Xie, F. 2009, *MNRAS*, 392, 325
Bu, D., Yuan, F., Gan, Z., & Yang, X. 2016a, *ApJ*, 818, 83
Bu, D., Yuan, F., Gan, Z., & Yang, X. 2016b, *ApJ*, 823, 90
Bu, D., & Mosallanezhad, A. 2018, *A&A* (arXiv:1805.03378)
Blandford, R., & Begelman, M. C. 1999, *MNRAS*, 303, L1
Cao, X. 2016, *ApJ*, 833, 30C
Chen, L., Cao, X., & Bai, J. 2012, *ApJ*, 748, 119C
Ciotti, L., & Ostriker, J. P. 2001, *ApJ*, 551, 131
Ciotti, L., & Ostriker, J. P. 2007, *ApJ*, 665, 1038
Gan, Z., Yuan, F., Ostriker, J. P., Ciotti, L, & Novak, G. S., 2014, *ApJ*, 789, 150
Gu, W. 2015, *ApJ*, 799, 71
Kormendy, J., & Ho, L. C. 2013, *ARA&A*, 51, 511
Li, S., & Cao, X. 2009, *MNRAS*, 400, 1734
Li, S., & Begelman, M. C. 2014, *ApJ*, 786, 6L
Narayan, R., & Yi, I. 1994, *ApJ*, 428, L13
Narayan, R., Igumenshchev I. V., & Abramowicz, M. A. 2000, *ApJ*, 539, 798
Quataert, E, & Gruzinov, A. 2000, *ApJ*, 539, 809
Stone, J. M., Pringle J. E., & Begelman, M. C. 1999, *MNRAS*, 310, 1002
Xie, F., & Yuan, F., 2008, *ApJ*, 681, 499
Yuan, F., Bu, D., & Wu, M. 2012, *ApJ*, 761, 130
Yuan, F., Gan, Z., Narayan, R. Sadowski, A., Bu, D., & Bai, X. 2015, *ApJ*, 804, 101
Yuan, F., Yoon, D., Li, Y., Gan, Z., Ho, L. C., & Guo, F. 2018, *ApJ*, 857, 121

Perseus in Sicily: From Black Hole to Cluster Outskirts
Proceedings IAU Symposium No. 342, 2018
K. Asada, E. de Gouveia Dal Pino, M. Giroletti,
H. Nagai & R. Nemmen, eds.

doi:10.1017/S1743921318005136

Radio-loud AGNs with peculiar shape of hard X-ray spectrum: figuring out the reasons

E. Fedorova, B. I. Hnatyk and V. I. Zhdanov

Astronomical Observatory of National Taras Shevchenko University of Kiev
email: efedorova@ukr.net

Abstract. We analyze the hard X-ray properties of five radio-loud active galactic nuclei with peculiar spectral shape. High-energy exponential cut-offs (HEC) in their hard X-ray 3-500 keV spectra are too high for radio loud AGNs (above 100 keV) or even absent.The probable reason of such visible spectral "peculiarity" can be due to the "jet contamination", i.e we see some mixture of the jet and nuclear emission, but not pure emission of the innermost nucleus. Here we try to estimate the jet and nuclear components of the spectra for a sample of "peculiar" RL AGNs to find out whether these are real or fake features.

Keywords. active galactic nuclei

1. Introduction

One of the prominent physical differences between AGNs of various classes is related with jets and RL/RQ dichotomy. The most known scheme of it referred to as "spin/gap-paradigm" associates jet activity with high values of a central sup er massive black hole (SMBH) spin and with accretion disk (AD) rotation direction (Garofalo *et al.* (2010)). Prograde disks with high SMBH spin ($a \geqslant 0.75$) and retrograde ones correspond to RL AGNs with jets, whereas low SMBH spin in prograde systems ($a \leqslant 0.5$) corresponds to RQ ones with no jets. Thus AD in RQ AGNs fit to geometrically thin, radiatively efficient, steady state Shakura & Sunyaev (1973) disk with zero viscosity at the sonic point and radial infall inside the innermost stable orbit; whereas in RL AGNs they are "torqued" magnetized disks by Agol & Krolik (2000). In such systems SMBH spin energy can dissipate through the sonic point into the disk due to magnetic reconnection between the inner part of AD and SMBH horizon. One of the consequences of this model is that the innermost part of AD in RL AGN is disrupted and smeared away by centrifugally-driven jet outflows, leading to low HEC values in the primary emission spectrum ($\leqslant$100 keV), whereas high HEC values are prescribed to RQ AGN spectrum. However, some objects were found to contradict this in Fedorova *et al.* (2017). The most probable reason of this "peculiarity" can be due to the "jet contamination". Following the results of Miller-Jones *et al.* (2008); Zdziarski *et al.* (2003) it is reasonable to suppose that jet emission spectra have the same photon indices from radio to X-rays. Here we test this possibility and use it to separate the jet emisison from primary nuclear one in RL AGN X-ray spectra.

2. X-rays vs. radio: spectral modeling

Here we try to estimate the jet and nuclear components in the spectra of five RL AGNs to clear out whether these are real or fake features. For this purpose we have modelled the data of XMM-Newton, Suzaku, Swift and INTEGRAL on five AGNS (3C 382, 3C 111, 3C 390.3, Pictor A, 3C 120) in the energy range 3-300 keV comparing them with the

Table 1. Best-fit results for the *INTEGRAL* spectra during the different phases.

Object	Γ_{Planck}	Γ_{prim}	R	$\Gamma_{jet\text{-}X}$	E_C, keV	k
3C 382	1.56±0.08	1.76±0.05	0.68±0.19	1.57±0.07	30±4	line not detected
3C 111	1.6±0.06	1.76±0.04	0.29±0.11	1.63±0.05	112^{+450}_{-65}	83.75
3C 390.3	1.66±0.14	1.41±0.03	1.28±0.17	1.66±0.03	36±4	117.0
Pictor A	2.0±0.02*	1.61±0.38	0.99±0.23	2.00±0.12	⩾400	24.1
3C 120	1.24±0.04	1.91*pm*0.09	⩾ 0.14	1.42±0.22	⩾150	25.9

* it's au upper index of bknpow, the lower one is 1.75±0.04

PLANCK data in 27 GHz - 1 THz range. To test whether is reasonable to use the same photon index for radio and X-ray spectra of a jet, we used two recipe. First one is based on the simple extrapolation of the PLANCK radio spectra of an AGN to X-rays, supposing that the photon index of the spectra of the jet base is the same in radio and in X-rays, i.e. $\Gamma_{Planck} = \Gamma_{jet-X}$. The second one uses the connection between the intensity of Fe-K emission line near 6.4 keV and primary continuum. Afterwards, the results obtained by both methods are compaerd to undestand, how the $\Gamma_{Planck} = \Gamma_{jet-X}$ equality reasonble is. To perform the spectra modelling, we use the model consisting of: reflected power-law comptonized emission of the corona (photon index Γ_{prim}, reflection coefficient R and HEC E_c), power-law jet emission (photon index Γ_{jet-X}), Gaussian Fe-Kα emission line frozen at 6.4 keV (line energy E_{line}, width σ and equivalent width EW). Using the first method, we had frozen the Γ_{jet-X} value to Γ_{Planck} one, obtained with simple or broken (for Pictor A) power-law (po or bknpo) fitting to the PLANCK spectra. In the second method this parameter was free, and we suppoe a linear dependence between 5-7 keV perimary nuclear emission flux and Fe-K line equivalent width $F_{5-7,prim} = k * EW$. The coefficient k can be estimated as $min(F_{5-7,total}/EW)$. Note that this give us possibility to estimate the upper limit on the nuclear contribution to the spectrum only.

3. Conclusions

• The photon indices of jet power-law emission obtained in a free-to-vary mode (second method) are compatible with the values obtained for the PLANCK spectra. This give us possibility to separate the jet contributions to the X-ray spectrum using the extrapolation of the jet base synchrotron emission from the radio to X-rays.

• HECs in the primary nuclear spectra, obtained using both methods, are below 100 keV or compatible with this within the error limits for 3C 382, 3C 111 and 3C 390.3, what means that in fact they do not contradict to the predictions of the spin/gap-paradigm. But in the same time HECS are essentially higher than 100 keV for Pictor A and 3C 120, indicating that thir nuclear spectra are truly peculiar in this sense.

References

Garofalo D., Evans D. A., & Sambruna R. M. *MNRAS*, 2010, Vol. 406, P.975–986
Garofalo D. Advances in Astronomy 2013, Vol. 2013, ID 213105, 11 pages
Shakura N. I., & Sunyaev R. A. *A&A*, 1973, Vol. 24, P.337–355
Agol E., & Krolik J.H. Astron.Journ., 2000, Vol. 528, p.161–170
Fedorova E., Vasylenko A., & Zhdanov V. Bull. KNU. Astronomy, 2017, no. 55, p. 29–34
Zdziarski A., Lubiski P., Gilfanov M., & Revnivtsev M. *MNRAS*, 2003, Vol. 342, Is. 2, P. 355–372
Miller-Jones J. C. A., Migliari S., Fender R. P. *et al.* Astroph. Journ., 2008, Vol. 682, P.1141–1151

Perseus in Sicily: From Black Hole to Cluster Outskirts
Proceedings IAU Symposium No. 342, 2018
K. Asada, E. de Gouveia Dal Pino, M. Giroletti,
H. Nagai & R. Nemmen, eds.

doi:10.1017/S1743921318008475

Expanding Radio Lobe associated with 3C 84

Keiichi Asada[1], Makoto Inoue[1], Hiroshi Nagai[2] and Seiji Kameno[2]

[1]Institute of Astronomy & Astrophysics, Academia Sinica, P.O. Box 23-141, Taipei 10617, Taiwan
email: asada@asiaa.sinica.edu.tw

[2]National Astronomical Observatory of Japan, 2-21-1, Osawa, Tokyo, 181-8588, Japan

Abstract. Following the discovery of the expansion of a young radio lobe associated with the bright radio source 3C 84 in the Seyfert galaxy NGC 1275 using VSOP monitoring observation with duration of 3 years, we further investigated its evolution by utilizing VLBA monitoring data taken by MOJAVE project. As the results, we confirmed the expanding motion of both the southern and northern lobes, and detected side-way expanding motions for the first time. In meantime, the total flux density of the lobes are decreasing, while total flux density associated with the nuclei is significantly increasing. This cooling of the radio lobe can be well-explained with the energy loss through the adiabatic cooling.

Keywords. 3C 84, VLBI, radio lobe, CSO

1. Introduction

Compact Symmetric Objects (CSOs), categorized by Wilkinson *et al.* (1994) and Readhead *et al.* (1996), are compact radio sources with a double lobe structure. Their structures are extremely compact < 1 kpc, compared to ∼ Mpc, which is usual for Large Symmetric Objects (LSOs), but are similar to those of LSOs, such as Cyg A. The compact structure of CSOs has led to a hypothesis that they are in a young stage of their evolution into LSOs. Measurements of kinematic aging Polatidis & Conway (2003) and spectral aging Nagai *et al.* (2006) have confirmed the ages of some CSOs < a few 1000 yrs.

The compact radio source 3C 84, hosted by the Seyfert galaxy NGC 1275, is one of the strongest radio sources and also a nearby (z =0.018) active galactic nucleus (AGN). Assuming that Hubble constant $H_0 = 65\,\mathrm{km\,s^{-1}\,Mpc^{-1}}$, 1 milli-arcsecond (mas) corresponds to 0.4 parsec (pc). Previous VLBI observations revealed that 3C 84 had a complicated radio structure on parsec scales, consisting of a bright core with a flat spectrum and radio lobe like southern and northern structures with a steep spectrum Walker *et al.* (2000). Multi epoch VSOP observations detected expanding motion of the southern radio lobe, and the age of radio lobe is estimated to be 45.7±8.9 years in 2001 by extrapolating the source size back to zero Asada *et al.* (2006). Interestingly, single dish observations of 3C 84 show that the outburst started at 1959 Nesterov *et al.* (1995), which is coincident well with the beginning of the expansion within the errors. In this paper, we report the results of our further investigation on the evolution of the radio lobe for much longer period by utilizing the archival data taken by VLBA through the MOJAVE project.

2. Data

We used data taken by VLBA observed at 15 GHz through MOJAVE project. Total number of the images were 45, those were taken during 1996 Nov. 21 to 2017 Aug. 25. Image were obtained by standard CLEAN method using Difmap.

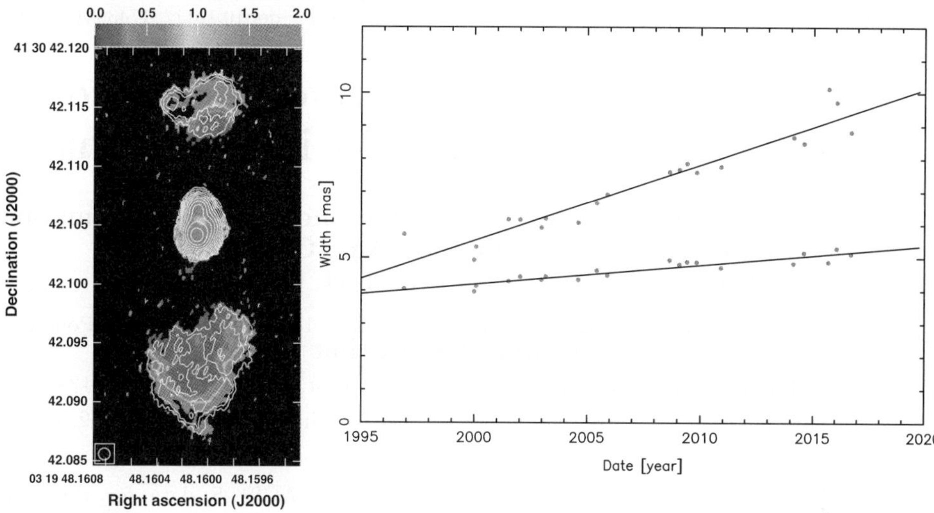

Figure 1. (left) Brightness ratio of the radio lobe at the epochs in 2000 and 2014. It is clearly shown that flux density increases at the edges of the radio lobes, while that at the inside of the radio lobe significantly decreased. Expansion towards side-way (i.e., towards east-west direction) is also clearly seen. (right) Deconvolved size of the southern radio lobe in both elongated (green) and side-way (red) directions as a function of time.

3. Results and Discussion

We show the brightness ration of the radio lobes between at the epochs in 2000 and 2014 in Fig. 1. Brightness at the nuclei region were significantly increased, while that at most of the region associated with the northern and southern radio lobes are decreased. However, the brightness at the edge of the radio lobes were clearly enhanced. It is an evidence that the radio lobe is expanding. We also show the size of the southern lobe as a function of the time Fig. 1. Southern radio lobe was expanded in both elongated and side-way directions, and its expanding speeds were estimated to be 0.3 and 0.08 c, respectively.

References

Asada, K., Kameno, S., & Shen, Z.-Q., *et al.*. 2006, *PASJ*, 58, 261

Nagai, H., Inoue, M., Asada, K., Kameno, S., & Doi, A. 2006, *ApJ*, 648, 148

Nesterov, N. S., Lyuty, V. M., & Valtaoja, E. 1995, *A&A*, 296, 628

Polatidis, A. G., & Conway, J. E. 2003, *PASA*, 20, 69

Walker, R. C., Dhawan, V., Romney, J. D., Kellermann, K. I., & Vermeulen, R. C. 2000, *ApJ*, 530, 233

Wilkinson, P. N., Polatidis, A. G., Readhead, A. C. S., Xu, W., & Pearson, T. J. 1994, *ApJL*, 432, L87

Perseus in Sicily: From Black Hole to Cluster Outskirts
Proceedings IAU Symposium No. 342, 2018
K. Asada, E. de Gouveia Dal Pino, M. Giroletti,
H. Nagai & R. Nemmen, eds.

doi:10.1017/S1743921318007445

Chandra view on the active nucleus of CGCG 292−057: Jet-ISM interactions

K. Balasubramaniam[1], Ł. Stawarz[1], V. Marchenko[1], R. Thimmappa[1], M. Sobolewska[2], A. Siemiginowska[2], C. C. Cheung[3], D. Kozieł-Wierzbowska[1] and M. Jamrozy[1]

[1]Astronomical Observatory, Jagiellonian University, Orla 171, 30-244 Krakow, Poland
email: karthik@oa.uj.edu.pl

[2]Harvard Smithsonian Center for Astrophysics, 60 Garden St, Cambridge, MA 02138, USA

[3]Space Science Division, Naval Research Laboratory, Washington, DC 20375-5352, USA

Abstract. We present the analysis of the 93 ksec *Chandra* ACIS−S data for the galaxy CGCG 292−057 ($z = 0.054$), with complex radio structure indicative of the intermittent jet activity. In order to characterize precisely the spectrum of the unresolved low-luminosity active nucleus in the source, we performed detailed MARX/PSF simulations and studied the radial profile of the source region surface brightness. In this way, we have detected an additional X-ray component extending from a few up to ~ 10 kpc from the unresolved core, which could be associated with the hot gaseous medium compressed and heated (up to 0.9 keV) by the expanding inner lobes of the radio galaxy. We modeled the X-ray spectrum of the unresolved nucleus assuming various emission models, including an absorbed power-law, a power-law plus thermal emission component, and a two-temperature thermal plasma. The best fit was however obtained assuming a power-law emission scattered by a hot ionized gas, giving rise to the 6.7 keV iron line.

Keywords. galaxies: active — X-rays: galaxies — X-rays: Individual (CGCG 292−057)

1. Introduction

CGCG 292−057 ($z = 0.054$) is a particularly interesting example of a post-merger starforming galaxy (Singh *et al.* (2015)), hosting two pair of radio lobes: the outer lobes are believed to have formed during the previous, long-terminated cycle of the jet activity, while the inner coaxial lobes are considered as a manifestation of a new episode of the enhanced jet production in the system, triggered by a sudden increase in the central black hole accretion rate (Kozieł-Wierzbowska *et al.* (2012)). We have imaged CGCG 292−057 in X-rays with the ACIS instrument onboard the *Chandra* X-ray Observatory for a total of 93 ks, and here we highlight some selected results of the data analysis, which will be presented in full in Balasubramaniam et al. (2018, in prep.).

2. Chandra data analysis

The *Chandra* data analysis was carried out with CIAO version 4.9 software, CALDB version 4.7.7, using standard procedure. Spectral fitting was carried out with Sherpa using cstat, due to the low photon statistics. First we performed 100 MARX simulations of the PSF at the position of the active nucleus in CGCG 292−057, and found out that the 2σ fraction of counts radius for the simulated PSFs ranges from 2.5 px up to 5 px. For this reason, for the spectral analysis of the nucleus described below, we adopted the circular source (AGN) extraction region with the radius of $5\,\text{px} \simeq 2.5''$; at the source redshift, this correspond to the physical scale of $\simeq 2.5$ kpc.

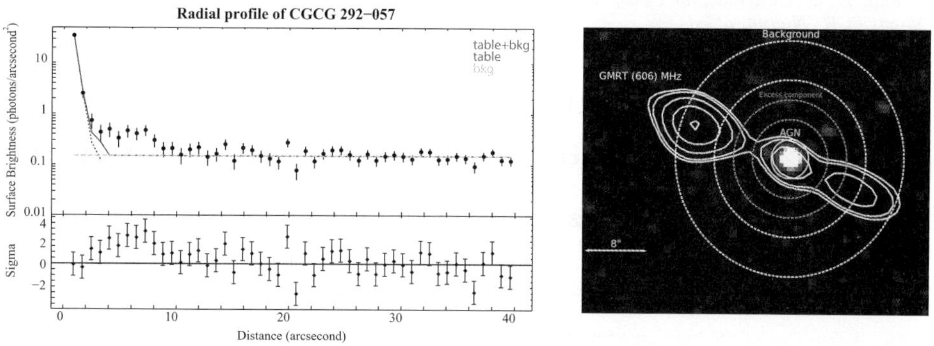

Figure 1. Left: the X-ray surface brightness profile centred on the CGCG 292−057 nucleus, along with fit including the table model for the PSF (based on the merged simulated PSF images) and a constant background. Right: ACIS-S image of CGCG 292-057 system, with the overlaid 606 MHz GMRT contours.

The X-ray surface brightness profile of the source, centred on the CGCG 292−057 nucleus, along with fit including the table model for the PSF (based on the merged simulated PSF images) and a constant background, is presented in Figure 1, left panel. As shown, positive residuals are seen within the range from about $2.5''$ up to roughly $10''$, signalling an additional X-ray emission component located at distances between a few kpc up to about 10 kpc from the core. In the following spectral analysis of this excess component, we adopted the annular region from $5''$ to $7.5''$, and the annular region from $10''$ to $15''$ for the constant background. The right panel in Figure 1 presents the *Chandra* image of the central parts of the CGCG 292−057 galaxy, with the 606 MHz GMRT contours of the inner radio lobes superimposed, and the source extraction regions as described above. As shown, the excess X-ray emission component is seen around the edges of the innermost parts of the radio structure, and hence we identify this feature with the diffuse phase of the ISM compressed and heated by the expanding jets/lobes.

First, we modelled the spectrum of the excess ISM component ($5'' - 7.5''$) simultaneously with the background ($10'' - 15''$), assuming an absorbed Apec model (with solar metallicity) for the former, and a single power-law (PL) for the latter. The best-fit parameters are $kT_{\rm ism} \simeq 0.88\,{\rm keV}$, $N_{\rm H,\,ism} \simeq 0.45 \times 10^{22}\,{\rm cm}^{-2}$, and $\Gamma_{\rm bcg} \simeq 0.9$. Next we modelled the spectrum of the AGN ($\leq 2.5''$), considering a variety of models. Our fits favour however the model consisting of a power-law emission scattered by a hot ionized gas giving rise to the 6.7 keV iron line. For such, the implied power-law photon index $\Gamma = 1.73 \pm 0.16$ and the 0.5–7.0 keV luminosity of 5×10^{41} erg/s, are consistent with the LINER classification of the active nucleus in CGCG 292-057, while the position of the line, $E_{line} = 6.75 \pm 0.09$, and the calculated $EW_{line} \simeq 241 \pm 77\,{\rm eV}$, are both, in principle, consistent with the FeXXV emission arising due to an X-ray reflection on a photoionized circumnuclear matter with the column density of $\sim 6 \times 10^{22}\,{\rm cm}^{-2}$.

Recently, ionized iron lines have been also detected in the X-ray spectra of the two other compact radio galaxies, namely 0710+439 (Siemiginowska *et al.* (2016)), and PMN J1603-4904 (Krauss *et al.* (2018)). This may suggest that such features are common in the systems hosting newly-born jets propagating through, and interacting with the multiphase ISM.

References

Kozieł-Wierzbowska, D., *et al.* 2012, *MNRAS*, 422, 1546
Krauss, F., *et al.* 2018, *A&A*, 610, 8
Siemiginowska, A., *et al.* 2016, *ApJ*, 823, 57
Singh, V., *et al.* 2015, *MNRAS*, 454, 1556

Perseus in Sicily: From Black Hole to Cluster Outskirts
Proceedings IAU Symposium No. 342, 2018
K. Asada, E. de Gouveia Dal Pino, M. Giroletti,
H. Nagai & R. Nemmen, eds.

doi:10.1017/S1743921318005148

Preliminary analysis of the X-ray emission from the central regions of the Pictor A

R. Thimmappa, Ł. Stawarz, K. Balasubramaniam and V. Marchenko

Astronomical Observatory, Jagiellonian University, 30-244 Krakow, Poland
email: `rameshan@oa.uj.edu.pl`

Abstract. Here we present some preliminary results of our analysis of the combined *Chandra* observations of the Pictor A radio galaxy. All the available *Chandra* data for the target, consisting of multiple pointings spanning over 15 years and amounting to the total exposure time of 464 ks, have been included in the analysis. We studied in detail the PSFs of the core region in the individual pointings, as well as the radial profile of the X-ray surface brightness of the source in the combined dataset, in order to discriminate between the radiative output of the unresolved core and the host galaxy. Based on these, we have performed spectral modeling of the active nucleus, constraining its variability.

Keywords. galaxies: active — X-rays: galaxies — X-rays: Individual (Pictor A)

1. Introduction

The radio galaxy Pictor A is one of the most prominent radio source in the Southern sky. It is an excellent candidate for a comprehensive study of the physics of relativistic jets, due to its large size and high luminosity of the extended lobes. Pictor A is also a natural laboratory for investigating the physics of Active Galactic Nuclei (AGN), for its strong emission-line spectrum, including broad forbidden and permitted lines.

2. The Analysis

Chandra has observed Pictor A on 14 separate occasions over the past 15 years, for a total of 464 ks of the observing time. The pointings of the observations differ, and this affects the quality of the available data for various regions of the source, because the effective point spread function of the combined dataset is a complicated function of the position. The other complication is that, due to a high flux of the Pictor A nucleus, the Chandra spectra for the central parts of the source suffer from a significant pile-up problem. Due to this fact, in their analysis of the AGN variability in Pictor A, Hardcastle *et al.* (2016) has considered only the wings of the central PSF (from 6.3 px up to 29 px), excluding in this way the pixels piled up at more than the 1% level at any epoch. Still, due to the dependence of the PSF shape on the photon energy, the core spectra extracted in this way had to be corrected further via simulations.

We have anayzed all the Chandra data for the target using software CIAO 4.9 and CALDB 4.6.7. The readout streaks were removed for each observation and the events files were then reprojected to a single physical coordinate system. The merge_obs script was used to produce merged events files, exposure maps and exposure-corrected ("fluxed") images after masking out point sources detected with wavdetect with minimum psf. The resulting X-ray surface brightness profile of the central parts of Pictor A (up to

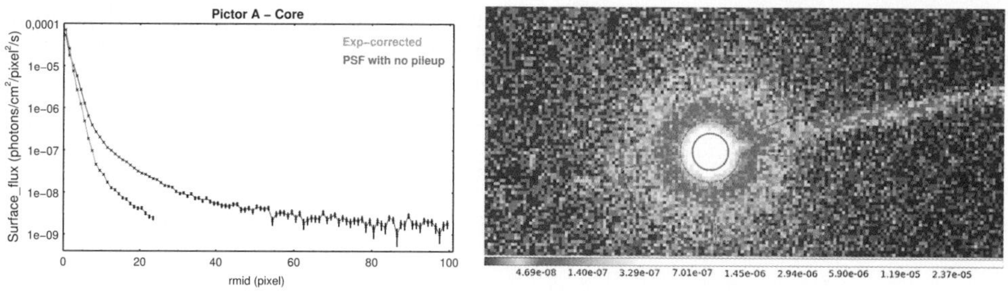

Figure 1. Left: The surface brightness profile of the Pictor A central region, obtained for the merged event file (red); the profile of the corresponding central PSF (not corrected for the pile-up) is shown for a comparison (blue). Right: ACIS-S image of the central parts of Pictor A, within the energy range 0.5–7 keV. AGN source extraction region (< 6 px) and the background region (10 px–30 px, omitting the jet), are denoted by green contours.

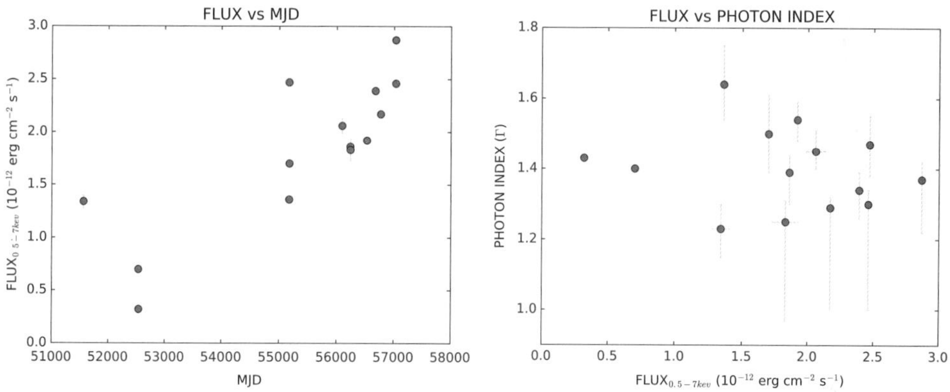

Figure 2. Left: Fluxes calculated for individual observations with 0.5–7.0 kev energy range, vs. observation time. Right: Relation between the 0.5–7.0 keV flux of the core and the photon index of the X-ray continuum.

100 px $\simeq 50''$ from the core, i.e. ~ 35 kpc at the source distance), is shown in Figure 1, left panel. There we also present the profile of the central PSF, obtained by merging the exposure-corrected MARX-simulated PSFs (50 simulated images for each pointings, totaling to 700 PSFs). As the simulated PSFs have not been corrected for the pile-up, a discrepancy can be seen in the figure between the observed surface brightness profile and the PSF profile, within the innermost parts of the source.

For the AGN spectral analysis, we have extracted the source spectra from the circular region with radius 6 px for each pointing; the backgrounds were taken from the annual region (omitting the jet) with radii 10 px and 30 px, as shown in Figure 1, right panel. Spectral fitting was done in SHERPA, assuming the simplest model consisting of the absorbed power-law. In all of the cases, the intrinsic absorption was consistent with hydrogen column densities much below the Galactic value in the direction of the source. The results of the fitting are shown in Figure 2.

3. Conclusions & Future work

The next step of the analysis will include updated PSF simulations including the pileup effect. With the properly characterized PSF shape, we will update also the source

(AGN) and the background (host galaxy, plus extended lobes) regions, and perform a more detailed spectral modeling, constraining also the presence of the iron line in the source spectrum. The image deconvolution and the spectral modeling for the large-scale jet and the extended lobes will also be performed.

Reference

Hardcastle, M. J., Lenc, E., & Birkinshaw, M., *et al.* 2016, MNRAS, 455, 3526

Perseus in Sicily: From Black Hole to Cluster Outskirts
Proceedings IAU Symposium No. 342, 2018
K. Asada, E. de Gouveia Dal Pino, M. Giroletti,
H. Nagai & R. Nemmen, eds.

doi:10.1017/S1743921318006075

Flip of the jet head position of 3C 84 in 2015

M. Kino[1,2], H. Nagai[2], K. Wajima[3], N. Kawakatu[4], M. Orienti[5], G. Giovannini[5,6], K. Hada[2], K. Niinuma[7] and M. Giroletti[5]

[1]Kogakuin University of Technology & Engineering, Academic Support Center, 2665-1 Nakano, Hachioji, Tokyo 192-0015, Japan

[2]National Astronomical Observatory of Japan, 2-21-1 Osawa, Mitaka, Tokyo, 181-8588, Japan

[3]Korea Astronomy and Space Science Institute, 776 Daedeokdae-ro, Yuseong-gu, Daejeon 34055, Republic of Korea

[4]National Institute of Technology, Kure College, 2-2-11, Agaminami, Kure, Hiroshima, 737-8506, Japan

[5]INAF - Istituto di Radioastronomia, via Gobetti 101, 40129 Bologna, Italy

[6]Dipartimento di Astronomia, Universita' di Bologna, via Gobetti 103/2, I-40129, Bologna, Italy

[7]Graduate School of Sciences and Technology for Innovation, Yamaguchi University, 1677-1 Yoshida, Yamaguchi, Yamaguchi 753-8512, Japan

Abstract. By performing the multi-epoch monitoring observation with KaVA at 43 GHz, we investigate the kinematics of the notable newborn bright component C3 located at the tip of the recurrent jet of 3C 84. During 2015 August-September,we discover the positional flip of the C3 component about 0.4 milli-arcsecond in angular scale. The flux density of the C3 component coherently showed the monotonic increase after the flip during our monitoring period. These phenomena are in good agreement with characteristic behaviors of a jet propagation in clumpy ambient medium predicted in hydrodynamical simulations.

Keywords. galaxies: active, galaxies: jets, radio continuum: galaxies

---

1. KaVA Observation of 3C 84

Jets in active galaxies are generally thought to have a strong impact on interstellar medium especially when they are young. Hydrodynamical simulations actually indicate flips of the jet head caused by the interaction with the clumps (Wagner & Bicknell 2011), although such a phenomenon has not yet been observed so far. NGC 1275 (a.k.a. 3C 84) is a notable nearby giant elliptical galaxy at the center of the Perseus cluster. Nagai *et al.* (2010) discovered the emergence of newborn bright component so-called C3 in the radio jet 3C 84. Hence, 3C 84 is regarded as a unique laboratory to explore interactions between the jet and the circum-nuclear environment on parsec scale.

From 2015 August to 2016 June, we conducted full- track 12 epochs monitoring of 3C 84 at 43 GHz with KaVA, the combination of the Korean VLBI Network (KVN) and the Japanese VLBI Exploration of Radio Astrometry (VERA) radio arrays. The joint array, KaVA, has a total of seven antennas at frequencies of 22 and 43 GHz, resulting in angular resolutions of about 1.2 and 0.6 mas, respectively. Basic imaging capabilities of KaVA has been summarized in detail by Niinuma *et al.* (2014) and (Hada 2017). Left-hand circular polarization was received and sampled with a 2 bit quantization. All the data were recorded at 1 Gbps (256 MHz bandwidth, 16 MHz, 16 sub-bands) and correlated by the Daejeon correlator. As a technical note, we note that in the flux calibration of

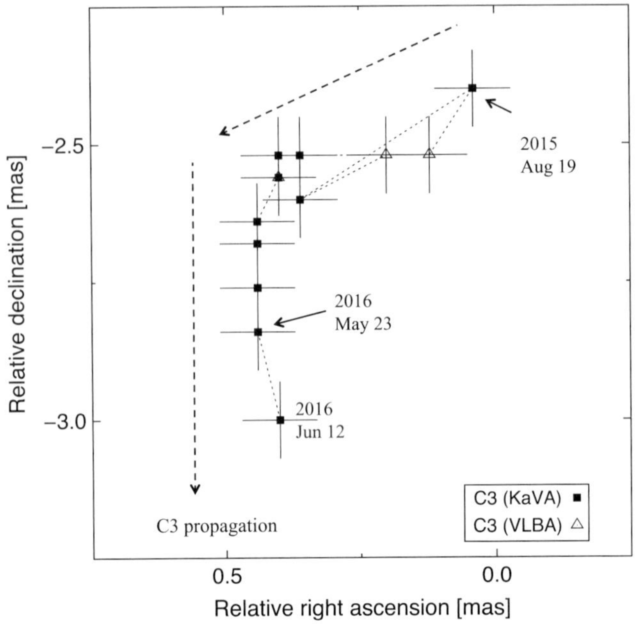

Figure 1. Relative peak intensity position of the component C3 with respect to C1 assuming as an origin (0, 0). The filled squares and open triangles represent the peak intensity position obtained with KaVA and VLBA observations, respectively.

KaVA one visibility amplitudes of KaVA correlated by the Daejeon hardware correlator (Lee *et al.* 2015a) should be corrected by multiplying the factor of 1.35 (Lee *et al.* 2015b).

2. Summary

Here we summarize our main findings following our recent work (Kino 2018).

• *Discovery of the position flip of C3.* We discover the significant flip of C3 with the angular scale of about 0.4 mas between Aug and Sep in 2015. As shown in Figure 1, the motion of the peak position of C3 was decelerated and C3 wobbled for a few months. After that, the peak position of C3 resume to propagate to the south direction.

• *Brightening of C3.* We find the clear monotonic increase of the peak intensity of the C3 component during the observation period. The peak intensity of C3 increased more than a factor 4 at the end of the observation period. The flip of C3 together with this monotonic enhancement of its flux density can be explained by a collision between the jet and a dense clump.

• *Estimate of the clump density.* We estimate the number density of the presumable clump based on the momentum balance between the jet thrust and the ram pressure from the clump and we find that the clump number density is estimated about $4 \times 10^3\ \mathrm{cm}^{-3} \lesssim n_{\rm cl} \lesssim 2 \times 10^5\ \mathrm{cm}^{-3}$. The clump's location and estimated number density are in good agreement with a cloud located deep side the NLR or a dense portion of an intergalactic molecular cloud.

References

Hada, K., Park, J., Kino, M., *et al.* 2017, *PASJ*, 69, 71
Kino, M., Wajima, K., Kawakatu., *et al.* 2018, *ApJ*, in press
Lee, S.-S., Oh, C. S., Roh, D.-G., *et al.* 2015a, *JKAS*, 48, 125
Lee, S.-S., Byun, D.-Y., Oh, C. S., *et al.* 2015b, *JKAS*, 48, 229
Nagai, H., Suzuki, K., Asada, K., *et al.* 2010, *PASJ*, 62, L11
Niinuma, K., Lee, S.-S., Kino, M., *et al.* 2014, *PASJ*, 66, 103
Wagner, A. Y., & Bicknell, G. V. 2011, *ApJ*, 728, 29

Perseus in Sicily: From Black Hole to Cluster Outskirts
Proceedings IAU Symposium No. 342, 2018
K. Asada, E. de Gouveia Dal Pino, M. Giroletti,
H. Nagai & R. Nemmen, eds.

doi:10.1017/S174392131800604X

AGN and Star Formation Feedback in Galaxy Outflows

Elisabete M. de Gouveia Dal Pino[1], **William Clavijo-Bohórquez**[1] **and Claudio Melioli**[2]

[1]Instituto de Astronomia, Geofísica e Ciências Atmosféricas (IAG), Universidade de São Paulo, CEP 05508-090, São Paulo, Brazil
emails: dalpino@iag.usp.br, weclavijob@usp.br

[2]University of Modena, Italy

Abstract. Large-scale, broad outflows are common in active galaxies. In systems where star formation coexists with an AGN, it is unclear yet the role that both play on driving the outflows. In this work we present three-dimensional radiative-cooling MHD simulations of the formation of these outflows, considering the feedback from both the AGN and supernovae-driven winds. We find that a large-opening-angle AGN wind develops fountain structures that make the expanding gas to fallback. Furthermore, it exhausts the gas near the nuclear region, extinguishing star formation and accretion within a few 100.000 yr, which establishes the duty cycle of these outflows. The AGN wind accounts for the highest speed features in the outflow with velocities around 10.000 km s^{-1} (as observed in UFOs), but these are not as cold and dense as required by observations of molecular outflows. The SNe-driven wind is the main responsible for the observed mass-loading of the outflows.

Keywords. Galaxy: active, Star Formation, Supernova

1. Introduction

Fast massive, poorly collimated outflows of HI and molecular gas, with velocities $\sim$ 600-1500 km s^{-1}, have been recently observed in the central regions of AGN sources like Seyfert galaxies and ULIRGs (e.g, Morganti *et al.* 2005, 2013; see also R. Morganti in these procs.). Also, much hotter ultra-fast outflows (UFOs) with velocities $\sim$ 0.03-0.3c have been detected in X-rays (e.g. Tombesi *et al.* 2013). An interesting example of a molecular outflow is the ULIRG AGN 4C12.50. It exhibits HI and CO gas outflow with velocity $\sim$ 1000 km s^{-1} which is co-spatial with the radio jet hot-spot, at 100 pc scales (Morganti *et al.* 2013, 2014, Dasyra & Combes 2012). This is probably produced by radiative cooling of shocked gas due to jet interaction with the ISM. Typically, gas column densities $\sim 10^{21} - 10^{22}$ cm^{-2} are inferred for these outflows, mass loss rates up to 100 $M_\odot$ yr^{-1}, and HI clumps with mass larger than 600 $M_\odot$. A spectacular example of UFO is the one detected (in X-rays) in the Seyfert NGC 4151 central region, denominated Eye of Sauron, which is surrounded by ionized hydrogen (HII) associated to star formation regions observed in optical and is also connected with infalling HI gas observed in radio with VLA (Wang *et al.* 2010). UFOs have even larger column densities $\sim 10^{23} - 10^{24}$ cm^{-2} (Kraemer *et al.* 2018).

The investigation of these outflows is important as they transport energy and gas outwards from the central regions and may affect the evolution of the host galaxy and its surrounds as well. Their origin is not yet fully understood, but the accreting supermassive black hole (SMBH) in the center of the AGN is believed to be their main driving source. Injection mechanisms such as radiation pressure, or the kinetic interaction of the radio plasma with the ISM, or even magneto-centrifugal processes are often invoked (e.g. Kremer et al 2018 and references therein). A key point when investigating these outflows is the fact that they are generally associated with Seyferts, for which SF regions and starbursts (SB) coexist with the AGN, and the energy power emitted by both the nuclear AGN and the host galaxy is comparable, so that it is unclear the relative role of the SF and the AGN on the driving of the outflows (Melioli & de Gouveia Dal Pino 2015).

In this work, we explore the formation and evolution of these outflows in a Seyfert-like galaxy (within 1 kpc scale) by means of 3D MHD simulations at high resolution (3.9 pc) including gas radiative cooling, magnetic fields and the mechanical feedback from both SF (through the production of type Ia and II SNe) and AGN outflows.

2. Results of the MHD Simulations

For the numerical setup of our simulations, we have considered a total gravitational potential provided by dark matter, the galactic bulge and the disk. We adopted a black hole to bulge mass ratio 10^{-3} and considered a multi-phase, initially stratified gas disk supported both by rotation, and thermal and magnetic pressures. We have considered a SF rate of 1 $M_\odot$ yr^{-1}, type Ia SNe rate $= 0.01$ yr^{-1} and type II SNe rate 0.1 yr^{-1}. We have considered equilibrium radiative cooling for a gas able to achieve temperatures between $T = 100$ K to 10^8 K. We have taken initial thermal to magnetic pressures $\beta \equiv \infty$, =300, 30, 3, column densities 10^{22} to 10^{23} cm^{-2}, AGN wind initial aperture angles $= 0°$, $10°$, or spherical, with injection power 10^{43} erg s^{-1}, velocity $v = 0.06c$ and mass rate 5×10^{-4} $M_\odot$ yr^{-1}. (For more details see Melioli & de Gouveia Dal Pino 2015; Clavijo-Bohórquez, de Gouveia Dal Pino and Melioli, in prep.).

As an example, Figure compares four evolved models at $t = 5$ Myr. These diagrams clearly indicate that a larger AGN outflow opening angle has a major influence on the nuclear region evolution, sweeping the ISM matter, dragging the SN-driven winds, quenching accretion and star formation, and accelerating much more gas to velocities above 10.000 km s^{-1}. The comparison of the results for the two different initial magnetic field geometries do not reveal substantial differences, this in part because of the large value adopted for β here (which is compatible to the values expected from observations at these scales). Nevertheless, the comparison of the velocity-density diagrams with pure HD simulations indicate that the presence of B helps to preserve and increase the number of denser structures at higher (positive and negative) velocities.

Our simulations also show that the winds driven by the products of SF alone (i.e., by explosions of supernovae, SNe) can drive outflows with velocities $\sim 100 - 1000$ km s^{-1} and mass outflow rates of the order of 50 $M_\odot$ yr^{-1} (Figure, which resemble the properties of warm absorbers (WAs) and molecular outflows. However, the resulting densities ($\sim$ 1 – 10 cm^{-3}) and temperatures (between 10^4 and 10^5 K) are too low and too high, respectively, compared to the expected values of these outflows.

The presence of an AGN-wind, particularly with large opening angle, causes the formation of fountain-like structures (Fig. 1) that make part of the expanding gas (pushed also by the SF-wind) to fallback, producing a "positive" feedback on the host galaxy evolution. This effect is more pronounced in presence of magnetic fields, due to the action of extra magnetic forces by the AGN. This reduces the mass loss rate in the outflows by factors up to 10 when compared with HD models (see Figure 2).

Figure 3 depicts the early evolution ($t \sim 950.000$ yr) for a model with an AGN-wind injected spherically in the nucleus, after the SNe-driven wind is entirely developed. We note that the wind is quickly collimated by the surrounding gas forming dipolar outflows that resemble UFOs. The diagrams and the density-velocity histogram show

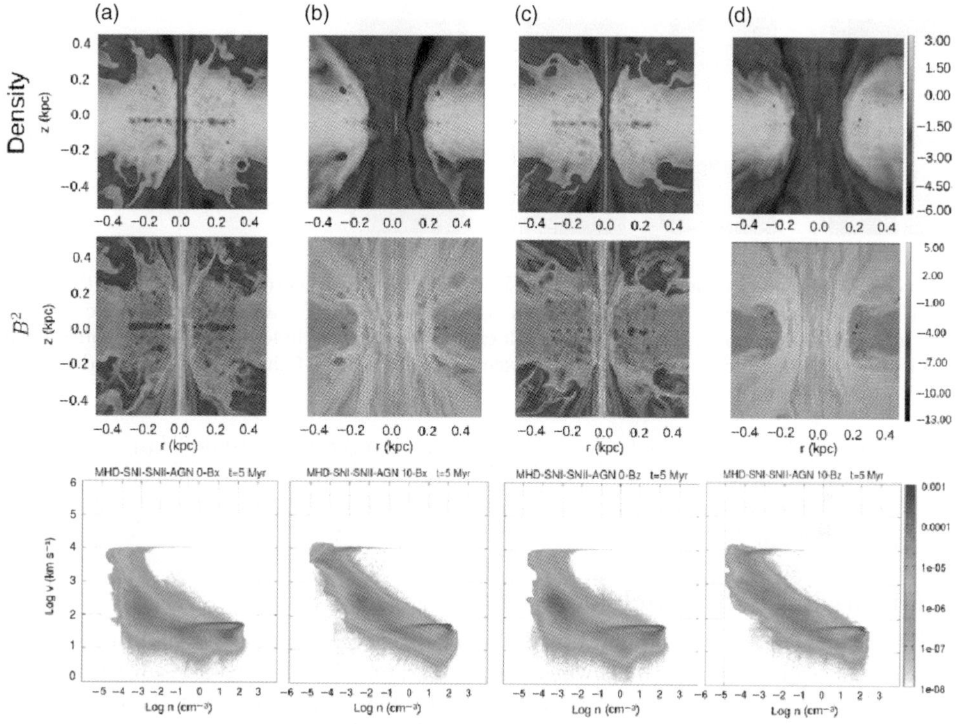

Figure 1. Top: Two-dimensional (2D) central cuts of the density distribution within one-kpc region of the evolved galaxy (at 5 Myr), for four different initial conditions. From left to right: (a) a system with initial homogeneous horizontal magnetic field ($\beta = 300$, and $B_x = 0.76\mu$G) and an AGN outflow with 0° opening angle (collimated); (b) the same as in the left panel, but with the AGN outflow with 10° opening angle; (c) the same as in (a), but with initial homogeneous vertical magnetic field (B_z); and (d) the same as in (c), but with the AGN outflow with 10° opening angle. Middle: 2D central cuts of the magnetic field strength (B^2), with the magnetic field vectors superposed to it. Bottom: 2D histograms of the vertical velocity versus density distribution, calculated considering all the cells of the systems. Velocity is in km s^{-1} and density in cm^{-3}, both in logarithmic scale. The color bar indicates the cell numbers normalized to their total number (see more details in Clavijo-Bohórquez, de Gouveia Dal Pino, Melioli in prep.).

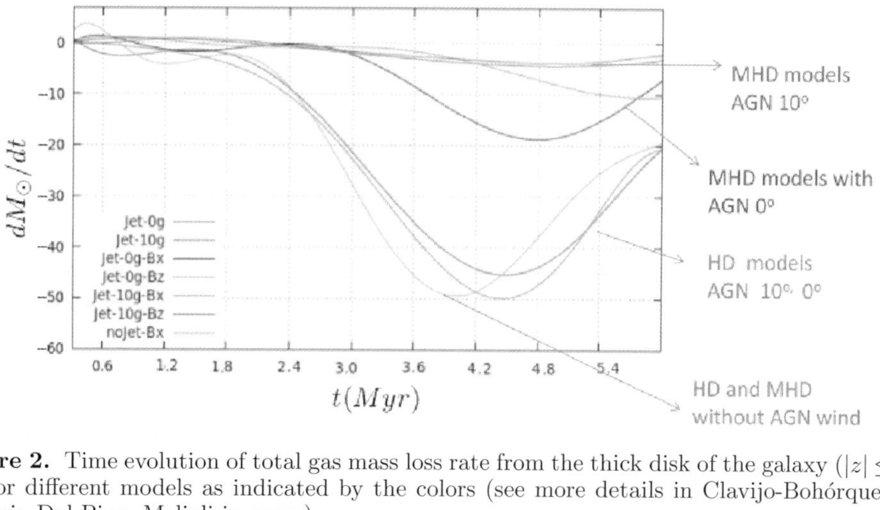

Figure 2. Time evolution of total gas mass loss rate from the thick disk of the galaxy ($|z| \leq 200$ pc) for different models as indicated by the colors (see more details in Clavijo-Bohórquez, de Gouveia Dal Pino, Melioli in prep.).

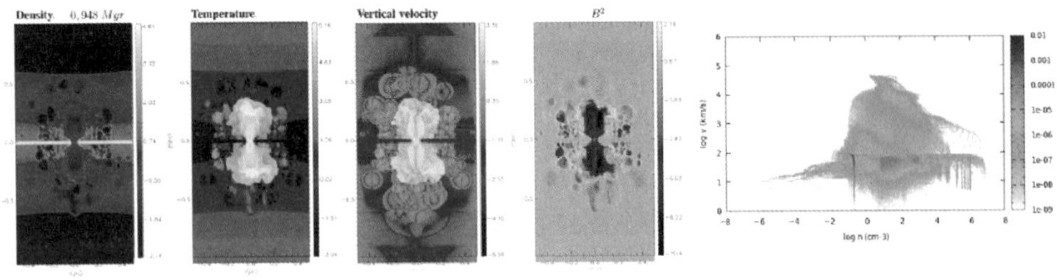

Figure 3. Early evolution for a model with an AGN wind injected spherically in the nuclear region with initial power 2×10^{45} erg s^{-1}, mass rate 1.5 M$_\odot$ yr^{-1}, and velocity 0.21 c. From left to right: 2D cuts of density, temperature, velocity and magnetic field strength distributions; and velocity-density histogram as in Figure 1 (see more details in Clavijo-Bohórquez, de Gouveia Dal Pino, Melioli in prep.).

that high speed, dense structures with velocities larger than 10.000 km s^{-1}, temperatures $> 10^7$ K, and densities > 100 cm^{-3} do develop. Besides, intermediate speed structures are also detected with velocities ~ 1000 km s^{-1}, temperatures > 1000 K, and densities $< 10^6$ cm^{-3}.

3. Discussion and Conclusions

MHD radiative cooling simulations of the nuclear region (1 kpc^3 scale) of Seyfert-like galaxies, taking into account both the SF and AGN outflow feedback, result the following:

The SNe can drive outflows with velocities of a few 1000 km/s, but the characteristic densities are too low and the temperatures too large in comparison to the observations of WA and molecular outflows.

A collimated AGN jet alone (without SNe-driven wind) is unable to drive massive outflows, but can accelerate structures to very high speeds (as observed in UFOs).

An AGN wind with large opening angle sweeps more ISM matter, accelerates more gas to $v >$ 10.000 km s^{-1}, but it exhausts gas fuel and quenches accretion onto the SMBH and star formation in the nuclear regions within a few 100.000 yr. This indicates that the duty cycle of these outflows is around a few 100.000 yr, which is compatible with the timescales inferred for the observed UFOs and molecular outflows.

Mass loss rates up to 50 M$_\odot$ yr^{-1} (which are compatible with the observations of molecular outflows) are driven mainly by the SNe-driven-wind component.

Large opening angle AGN-winds, specially in MHD flows, favor the fallback of gas to the galaxy (fountain) which decreases the mass loss rate.

An AGN spherical wind injected when the SNe wind is fully developed improves the results, producing UFO characteristics ($v >$ 10.000 km s^{-1}, $T > 10^7$ K, and $n >$ 100 cm^{-3}), but the slower outflow component with $v \sim 1000$ km s^{-1} is too hot to explain the molecular outflows.

Missing ingredient in our models, like the presence of non-equilibrium ionization radiative cooling to account for more efficient coolers (molecules and dust) that could survive to evaporation and destruction in interactions with high speed hot gas, will help to improve our results and ensure the survival of the cold, dense clumpy structures that are swept to very high speeds as observed in molecular outflows.

Acknowledgments

We acknowledge support from the Brazilian agencies FAPESP (2013/10559-5 grant) and CNPq (306598/2009-4 grant). The simulations presented in this lecture have made use of the computing facilities of the GAPAE group (IAG-USP) and the Laboratory of Astroinformatics IAG/USP, NAT/Unicsul (FAPESP grant 2009/54006-4).

References

Dasyra, K. M & Combes, F. 2012, *A&A*, 541, L7

Amari, S., Hoppe, P., Zinner, E., & Lewis R.S. 1995, *ApJ*, 30, 490

Kraemer, S. B., Tombesi, F., & Bottorff M. C. 2018, *ApJ*, 852, 35

Melioli, C., & de Gouveia Dal Pino, E. M. 2015, *ApJ*, 812, 90

Morganti, R., Oosterloo, T. A., Tadhunter, C. N., van Moorsel, G., & Emonts, B., 2005 *A&A*, 439, 521

Morganti, R., Frieswijk, W., Oonk, R. J. B., Oosterloo, T., & Tadhunter, C. 2013, *A&A*, 552, L4

Morganti, R., Oosterloo, T. A., Oonk, J. B. R., Frieswijk, W., & Tadhunter, C. N. 2015, *ASP-CS*, 499, 125

Oosterloo, T. A., Morganti, R., Tzioumis, A., Reynolds, J., King, E., McCulloch, P., & Tsvetanov, Z. 2000, *AJ*, 119, 2085

Tadhunter, C., Morganti, R., Rose, M., Oonk, J. B. R., & Oosterloo, T. 2014, *Nature*, 511, 440

Tombesi, F., Cappi, M., Reeves, J. N., Palumbo, G. G. C., Braito, V., & Dadina, M. 2011, *ApJ*, 742, 44

Tombesi, F., Cappi, M., Reeves, J. N., Nemmen, R. S., Braito, V., Gaspari, M., & Reynolds, S. 2013, *MNRAS*, 430, 1102

Tombesi, F., Meléndez, M., Veilleux, S., Reeves, J.M., González-Alfonso, C., & Reynolds, C. S. 2015, *Nature*, 519, 436

Wagner, A. Y., Umemura, M., & Bicknell, G. V. 2013, *ApJ*, 763, L18

Wang, J., Fabbiano, G., Risaliti, G., Elvis, M., Mundell, C. G., Dumas, G., Schinnerer, E., & Zezas, A. 2010, *ApJ*, 719, L208

Perseus in Sicily: From Black Hole to Cluster Outskirts
Proceedings IAU Symposium No. 342, 2018
K. Asada, E. de Gouveia Dal Pino, M. Giroletti,
H. Nagai & R. Nemmen, eds.

doi:10.1017/S1743921318007962

Update on the Multi-Frequency Monitoring of Blazars with Medicina and Noto

Simona Righini[1], Pietro Cassaro[1], Uwe Bach[2], Marcello Giroletti[1], Carla Buemi[1], Paolo Leto[1], Monica Orienti[1], Andrea Orlati[1], Claudia M. Raiteri[1], Corrado Trigilio[1], Grazia Umana[1] and Massimo Villata[1]

[1]INAF - Istituto Nazionale di Astrofisica, Italy
email: simona.righini@inaf.it

[2]MPIfR - Max Planck Institute für Radioastronomie, Germany

Abstract. The Medicina and Noto radiotelescopes have been employed for over 14 years to monitor the flux density variations of a vast sample of blazars at different radio frequencies. Radio data are essential components of blazar spectral energy distribution (SED, spanning from radio waves to gamma rays), whose trend with luminosity and shape changes provide decisive information on the physics of extra-galactic jets and, eventually, on the mechanism extracting energy from the central black hole in radio-loud AGN. Observations presently carried out at 5, 8 and 24 GHz have taken advantage of the continually evolving control system installed at the antennas. A new, batch-wise analysis tool was also produced, in order to easily handle and reduce the datasets acquired in monthly sessions. We here describe the latest developments and achievements.

Keywords. radio continuum: galaxies, galaxies: active

1. Overview

The name Blazars identifies a family of radio-loud Active Galactic Nuclei (AGN) showing a rather complex phenomenology: extreme variability at all wavelengths, polarization, strong gamma-ray emission and brightness temperatures exceeding the Compton limit. The large amount of work done in the last decade has led to a rather general consensus on the global mechanism responsible for the emission: a rotating black hole surrounded by a massive accretion disk with an intense relativistic plasma jet closely aligned to the line of sight. Relativistic electrons produce the soft photons through synchrotron emission, while hard photons are likely produced by inverse Compton scattering. This overall scenario, however, still presents a large number of poorly understood details which, in turn, lead to a wide variety of models and call for long term and multi-wavelength campaigns capable of providing the necessary observational constraints.

Variability measurements provide key information on the AGN structure, down to linear scales or flux density levels not accessible even with interferometric imaging. The mechanisms for variability are still not well understood. Possibilities discussed in the literature include shocks in jets, changes in the direction of forward beaming due, e.g., to helical trajectories of plasma, to a precessing binary black-hole system, or to a rotation of a helical jet, introducing flares due to the lighthouse effect and variability due to colliding relativistic plasma shells. Thus, variability furnishes important clues into size, structure, physics and dynamics of the radiating source region.

2. Our project

Since blazars are highly variable it is important to take simultaneous broad-band radio spectra. We proposed such simultaneous observations back in 2004, and in 2007 our project was approved as a key project at INAFs Institute of Radioastronomy. We have hence performed monthly multi-frequency (5, 8, 22/24 and 43 GHz) observations on a vast sample of blazars, which includes some of the brightest and most studied blazars detected in gamma rays by Fermi and AGILE. Details are provided on the project website (Bach *et al.*). Data from this project have produced about 40 refereed publications, mostly resulting from the participation to multi-wavelength campaigns by the Whole Earth Blazar Telescope (WEBT).

Since December 2011, all the sessions taking place in Medicina have been carried out using a new control software, developed to fully exploit the single-dish potential of the Italian radio telescopes. We achieve a good characterization of the source flux density with a few cross-scans on each target source. On the brightest targets, each scan provides an estimate of the flux density to within 10%, while for the faintest sources stacking of the individual scans yields a good signal-to-noise detection. Since 2017, also the Noto sessions are carried out using such system. As the 43 GHz receiver is now under maintenance, the only observations performed in Noto are the 5 GHz acquisitions. This constitutes a precious contribution, as such frequency band is highly polluted by RFI in Medicina.

3. Dealing with an evolving dataset: introducing CAP (Cross-scan Analysis Pipeline)

During the 12+ years of acquisitions, a large dataset was formed, evolving during the development of the overall control system. To handle it, we produced a new, comprehensive reduction pipeline able to perform a batch (re-)analysis on the data, if needed, of any epoch. Such tool has very recently been completed and made available on a GitHub repository, which also provides a basic wiki.

CAP is devoted to the integration, analysis and calibration of the total power cross-scans produced by the Italian radio telescopes, including the soon-to-be-operative Sardinia Radio Telescope. It is a set of IDL-programming-language (Exelis) tools, developed to process cross-scans (FITS files) acquired on point-like targets. It allows the users to perform the analysis through the following phases:

- rearrange the date-based datasets, separating acquisition in frequency-based folders;
- visually inspect and flag the data [optional];
- estimate the atmospheric opacity via the observed "skydips" [optional];
- retrieve the proper calibration factors from cross-scans acquired on flux density calibrators (considering Perley & Butler 2013);
- integrate scans on the target sources and obtain flux density measurements, taking into account both the atmospheric opacity (if provided) and the antenna gain curve, also correcting for pointing offset;
- reprocess the measurement list and average measurements recorded in different takes [optional].

The final output products comprise (but are not limited to): tables listing the measurements obtained on calibrators and the raw-counts-to-Jy conversion factors hence derived, tables with the raw amplitudes and calibrated flux densities measured on the targets, plots showing the single acquisitions and the integrated scans, with overlaid fitted data - both for flux density calibrators and targets. Two examples of the achieved results are provided in Fig. 1.

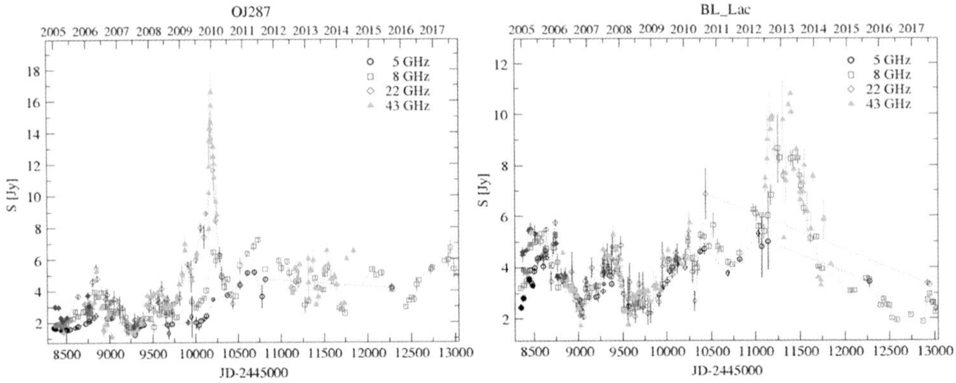

Figure 1. Radio flux density measurements of two of the monitored sources. The temporary unavailability of some of the receivers, the occurrence of stops due to HW and SW upgrades and the increasing incidence of RFI in low-frequency bands have affected the timeline. However, the visible lack of 22-GHz measurements since 2011 is mostly due to the analysis still being underway.

References

Bach, U. *et al.* www.mpifr-bonn.mpg.de/staff/ubach/bmonit/
WEBT, www.oato.inaf.it/blazars/webt/
CAP GitHub repository, https://github.com/discos/CAP
CAP wiki, https://github.com/discos/CAP/wiki

Perseus in Sicily: From Black Hole to Cluster Outskirts
Proceedings IAU Symposium No. 342, 2018
K. Asada, E. de Gouveia Dal Pino, M. Giroletti,
H. Nagai & R. Nemmen, eds.

doi:10.1017/S1743921318007196

Hints of radio sources evolution

Pietro Cassaro

INAF Istituto di Radioastronomia, CP 161, I97016 Noto (SR), Italy
email: pietro.cassaro@inaf.it

Abstract. An understanding of the mechanisms that regulate the Active Galactic Nuclei passes through the study of their early life stages, observable in Compact Symmetric Objects. To this purpose, a study was carried out on two compact radio sources, based on data from the VLBA archive at different times and frequencies. The results are compatible with an intermediate scenario between the two main theories about these objects: frustration and youth.

Keywords. Galaxies: active, galaxies: jets, radio continuum: galaxies

1. Introduction

A Compact Symmetric Object (CSO) is a milliarcsecond version of a radiogalaxy, and it is a widely shared opinion that such small dimensions are due to their young age, ranging from hundreds to thousands of years of activity. In order to check the ages both radiative and kinematic studies have been realized by many authors (Owsianik *et al.* (1998), Taylor *et al.* (2000), Conway (2002), Murgia (2003), Gugliucci *et al.* (2005), Britzen *et al.* (2008), An *et al.* (2012)).

To extend the studies on the age of some of these sources I used VLBA archive data at different frequencies and epochs of two CSOs, J1035+5628 and J1358+4737, in order to obtain information on expansion speed and break frequency in their radio spectrum.

2. Kinematic and radiative ages

The most updated analysis of the kinematics of J1035+5628 is in Britzen *et al.* (2008) and refers to C-band data from 1996 to 2000. I collected data for this source in X-band from 1994 to 2011 in the VLBA database. The data were calibrated and then self-calibrated in phase using the AIPS package. The position of the components has been determined by using the DIFMAP modelfit. Only the position of the main component (hot-spot) of each minilobe, i.e. the brightest, has been determined and the trend of the separation between hot-spots over time is shown in Fig. 1 (Left panel). The linear fit is represented by the continuous line and a sinusoidal fit is drawn in dot-dashed line. The sinusoid fits better the data than the line, with a χ^2 value of 0.040 against 0.258. From the linear fit we can determine an expansion motion of the hot-spots of 0.0074 mas/yr, which gives us a kinematic age of 4324 years. The radiative age has been calculated using the Eq.2 in Murgia (2003), assuming the source in equipartition and with simultaneous radio data. The total power spectrum has a break at 8.4 GHz and this leads to a radiative age of 9199 years, more than double of the kinematic age. If the radiative age were equal to the kinematic the source should have a break frequency of more than 32 GHz.

The source J1358+4737 has only two epochs in X-band, so I had to use the C-band data to obtain the position of the hot-spots. The plot in Fig. 1, right panel, clearly shows a contraction of the source in time and this behaviour is obviously not real. As an example, if the data of the previous source were stopped at 2007, we would also observe a contraction for it. The radiative age of this source, calculated in the same way

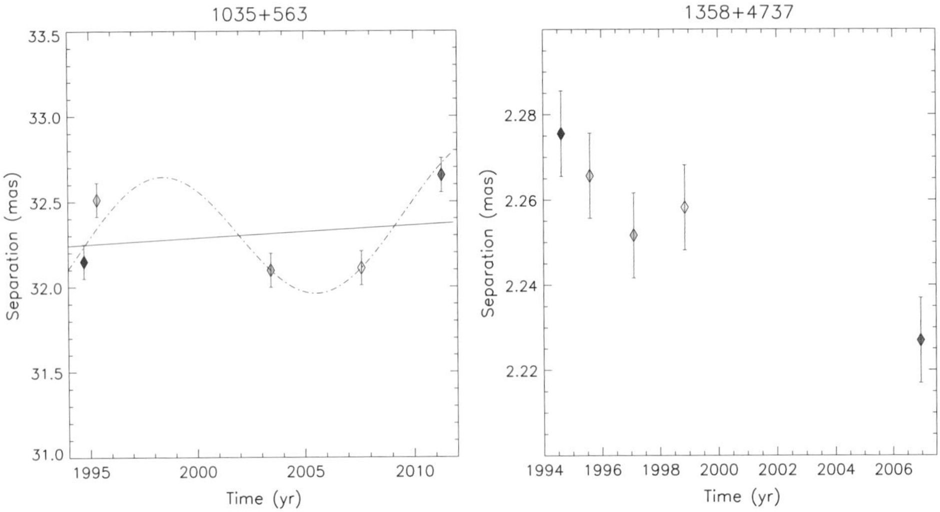

Figure 1. *Left panel* Separation versus time in J1035+5628. Data are taken in X-band. *Right panel* Separation versus time in J1358+4737. Data are taken in C-band

of J1035+5628, is of 4320 yr. If we suppose that this source had the same expansion speed of the previous, the kinematic age would be of only 310 yr, ten times less than the radiative one.

3. Conclusion

As already noticed by An *et al.* (2012), the motion of the components of a CSO is not so straightforward. In this work it is possible to recognize hints of periodic behaviour, as in J1035+5628, or a sharp contraction as we can see in J1358+4737. Moreover the radiative age tells a different story: the sources are more aged than the kinematics allow to infer. This, jointed with the observed motion suggests a more complicated evolution for this kind of sources: the motion appears composed by a linear and a periodic (or quasi-periodic) components, the second one possibly due to a precession of the inner jet. During the precession the hot-spot can encounter some gas clouds and, as a consequence, the expansion of the source slows. In other words, a combination of youth and frustration scenarios.

It appears necessary to deepen the study of the kinematics of the sources with observations closer in time and at high frequencies in order to trace the true hot-spots.

References

An T., Wu F., Yang J., Taylor G. B., Hong X., Baan W. A., Liu X., Wang M., Zhang H., Wang W., Chen X., Cui L., Hao L. & Zhu X. 2012, *ApJSS*, 198, 5

Britzen S., Vermeulen R. C., Campbell R. M., Taylor G. B., Pearson T. J., Readhead A. C. S., Xu W., Browne I. W., Henstock D. R., & P. Wilkinson 2008, *A&A*, 484, 119

Conway J. E. 2002, *New Astron. Revs*, 46, 263

Gugliucci N. E., Taylor G. B., PeckA. B., & Giroletti M., 2005, *ApJ*, 622, 136

Murgia M. 2003, *PASA*, 20, 19

Owsianik I., Conway J. E., & Polatidis A. G. 1998, *A&A*(Letters), 336, 37

G. B. Taylor, J. M. Marr, T. J. Pearson, & A. C. S. Readhead 2000, *ApJ*, 541, 112

Perseus in Sicily: From Black Hole to Cluster Outskirts
Proceedings IAU Symposium No. 342, 2018
K. Asada, E. de Gouveia Dal Pino, M. Giroletti,
H. Nagai & R. Nemmen, eds.

doi:10.1017/S1743921318008001

Bending of the pc scale jet in 3C84

A. Hirano[1], K. Fujisawa[2], K. Niinuma[1] and GENJI Collaboration

[1]Graduated School of Sciences and Technology for Innovation, Yamaguchi University, Yoshida 1677-1, Yamaguchi, 753-8512 Japan

[2]Research institute for Time Studies, Yamaguchi University

Abstract. We have conducted VLBI monitoring observations for a radio galaxy 3C 84 to investigate how the pc scale jet structure changes over a long period. VERA, a VLBI observation network in Japan, was used for the observation. The C3 component of the jet has continuously moved toward the south from the core. The motion was, however, not straight, but it showed a bending of about 0.3 mas (0.1 pc) with a time scale of 500-1000 days. Two models explaining the bending, local brightness distribution change or real change of the jet traveling direction, are discussed.

Keywords. galaxies: active, galaxies: individual(3C 84/NGC 1275), galaxies: jets, radio continuum: galaxies

---

1. Introduction

The bright radio source 3C 84 is associated with the giant elliptical galaxy NGC 1275 (z = 0.0176), which is a dominant member of the Perseus cluster.

Fig 1 shows a 22 GHz total intensity image of 3C 84 . In previous studies it was shown that the parsec-scale structure is consisted of four components. We have investigated the detailed kinematics of C3 with the GENJI programme data at 22 GHz for the period from 2015 January to 2016 December, 33 epochs. The GENJI programme [1] is a monitoring program of gamma-ray bright AGN with VERA array (VLBI Exploration of Radio Astrometry). Thanks to the high cadence monitoring of GENJI, we can observe the detailed motion of components in 3C 84 on subparsec scales. We also focus on the correlation between changes in radio structure (Fig 1) and the gamma-ray activity, that may provide a clue to identify of the gamma-ray radiation region of 3C84.

2. Results and Discussions

Hiura et al. (2018) made a monitoring of jet motion in 3C84 in the period before 2015. The position of the jet components relative to the core that we derived are shown in Fig 2 as overlapped to the result of Hiura et al. (2018). Although, all the components have moved to the south, the motion is not ballistic. A striking steep bending (flip motion) was observed for C3 in 2015. The position of C3 suddenly shifted 0.5 mas to the east with a time scale of 400 days, and is going back to the south-west in 2016. In addition, the flux density of C3 quickly increased from 10 Jy to 17 Jy during 2016, while the other components did not show prominent variability.

We considered the origin of this flip motion. A simple and plausible model is changing the local brightness distribution. Change in local brightness distribution may cause an apparent bending of the trajectory of C3. Since C3 has a spread structure, the peak position of the brightness distribution could be shifted by a slight brightness change. Such local brightening occur if the jet collide with an ambient matter and a shock occurred, and

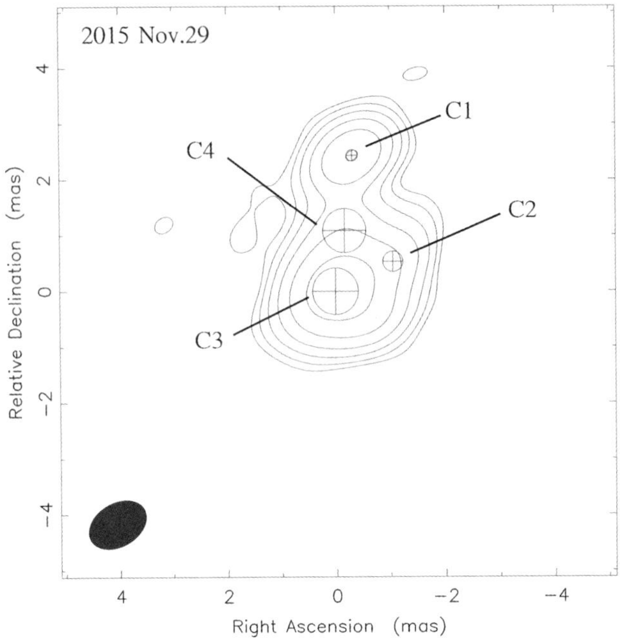

Figure 1. A 22 GHz total intensity image of 3C84 on Nov 29, 2015 by VERA with circular Gaussian components imposed.

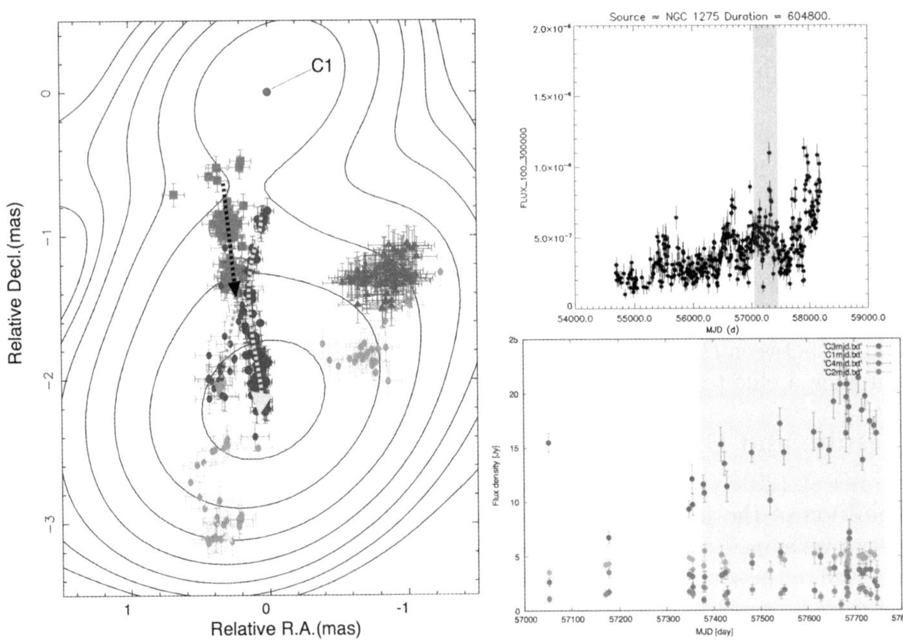

Figure 2. (Left) The positions of the jet components relative to the core (C1) for 33 epoch from 2015 to 2016 (C2: orange, C3: light blue, C4: purple) overlapped to the result by Hiura et al. (2018) from 2007 to 2013 (C2: green, C3: blue, C4: red). It superposed on the contours of the 22 GHz intensity distribution at January 6, 2014. Right) Radio (upper-panel) and gamma-ray (lower-panel) light curves. The gray zones in two panels indicate the period in which the flux density of C3 increases.

high energy particles are generated in the shock, consequently the synchrotron radiation is strengthened. As well as the radio emission, gamma-rays would increase in such shock region. The flux of the gamma-ray was at high-level during the bending period (Fig 2), and this may be caused by the shock.

References

Nagai *et al.* 2013, *PASJ*, 65, 24
Hiura *et al.* 2018, arXiv180605302H

Perseus in Sicily: From Black Hole to Cluster Outskirts
Proceedings IAU Symposium No. 342, 2018
K. Asada, E. de Gouveia Dal Pino, M. Giroletti,
H. Nagai & R. Nemmen, eds.

doi:10.1017/S174392131800594X

Chandra Early-Type Galaxy Atlas

Dong-Woo Kim[1], Craig Anderson[1], Douglas Burke[1], Raffaele D'Abrusco[1], Giuseppina Fabbiano[1], Antonella Fruscione[1], Jennifer Lauer[1], Michael McCollough[1], Douglas Morgan[1], Amy Mossman[1], Ewan O'Sullivan[1], Alessandro Paggi[1], Saeqa Vrtilek[1] and Ginevra Trinchieri[2]

[1]Smithsonian Astrophysical Observatory 60 Garden Street, Cambridge,MA 02138, USA
[2]INAF-Osservatorio Astronomico di Brera, Via Brera 28, 20121 Milan, Italy

Abstract. The hot ISM in early-type galaxies (ETGs) plays a crucial role in understanding their formation and evolution. The structural features of the hot gas identified by Chandra observations point to key evolutionary mechanisms, (e.g., Kim & Pellegrini 2012). In our Chandra Galaxy Atlas (CGA) project, taking full advantage of the Chandra capabilities, we systematically analyzed the archival Chandra data of 72 ETGs and produced uniform data products of the hot gas properties. The main data products include spatially resolved 2D spectral maps of the hot gas from individual galaxies. We emphasize that new features can be identified in the spectral maps which are not easily visible in the surface brightness maps. The high-level images can be viewed at the dedicated CGA website, and the CGA data products can be downloaded to compare with other wavelength data and to perform user-specific analyses. Utilizing our data products, we will further address focused science topics.

Keywords. galaxies: elliptical and lenticular, cD – X-rays: galaxies

---

72 E and S0 galaxies (type < 0, based on RC3) from the Chandra archive through AO15 (earlier work by Diehl & Statler 2007 used data only up to AO4) were selected and homogeneous data products of the spatially resolved 2D distribution of spectral properties of the hot gas were explored.

These 2D *spectral* maps (of the hot gas temperature, emission measure, projected pseudo-pressure and projected pseudo-entropy) can reveal unique features, which may not be visible in 1D radial profiles or the 2D surface brightness maps alone. We make use of newer and improved data analysis (and statistical) techniques (e.g., CIAO tools and scripts), instrument calibration data (in Chandra CALDB), and atomic data (in ATOMDB).

We have developed our own analysis pipelines to apply robust data reduction methods in three main steps: (1) merging multiple observations and imaging diffuse emission after point sources are excluded, (2) adaptively binning to determine optimal spectral extraction regions, (3) extracting spectra from each spatial bin, fitting and mapping spectral parameters.

We provide the Chandra Galaxy Atlas data products in a dedicated CGA website. The 2D images can be directly viewed on the browser and the necessary data products can be downloaded for further analysis. For those who want to take a quick look at the hot gas distribution and the thermal structure,the figures posted on the CGA website can be useful. A screenshot example is in Figure 1. The left panel shows a part of the CGA image gallery with postage stamp images of diffuse gas of individual galaxies and the right panel show the spectral maps of NGC 4649 (4th-row 2nd-column in the left panel) which is displayed simply by clicking the galaxy image on the left panel.

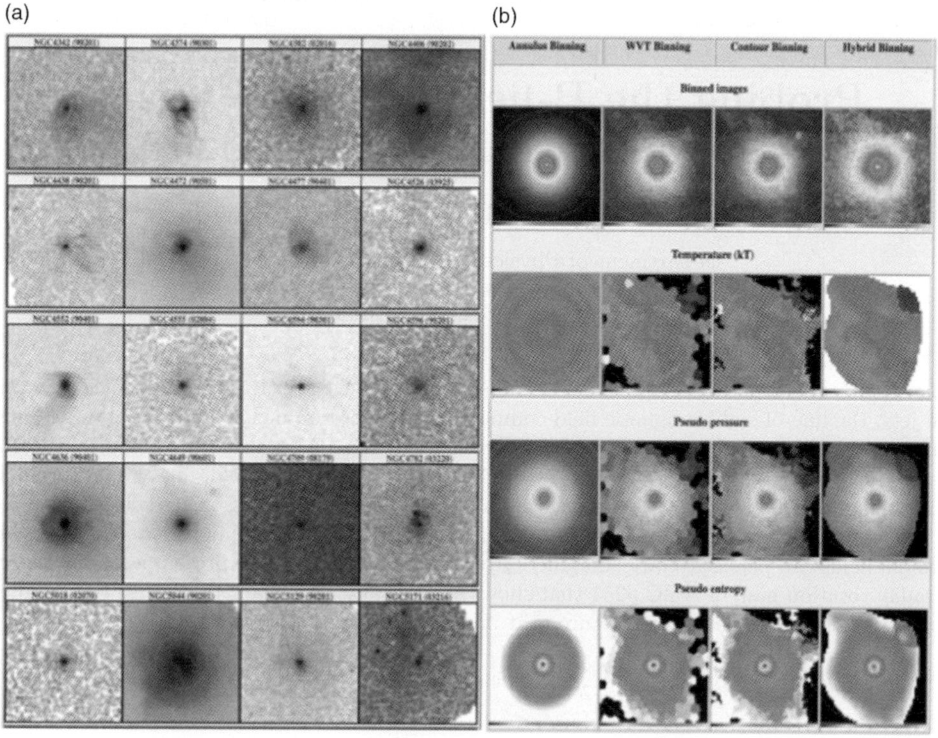

Figure 1. (a) The image gallery of diffuse gas from individual galaxies available in the CGA web site. (b) An example of a set of spectral maps (binned image, temperature map, pressure map and entropy map from top to bottom) of NGC 4649. (The color image is available in the online version.)

Data products are made available in two packages. The main package consists of the high-level data products, e.g., point source removed and filled, exposure corrected images (jpg/png and FITS format) in multiple energy bands to best illustrate the diffuse hot gas distribution, derived temperature maps (jpg/png and FITS format) in various adaptive binning methods to show the hot gas thermal structure. They can be used to get an overview, to obtain the necessary hot gas quantities and to compare them with data at other wavelengths with no additional X-ray data reduction. For more specific analyses, e.g., to extract and fit X-ray spectra from user-specified regions, we provide all the necessary data in the supplementary package, e.g., event files per obsid and per ccd.

Utilizing our CGA data products, we are investigating a range of focused scientific questions, including the hot gas morphology, global properties, radial profiles, X-ray based mass profiles (see Paggi *et al.* 2017 as an example of our CGA data usage), and LMXBs (see Kim *et al.* 2018 in prep).

References

Diehl, S. & Statler, T. S. 2007, *ApJ*, 668, 150
Kim, D.-W. & Pellegrini, S. 2012 “Hot ISM in Elliptical Galaxies” *ASAL* Vol. 378 (Springer)
Paggi, A. Kim, & D.-W. Anderson, C. *et al.* 2017, *ApJ*, 844, 5

Perseus in Sicily: From Black Hole to Cluster Outskirts
Proceedings IAU Symposium No. 342, 2018
K. Asada, E. de Gouveia Dal Pino, M. Giroletti,
H. Nagai & R. Nemmen, eds.

doi:10.1017/S1743921318005161

Probing the B-fields of AGN jets on kiloparsec scales - NGC 6251

Sebastian Knuettel and Denise Gabuzda

Department of Physics, University College Cork,
Cork, Ireland
email: s.knuettel@mars.ucc.ie

Abstract. By constructing images of the Faraday rotation measure (RM) of large scale astrophysical jets, the line-of-sight magnetic field component and electron density in the region of Farady rotation can be investigated. A significant gradient in the RM transverse to the jet direction may indicate a corresponding gradient in the line-of-sight magnetic field, implying a toroidal or helical magnetic field, which would, in turn, imply the presence of an associated electrical current in the jet. The detection of such large scale gradients can reliably demonstrate that helical or toroidal fields can persist to large distances from the central AGN. We present a kiloparsec-scale Faraday rotation map of NGC 6251 that shows statistically significant transverse RM gradients across its kiloparsec scale jet structure that imply an outward current.

Keywords. galaxies: active, magnetic fields, polarization, galaxies: individual (NGC 6251).

1. Introduction

The radio emission from astrophysical jets is primarily synchrotron and can be highly linearly polarised, making the behaviour of the polarised radiation a useful tool in examining the B field environment of these jets. Faraday rotation plays an important role in the detection such magnetic fields. This phenomenon occurs in regions with free electrons and magnetic field, where the polarization angle of an electromagnetic wave changes according to the equations: $\chi_{\rm obs} = \chi_0 + {\rm RM}\lambda^2$, ${\rm RM} \propto \int_0^d n_e \vec{B} \cdot {\rm d}\vec{l}$; where $\chi_{\rm obs}$ is the observed polarization angle, χ_0 is the intrinsic (zero wavelength) polarization angle, $\vec{B}$ is the magnetic field, ${\rm d}\vec{l}$ is a path element along the line of sight, n_e is the electron density, and λ is the observing wavelength. Gradients in the Faraday rotation measure (RM) as evidence for toroidal magnetic fields on parsec and kiloparsec scales have been discussed by Gabuzda *et al.* (2018) and Christodoulou *et al.* (2016). The presence of a transverse RM gradient across the kiloparsec-scale jet of NGC 6251 was suggested by Perley *et al.* (1984). To accurately determine the statistical significance of these gradients the original data used in Perley *et al.* (1984) were downloaded from the VLA archive and analysed again using modern standard methods in AIPS.

2. Overview

An approximately 25″ section of the jet shows monotonic transverse RM gradients with significances of over 3σ, see Figure 1. The plot to the right shows the significances of the gradients for a series of transverse RM slices along this section of the jet with the dashed lines representing the 2σ and 3σ thresholds. This significance was calculated using the end point RM values and their associated errors from the χ versus λ^2 fits:

$$\text{significance}(\sigma) = \frac{|{\rm RM}_1 - {\rm RM}_2|}{\sqrt{\Delta{\rm RM}_1^2 + \Delta{\rm RM}_2^2}} \tag{2.1}$$

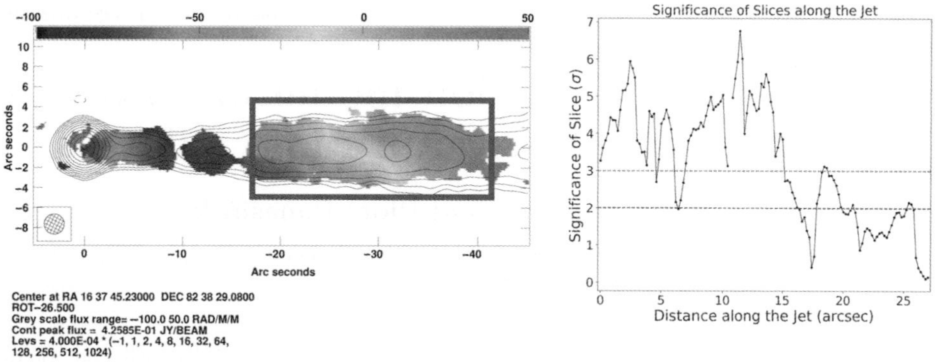

Figure 1. RM map of the inner part of the kiloparsec scale jet of NGC 6251 is given in the left panel. The colour scale represents the RM in rad m^{-2}, the contours of total intensity at 1.37 GHz are in black and start at 0.4 mJy and increase in factors of 2. A 2$''$ circular beam was used to produce the RM map and contour map. Note that the image has been rotated by 26.5° in the clockwise direction. The right panel shows a plot of the significances of the gradients shown by a series of transverse slices in the region bordered by the purple box in the left panel.

Here the subscripts indicate the endpoint values of the RM gradient. The Galactic component of the RM has been estimated to be $\simeq -49$ rad m^{-2} in Perley *et al.* (1984) and has been subtracted from the map; this reveals gradients where the RM changes from positive to negative, a clear indicator that the line-of-sight magnetic field is changing direction across the jet, rather than just the electron density.

3. Implications

NGC 6251 shows statistically significant transverse RM gradients across its kiloparsec scale jet that imply an outward current, as is indicated by the right-hand rule for the relationship between current and associated B field. While it has been suggested the jet of NGC 6251 is collimated due to the more dense intergalactic medium around it, X-ray observations suggest the pressure of the intergalactic medium is not suffficent and magnetic confinement seems more likely (Mack *et al.* (1997)). This result adds to previous detections of significant transverse Faraday rotation gradients across large-scale jets, which have all corresponded to outward currents (11 in Christodoulou *et al.* (2016) and Knuettel *et al.* (2017)). This is a highly statistically significant result, and may provide key information about the system of currents and magnetic fields generated in AGN jets. Combining this with results for AGN jets on parsec scales, which imply a predominance of inward currents, this suggests that the jets have a current/magnetic field configuration similar to that of a co-axial cable as dexcribed by Gabuzda *et al.* (2018).

References

Christodoulou, D. M., Gabuzda, D. C., Knuettel, S.,Contopoulos, I., Kazanas, D., & Coughlan, C. P. 2016, *A&A*, 591, A61

Gabuzda, D. C., Nagle, M. & Roche, N. 2017, *A&A*, 612, A67

Knuettel, S., Gabuzda, D., & O'Sullivan, S., 2017, in: E. Angelakis, M. Boettcher & J. L. Gómez (eds), *Polarized Emission from Astrophysical Jets*, Galaxies 2017, 5, p. 61

Mack1, K.-H., Kerp, J., & Klein, U., 1997, *A&A*, 324, 87087

Perley, R. A., Bridle, A. H., & Willis, A. G. 1984, *ApJS*, 54, 291

Perseus in Sicily: From Black Hole to Cluster Outskirts
Proceedings IAU Symposium No. 342, 2018
K. Asada, E. de Gouveia Dal Pino, M. Giroletti,
H. Nagai & R. Nemmen, eds.

doi:10.1017/S1743921318008141

External blob radiation model for the TeV gamma-ray emission in radio galaxies

Wlodek Bednarek and Piotr Banasiński

Department of Astrophysics, University of Lodz, 90-236 Lodz, ul. Pomorska 149/153
email: bednar@uni.lodz.pl

Abstract. We propose that two emission regions are present in the decelerating jet of radio galaxy at this same moment. Isotropic electrons, in the farther blob as counted along the jet, up-scatter mono-directional soft radiation, from the closer blob, at a relatively large angles to the jet axis. Such model can naturally explain the large angle gamma-ray emission observed from radio galaxies.

Keywords. Galaxies: active, galaxies: jets, radiation mechanisms: non-thermal, gamma-rays: galaxies

---

The inner radio jets in radio active galaxies observed at TeV γ-ray energies are inclined at large angles to the line of sight. We consider the standard blob in a jet scenario for the variable, high energy emission from the inner jet in radio galaxies. It is assumed that two emission regions are present within the jet at this same moment. The inner region (called blob I) moves fast and emits highly collimated radiation towards the outer, slowly moving region (called blob II). Isotropically distributed relativistic electrons, in the blob II reference frame, up-scatter collimated radiation from the inner blob I, preferentially in the direction opposite to the jet propagation. Therefore, γ-rays, produced in the blob II, are emitted at larger angles to the observer's line of sight than expected in the case of the isotropic emission region. On the other hand, radiation produced in blob I is highly collimated along the jet axis. This emission is characteristic for the BL Lac type objects observed at a small viewing angles. We discuss the basic emission features of such external blob radiation model. We argue that the model can describe the high energy emission from radio galaxies viewed at large angles. The details of the model can be found in Banasiński & Bednarek (2019).

Our calculations show that the large angle emission can be significantly enhanced if the relativistic electrons in the blob II inverse Compton up-scatter mono-directional soft radiation produced in the blob I. Therefore, the existence of such two regions in the jet, either well localized or quite extended, seems to provide natural mechanism for the γ-ray emission observed from radio galaxies.

We calculate the γ-ray spectra produced by isotropic electrons with the equilibrium spectrum, $dN_{\rm e}/dE_{\rm e}$, which scatter mono-directional soft radiation with the spectrum, $n(\varepsilon)$, in the blob II reference frame, by integration the following formula,

$$\frac{dN}{d\varepsilon_\gamma dt' d\Omega'} = \int_{E_{\rm L}}^{E_{\rm U}} \frac{dN_{\rm e}}{dE_{\rm e}} \int_{\varepsilon_{\rm L}}^{\varepsilon_{\rm U}} \frac{dN(\theta', \varepsilon, E_{\rm e})}{d\varepsilon_\gamma d\varepsilon dt' d\Omega'} d\varepsilon dE_{\rm e}, \tag{1}$$

where $dN(\theta', \varepsilon, E_{\rm e})/(d\varepsilon_\gamma d\varepsilon dt' d\Omega')$ is the γ-ray spectrum produced by the mono-energetic electrons by scattering soft photons with energy ε. It is expressed by the convenient formulas obtained by Aharonian & Atoyan (1981). These spectra depend on the angle, θ', which is the angle between the directions of the soft photon and the γ-ray as measured in the blob II reference frame. In our geometrical case this angle is measured from the jet axis. In order to produce γ-ray photon with energy, ε_γ, the electron, soft photon, and γ-ray has to fulfill the following minimum kinetic condition given by, $\varepsilon_{\rm min} = 0.5\varepsilon_\gamma m_{\rm e}c^2/[(1-\cos\theta)E_{\rm e}^2(1-\varepsilon_\gamma/E_{\rm e})]$ (see Appendix A in Moderski *et al.* 2005). If $\varepsilon_{\rm min}$ is greater than the lower range of the soft photon spectrum, $\varepsilon_{\rm L}$, the integration is performed only over the energy range of soft photons above $\varepsilon_{\rm min}$. $\varepsilon_{\rm U}$ is the upper energy range of the soft radiation. We assume that electrons in the blob have the energies in the range $E_{\rm min}$ and $E_{\rm max}$. The lower limit on the energy of the electron, $E_{\rm L}$, which is able to produce γ-ray with energy, ε_γ, is also obtained from the above kinetic condition by introducing in the above formula $\varepsilon = \varepsilon_{\rm U}$. This lower limit has to be below the maximum energy of available electrons in the blob, $E_{\rm max}$. If $E_{\rm L}$ is above $E_{\rm min}$ then we integrate the electron spectrum only from $E_{\rm L}$. We assume that electrons in both blobs have the equilibrium spectra well described by the power law function, i.e. $dN_{\rm e}/dE_{\rm e} = AE_{\rm e}^{-\beta}$.

The model has been applied to the radio galaxy, NGC 1275. We perform the example calculations of the γ-ray spectra assuming two limiting values for the observation angles of the jet in NGC 1275, i.e. $\theta = 30°$ and $50°$. These γ-rays have to propagate through the synchrotron radiation field produced by electrons within the blob II. To find out whether their absorption is important, we calculate the optical depths for the γ-rays in the radiation field of the homogeneous blob II. The absorption effects in the soft radiation field of the blob II has to be taken into account for the case of the blob II. We include them by introducing the reduction factor equal to $R(E_\gamma) = \{1-\exp[-\tau(E_\gamma)]\}/\tau(E_\gamma)$.

We conclude that the spectrum observed simultaneously by the *Fermi*-LAT and the MAGIC telescopes, in the GeV-TeV energy range (Aleksić *et al.* 2014), can be well described for the reasonable power in the low energy soft photons emitted from the blob I, equal to the isotropic equivalent synchrotron power of the order of a few 10^{45} erg s^{-1}. Such emission power is within the range of the powers (a few $\sim 10^{45} - 10^{46}$ erg s^{-1}) observed from the BL Lac type objects, e.g. Mrk 501 (Ahnen *et al.* 2017), Mrk 421 (Aleksić et al. 2012b) or PKS 2155-304 (Aharonian *et al.* 2009, Abramowski *et al.* 2012). In the case of a relatively small optically thick blob ($R = 3\times10^{15}$ cm), the total energy in the relativistic electrons in the blob II, able to explain the γ-ray emission from NGC 1275, should be equal to $\sim 10^{48}$ ergs, for the viewing angles not far from $\sim 30°$. However, the absorption effects for the small blob II viewed at the angle $50°$ are too strong to provide correct description of the NGC 1275 spectrum. In the large blob II model ($R = 3\times10^{16}$ cm), the absorption effects are small enough to provide correct description of the observed spectrum for the whole range of the observation angles $30° - 50°$. However, in this case the energy in relativistic electrons should be much larger, i.e. of the order of a few 10^{49} erg, which is still acceptable in the case of blobs in jets of AGNs.

This work is supported by the grant through the Polish Narodowe Centrum Nauki No. 2014/15/B/ ST9/04043.

References

Abramowski, A. *et al.* 2012 *A&A* 539 149A
Aharonian, F. A., & Atoyan A.M. 1981 *Ap&SS*, 79, 321, 336
Aharonian, F. *et al.* 2009 *A&A* 502, 749
Ahnen, M. L. *et al.* 2017, *A&A* 603, 31A
Aleksić, J. *et al.* 2012 *A&A* 542, 100A Aleksić *et al.* (2014)
Aleksić, J. *et al.* 2014a *A&A* 564, A5
Banasiński, P., Bednarek, W. *et al.* 2018 submitted to *ApJ*
Moderski, R., Sikora, M., Coppi, P. S., & Aharonian, F. 2005 *MNRAS* 363, 954

Perseus in Sicily: From Black Hole to Cluster Outskirts
Proceedings IAU Symposium No. 342, 2018
K. Asada, E. de Gouveia Dal Pino, M. Giroletti,
H. Nagai & R. Nemmen, eds.

doi:10.1017/S1743921318005112

A negative correlation between $R_{\rm UV}$ and $\alpha_{\rm ox}$ in low-luminosity AGNs

Shuang-Liang Li

Key Laboratory for Research in Galaxies and Cosmology, Shanghai Astronomical Observatory, Chinese Academy of Sciences, 80 Nandan Road, Shanghai 200030, China
email: lisl@shao.ac.cn

Abstract. It has been controversial for years that the accretion mode is different for bright active galactic nuclei (AGNs) and low-luminosity AGNs (LLAGNs). In this work, we compile from literature a sample of 32 LLAGNs, consisting 18 LINERs and 14 low Eddington ratio (λ) Seyfert galaxies. A strong negative correlation between the radio loudness $R_{\rm UV}$ and the optical to X-ray spectral index $\alpha_{\rm ox}$ is reported for the first time. We further demonstrate that this negative correlation can be understood consistently and comprehensively under the truncated accretion–jet model, a model that has been applied successfully to LLAGNs. We argue that the scatter in the observations is mainly due to the spread in the viscosity parameter α of a hot accretion flow, a parameter that can potentially serve as a diagnosis of the strength and/or configuration of magnetic fields in accretion flows.

Keywords. accretion, accretion discs - black hole physics - galaxies: active - galaxies:Seyfert

1. Introduction

It has been known for years that, statistically, LLAGNs are distinct in various aspects from bright AGNs. A hot accretion flow is believed to be present in LLAGNs (Narayan & Yi 1994, Bu *et al.* 2016). However, in observations, apart from the global trends observed, large dispersion is also witnessed in both the $\Gamma - \lambda$ correlation (e.g. Gu & Cao 2009) and the $\alpha_{\rm ox} - \lambda$ correlation (e.g. Xu 2011). Obviously these large scatter trends indicate that, besides the accretion rate that controls the luminosities in different wavebands, additional parameters/factors waiting to be identified should also play certain roles. One possible factor is the difference in the viscosity parameter α of the hot accretion flow among different systems.

2. The LLAGN sample

We gather from the literature a sample of LLAGNs, which consist of both LINERs and low-λ Seyfert galaxies. After excluding the LLAGNs whose X-rays may be dominated by emission from relativistic jets, our final LLAGN sample includes 14 low-λ Seyfert galaxies and 18 LINERs (Li & Xie 2017).

3. Main results and theoretical interpretation

For the first time, a very strong negative correlation between $R_{\rm UV}$ and $\alpha_{\rm ox}$, i.e., $\log R_{\rm UV} = (-2.39 \pm 0.48)\,\alpha_{\rm ox} + 5.61 \pm 0.49$, is found with the confidence level larger than 99.9 % (Fig. 1). We suggest that this correlation can be well fitted by the truncated accretion-jet model. The scatter in this correlation is argued to relate to the viscosity parameter α of the hot accretion flow component. A theoretical correlation between $R_{\rm UV}$

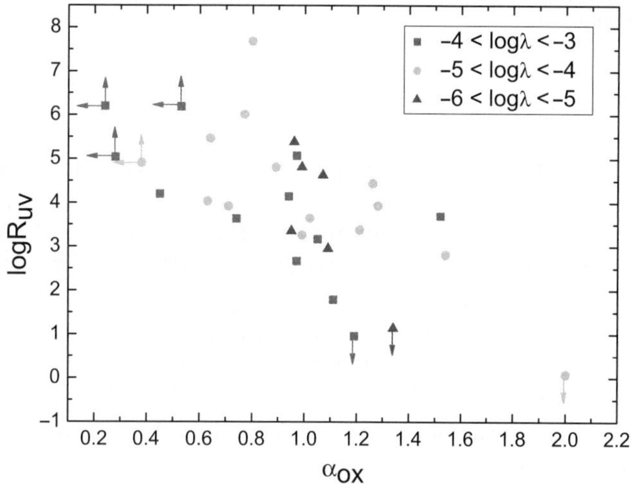

Figure 1. The relationship between radio loudness R_{UV} and the optical to X-ray spectral index α_{ox} for bright AGNs with $\lambda > 10^{-2}$.

and α_{ox} ($\log R_{UV} \approx -3.25\ \alpha_{ox} - 0.95\ \log\lambda + const.$) can provide very well fitting for the observations (see Li & Xie 2017 for details).

4. Summary

In this work, we gather from the literature a sample of 32 LLAGNs, including 14 low-λ Seyfert galaxies and 18 LINERs. From this sample, we observe a very strong negative correlation between R_{UV} and α_{ox}. Then, based on the truncated accretion-jet model, we provide a comprehensive understanding of this new observation.

Acknowledgements

This work is supported by the NSFC (grants 11773056).

References

Bu D. F., Yuan F., Gan Z. M., & Yang X. H., 2016, *ApJ*, 818, 83
Gu, M., & Cao, X., 2009, *MNRAS*, 399, 349
Li, S.-L., & Xie, F.-G., 2017, *MNRAS*, 471, 2848
Narayan, R., & Yi, I. 1994, *ApJ*, 428, L13
Xu, Y.-D., 2011, *ApJ*, 739, 64

Perseus in Sicily: From Black Hole to Cluster Outskirts
Proceedings IAU Symposium No. 342, 2018
K. Asada, E. de Gouveia Dal Pino, M. Giroletti,
H. Nagai & R. Nemmen, eds.

doi:10.1017/S1743921318008049

A precessing and nutating jet in OJ287

Silke Britzen[1], C. Fendt[2], G. Witzel[3], S.-J. Qian[4], I. N. Pashchenko[5], O. Kurtanidze[6,7], M. Zajacek[1,8,10], G. Martinez[3], V. Karas[8], M. Aller[9], H. Aller[9], A. Eckart[1,10], K. Nilsson[11], P. Arévalo[12], J. Cuadra[13], M. Subroweit[10] and A. Witzel[1]

[1]Max-Planck-Institut für Radioastronomie, Auf dem Huegel 69
53121 Bonn, Germany
email: sbritzen@mpifr.de

[2]Max-Planck-Institut für Astronomie, Königstuhl, Heidelberg, Germany

[3]UCLA, Department of Physics and Astronomy, LA, CA 90095, USA

[4]National Astronomical Observatories, Chinese Academy of Sciences, Beijing 100012, China

[5]Astro Space Center, Lebedev Physical Institute, Russian Academy of Sciences

[6]Abastumani Observatory, Mt Kanobili, 0301 Abastumani, Georgia

[7]Engelhardt Astronomical Observatory, Kazan Federal University, Tatarstan, Russia

[8]Astronomical Institute, Academy of Sciences, Boční II 1401, CZ-14131 Prague, Czech Republic

[9]University of Michigan, Ann Arbor, MI 48109, USA

[10]I. Physikalisches Institut der Universität zu Köln, Zülpicher Str. 77, 50937 Köln, Germany

[11]Tuorla Observatory, Department of Physics and Astronomy, University of Turku, 20500, Turku, Finland

[12]Instituto de Física y Astronomía, Facultad de Ciencias, Universidad de Valparaíso, Gran Bretaña No. 1111, Playa Ancha, 2360102 Valparaíso, Chile

[13]Instituto de Astrofísica, Pontificia Universidad Católica de Chile, 782-0436 Santiago, Chile

Abstract. We re-analyzed OJ287 in 120 Very Long Baseline Array (VLBA, MOJAVE) observations (at 15 GHz) covering the time span between Apr. 1995 and Apr. 2017. We find that the radio jet motion over the sky is consistent with a precessing and nutating jet source. The variability of the radio flux-density can be explained by Doppler beaming due to a change in the viewing angle. We suggest that part of the optical emission is due to synchrotron emission related to the jet radiation. We find a strikingly similar scaling for the timescales for precession and nutation as indicated for SS433 with a factor of roughly 50 times longer in OJ287.

Keywords. techniques: interferometric, galaxies: jets, BL Lacertae objects: individual (OJ 287)

1. Introduction

OJ287 (z=0.306 Stickel *et al.* (1989), a low synchrotron peaked (LSP) BL Lac Object) has been named the Rosetta stone of blazars by Takalo (1994), expressing the hope, that deciphering this particular blazar will lead to a better understanding of the active galactic nucleus (AGN) phenomenon in general. A light-curve observed in the optical V-band since 1890 shows repeated outbursts at ~11.65 yr intervals of this object (Sillanpää *et al.* (1988)). Sillanpää *et al.* (1988) proposed that OJ287 is a supermassive binary black hole (SMBBH) with an orbital period of 9 yr (in the rest frame of OJ287). Further works by Lehto & Valtonen (1996), Valtonen & Wiik (2012), and Valtonen *et al.* (2016) have explored the binary BH nature of this AGN in a scenario where disturbances of its accretion disc are caused by a plunging black hole (BH).

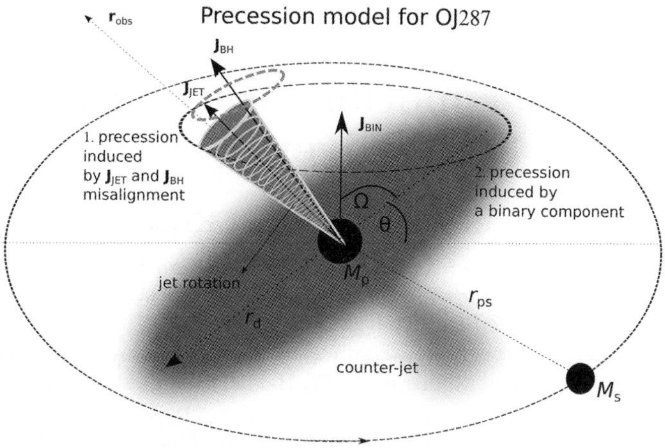

Figure 1. Sketch to illustrate two of the scenarios discussed in more detail in Britzen *et al.* (2018)): 1. The Lense-Thirring precession due to the misalignment of angular momenta of the Kerr black hole and that of the accretion disc. 2. The disc-jet precession due to the torques induced by a secondary black hole.

2. Results

We briefly list the most important results of the findings presented in detail in Britzen *et al.* (2018).

Jet stability. Despite several claims, the jet is not wobbling in an erratic way but, instead, it reveals smooth and continuous motion in the plane of the sky.

Jet precession. The jet ridge line connects all jet features modeled per Gaussian components per epoch, which reveals a precession of the radio jet on the timescale of about 22 yr. This appears to be consistent with (double of) the previously claimed OJ287 perodicity of $\sim$ 11 yr observed in the optical data.

Jet nutation. In addition to the dominant motion of the jet - the precession - a yearly motion is superimposed. This can best be explained and modeled as the jet nutation.

In summary, the radio variability on long time scales ($\sim$ two decades) can best be explained and modeled via jet precession, whereas the short term variability on timescales of a year can best be explained and modeled by jet nutation.

Physically, the *precession can be caused by a SMBBH or Lense-Thirring precession (a misaligned disc around a single black hole).* A plunging black hole in a SMBBH scenario does not seem to be required to explain the observed jet motion phenomena. A sketch is provided to illustrate both options (Fig. 1).

References

Britzen, S., Fendt, C., Witzel, G., Qian, S.-J., *et al.* 2018, *MNRAS*, 478, 3199

Lehto H. J., & Valtonen M. J. 1996, *ApJ*, 460, 207

Sillanpää A., Haarala S., Valtonen M. J., *et al.*, 1988, *ApJ*, 325, 628

Stickel, M., Fried, J. W., & Kühr, H. 1989, *A&AS*, 80, 103

Takalo L. 1994, *Vistas in Astronomy*, 38, 77

Valtonen M. J., & Wiik K. 2012, *MNRAS*, 421, 1861

Valtonen M. J., Zola S., Ciprini S., *et al.* 2016, *ApJL*, 819, L37

Perseus in Sicily: From Black Hole to Cluster Outskirts
Proceedings IAU Symposium No. 342, 2018
K. Asada, E. de Gouveia Dal Pino, M. Giroletti,
H. Nagai & R. Nemmen, eds.

doi:10.1017/S174392131800385X

A jet proper motion study in the early Universe

Krisztina Perger[1,2], Sándor Frey[2] and Krisztina É. Gabányi[2,3]

[1]Department of Astronomy, Eötvös Loránd University
Pázmány Péter sétány 1/A, H-1117 Budapest, Hungary
email: k.perger@astro.elte.hu

[2]Konkoly Observatory, MTA Research Centre for Astronomy and Earth Sciences
Konkoly Thege Miklós út 15-17, H-1121 Budapest, Hungary

[3]MTA-ELTE Extragalactic Astrophysics Research Group
Pázmány Péter sétány 1/A, H-1117 Budapest, Hungary

Abstract. We constrained the jet proper motion of PMN J2134–0419 at $z = 4.33$ to $\mu = 0.035 \pm 0.023$ mas yr^{-1} using very long baseline interferometry (VLBI). This is the second most distant source where such a measurement could be made. Another distant blazar SDSS J1026+2542 ($z = 5.27$) shows jet component proper motions up to $\mu = 0.112 \pm 0.031$ mas yr^{-1}. In both cases, the measured values are consistent with the expectations of current ΛCDM cosmological models.

Keywords. galaxies: active, galaxies: jets, galaxies: high-redshift, quasars: individual (PMN J2134–0419), quasars: individual (SDSS J1026+2542)

---

1. Introduction

Observing high-redshift active galactic nuclei (AGN) at radio wavelengths provides crucial information about the early epochs of our Universe. Using the very long baseline interferometry (VLBI) technique, the structure of radio jets can be studied with milliarcsecond (mas) resolution at cm wavelengths. With observations at two suitably distant epochs, the proper motion of the jet components and the physical and geometric properties of the plasma flow can be determined.

So far there are only two AGN at the highest redshifts (PMN J2134–0419 at $z = 4.33$ and SDSS J1026+2542 at $z = 5.27$) that show suitably extended mas-scale jet structure and have VLBI imaging data at two epochs so that a jet proper motion study could be performed (Frey *et al.*, 2015; Perger *et al.*, 2018).

2. Radio jet proper motion at $z > 4$

J1026+2542 was observed with the Very Long Baseline Array (VLBA) and with the European VLBI Network (EVN), in 2006 and 2013, respectively, while J2134–0419 was observed by the EVN in 1999 and 2015. All observations were carried out at 5 GHz.

The two-epoch analysis of the radio emission in the parsec-scale jet of both AGN indicates an outward motion. J1026+2542 has multiple jet components, while J2134–0419 exhibits a more simple structure. The change in the angular distance of the core and jet components yields proper motion values less than 0.1 mas yr^{-1} for both sources.

Applying standard ΛCDM cosmology ($H_0 = 70$ km s^{-1} Mpc^{-1}, $\Omega_M = 0.3$, and $\Omega_\Lambda = 0.7$), the proper motions can be translated to apparent superluminal speeds, which span the range of $[3.3 \pm 3.9, 14.0 \pm 3.9]\,c$ for J1026+2542, and has the value of $4.1 \pm 2.7\,c$ -for J2134–0419. The calculated values of the bulk Lorentz factor and the jet

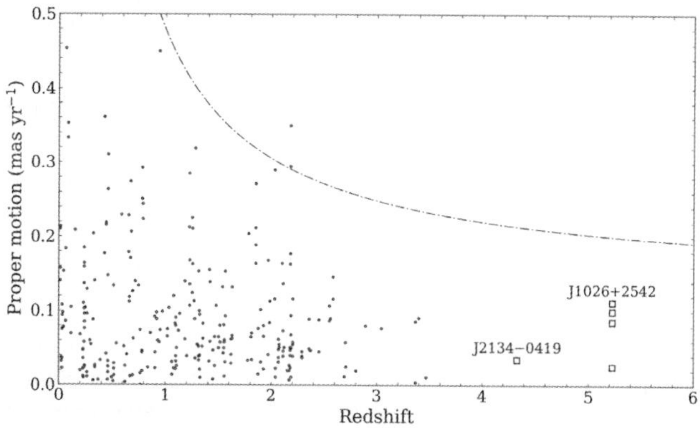

Figure 1. Distribution of jet proper motions in AGN as a function of redshift. Circles denote the best quality flag sources from Britzen *et al.* (2008) for comparison at lower redshifts. The curve represents the upper limit based on the applied cosmology, assuming a Lorentz-factor $\Gamma = 25$.

inclination angle to the line of sight span the ranges $\Gamma = 12.5 - 15$ and $\theta = 4.6° - 10°$ for J1026+2542, and the ranges $\Gamma = 2 - 7$ and $\theta = 3° - 20°$ for J2134–0419.

3. Conclusions

The constraints obtained on the jet proper motion of the two high-redshift AGN are consistent with the upper limit estimated by the current cosmological models (e.g. Kellermann *et al.* 2004). The determined Lorentz factor and viewing angle are consistent with those obtained from models fitted to the broad-band spectral energy distribution (SED, Sbarrato *et al.* 2013) for J1026+2542. The radio interferometric data show slightly lower Lorentz factor and higher viewing angle for J2134–0419 than the SED study (Sbarrato *et al.* 2015).

Acknowledgements

KP is grateful for the IAU grant supporting her to attend the Symposium. KÉG was supported by the János Bolyai Research Scholarship of the Hungarian Academy of Sciences. We thank the Hungarian National Research, Development and Innovation Office (OTKA NN110333) for support.

References

Britzen, S., Vermeulen, R. C., Campbell, R. M., *et al.* 2008, *A&A*, 484, 119
Frey, S., Paragi, Z., Fogasy, J. O. & Gurvits, L. I. 2015, *MNRAS*, 446, 2921
Kellermann K. I., Lister, M. L., Homan, D. C. *et al.* 2004, *ApJ*, 609, 539
Perger, K., Frey, S., Gabányi, K. É. *et al.* 2018, *MNRAS* 477, 1065
Sbarrato, T., Tagliaferri, G., Ghisellini, G. *et al.* 2013, *ApJ*, 777, 147
Sbarrato, T., Ghisellini, G., Tagliaferri, G. *et al.* 2015 *MNRAS*, 446, 2483

Perseus in Sicily: From Black Hole to Cluster Outskirts
Proceedings IAU Symposium No. 342, 2018
K. Asada, E. de Gouveia Dal Pino, M. Giroletti,
H. Nagai & R. Nemmen, eds.

doi:10.1017/S1743921318007251

Evolution of the $M_\bullet$-σ relation

D. Bhattacharyya and A. Mangalam

Indian Institute of Astrophysics, Sarjapur Road, Koramangala, Bangalore, 560034, India
emails: dipanweeta@iiap.res.in, mangalam@iiap.res.in

Abstract. Black holes at the centers of the galaxies grow mainly by the processes of accretion, mergers, and consumption of stars. In the case of gas accretion with cooling sources, the flow is momentum driven, after which the black hole reaches a saturated mass, and subsequently, it grows only by consumption of stars. In addition, we include the effect of mergers on the growth of black hole spin and mass and study its evolution as a function of redshift in a ΛCDM cosmology using an initial seed mass and spin distribution functions that we have derived. For the stellar ingestion, we have assumed a power-law density profile for the galaxy in our framework of a new relativistic loss cone theory that includes the effect of the black hole spin. We predict the impact of the evolution on the $M_\bullet - \sigma$ relation and compare it with available observations.

Keywords. spin, mass of black hole —$M_\bullet - \sigma$ relation — redshift — evolution.

1. Introduction

The $M_\bullet - \sigma$ relation, $M_\bullet \propto \sigma^p$, is a very astonishing relation and its origin is still a topic of debate. Bhattacharyya & Mangalam (2018) [BM18 hereafter] have constructed a static model of deriving the $M_\bullet - \sigma$ relation and $M_\bullet - M_b$ relation together from intensity profiles of 12 elliptical galaxies. Here we discuss the dynamical aspect of the $M_\bullet - \sigma$ relation, i. e., the evolution of the relation with redshift and time in ΛCDM cosmology. From the basic equations of mass and spin evolution of the black hole [c. f. Mangalam (2015)] we have derived the evolution of this relation. For the rate of growth of mass, $\dot{M}_\bullet$, we have considered the accretion process, stellar ingestion process and mergers, where, the accretion rate is considered to be a fraction of the Eddington accretion rate, the mass growth by stellar consumption is calculated by using both full and steady loss cone theory [Mageshwaran & Mangalam (2015)] for a galaxy cusp following power law mass density (power law index is γ) and in case of mergers we have considered both major and minor mergers [Stewart *et al.*, (2009)]. We have also considered the BZ torque which contributes to spinning down the black hole while the accretion process spins it up. For spin evolution we have considered only the effect of minor mergers which also contributes in spinning the hole down [Gammie *et al.*, (2004)], while we neglect major mergers because the accretion process is the dominant one for spinning up the black hole. We have also incorporated the prescription of saturated mass, M_{sat}, given by King (2003) which is the critical mass after which the accretion stops and the black hole grows only by stellar ingestion and mergers. We have considered the mergers to be effective from $z \leqslant 4$ considering the peak of merger activity. Starting from an initial spin, j_s, an initial mass, $M_{\bullet s}$ we have derived the evolution of the mass and spin of the black hole and its effect on the $M_\bullet - \sigma$ relation. For a few examples, the resulting mass and spin evolution of the black hole is shown in Figures [1(a), 1(c)] and the impact on the index of the $M_\bullet - \sigma$ relation which is obtained from the slope of $\log M_\bullet$ vs $\log \sigma$ plot is shown in Figure 1(b). We show how our results matches with observations in Figure 1(d), where the data of

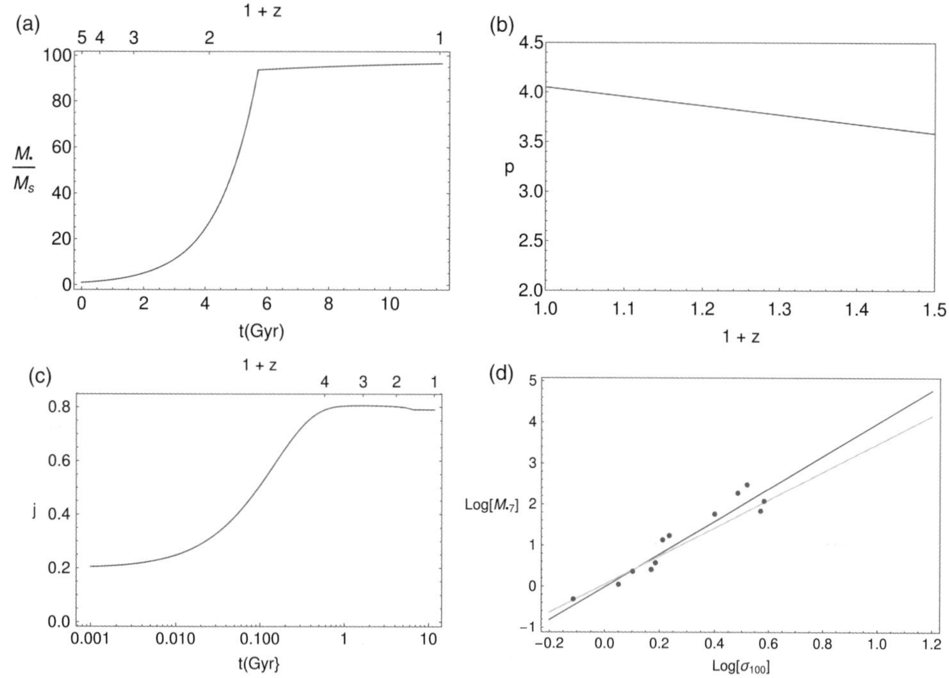

Figure 1. Comic evolution of the mass and the spin of the black hole and the $M_\bullet - \sigma$ relation. *(1a, 1c)* Evolution of the mass of the black hole in units of $M_{\bullet s}$ and spin of the black hole as function of time and redshift with merger mass ratio, $q = 0.1$, $j_s = 0.2$, magnetic field of BZ torque, $B_4 = B/(10^4$ Gauss$) = 5$, formation redshift, $z_f = 4$ and $\gamma = 1.5$ and $M_{\bullet s} = 10^5 M_\odot$. *(1b)* Evolution of the index p of the $M_\bullet \propto \sigma^p$ relation. *(1d)* shows $\log(M_{\bullet 7})$ vs $\log(\sigma_{100})$ for two different redshifts ($z = 0.003$, red and $z = 0.23$, green) along with the data obtained from our calculation in BM18 for 12 elliptical galaxies (their redshift lies in the range 0.004 - 0.002).

$M_\bullet$ and σ of 12 elliptical galaxies was obtained from their observed intensity profile in BM18. A more detailed study of the models is discussed in the paper by Bhattacharyya & Mangalam (2019, submitted).

2. Summary and conclusions

We have obtained the relativistic as well as the non-relativistic evolution of black hole mass and spin as a function of redshift using cosmological ΛCDM model for different values of the spin parameter, seed masses and different formation redshifts. For calculation of stellar consumption rate, we use the steady loss cone model. We have compared our model with the available observations of z, $M_\bullet$ and $\sigma_{||}$ of 12 elliptical galaxies which follow the $M_\bullet - \sigma$ relation and we are able to explain the observations from our model. We have computed the evolution of the $M_\bullet - \sigma$ relation with redshift by deriving the evolution of the slope and intercept of $\log[M_{\bullet 7}]$ vs $\log[\sigma_{100}]$ plot. Using formula given by Shankar *et al.* (2009), for renormalization, $p \propto (1+z)^{-\alpha}$, we show that the $M_\bullet - \sigma$ relation holds with $\alpha \simeq 0.24$ - 0.34 upto $z \simeq 1$ for the index, p lying between 4 - 5. The data from future surveys at high redshift, for example from TMT, can be used to probe the $M_\bullet - \sigma$ evolution.

References

Stewart *et al.* 2009, *ApJ*, 702, 1005
Mangalam, A., 2015, *ASI conference Series*, 12, 51–56

King 2003, *ApJ*, 596, L27
Bhattacharyya, D. & Mangalam, A., 2018, *JAA*, 39, Issue 1, article id 4 (2018a)
Gammie *et al.* 2004, *ApJ*, 602, 312
Mageshwaran, T. & Mangalam, A., 2015, *ApJ*, 814, 141
Shankar, F., *et al.* 2009, *ApJ*, 694, 867

Perseus in Sicily: From Black Hole to Cluster Outskirts
Proceedings IAU Symposium No. 342, 2018
K. Asada, E. de Gouveia Dal Pino, M. Giroletti,
H. Nagai & R. Nemmen, eds.

doi:10.1017/S1743921318005951

Simulations of the W50-SS433 system

Dimitrios Millas[1], **Oliver Porth**[2] **and Rony Keppens**[1]

[1]Centre for mathematical Plasma Astrophysics, KU Leuven,
Celestijnenlaan 200B, 3001, Leuven, Belgium
email: dimitrios.millas@kuleuven.be

[2]Institut für Theoretische Physik, Goethe-Universität Frankfurt am Main,
Max-von-Laue-Str. 1, 60438, Frankfurt am Main, Germany

Abstract. Supernovae and astrophysical jets are two of the most energetic and intriguing objects in the universe. We examine an interesting scenario that involves the interaction of these two extreme phenomena, motivated by observations of the W50-SS433 system: a jet launched from the microquasar SS433 (an X-ray binary) located inside a supernova remnant, W50. These observations revealed a unique morphology of the remnant, attributed to the presence of the jet. We performed full 3D relativistic hydrodynamic simulations to better capture the interaction between the remnant and the jet and post-processed the data with a radiative transfer code to create emission maps.

Keywords. hydrodynamics, methods: numerical, (stars:) supernovae: general, ISM: jets and outflows, (ISM:) supernova remnants

---

1. Introduction

W50 is a quite peculiar supernova remnant (SNR), well known for its characteristic "manatee" shape, with a profound "east - west" asymmetry. SS433 is an X-ray binary, know to host a precessing, relativistic jet (Dubner *et al.* 1998; Monceau-Baroux *et al.* 2014, 2015), with its mean axis almost parallel to the elongated part of the SNR. This elongated shape of the remnant is often associated with the presence of the jet and its interaction with the shell and has been examined in 2D simulations, e.g. Goodall *et al.* 2011. We now performed full 3D relativistic hydrodynamic simulations, aiming to capture the interaction between the SNR and the jet.

2. Simulations & Results

For the simulations we used the relativistic hydrodynamic module from the open source, parallel, grid adaptive, MPI-AMRVAC code (Keppens *et al.* 2012, Porth *et al.* 2014). The domain is a Cartesian grid with dimensions $-100pc \leq x, y, z \leq 100pc$, a base resolution 200^3 and 5 adaptive mesh levels (AMR), leading to an effective resolution of 3200^3. This feature is necessary to capture the initial supernova blast and the injection of the jet at a later time (treated as an internal boundary), while keeping the computational cost as low as possible. This is partially achieved by scaling-up the injection region of the jet. The mean jet axis is considered to be parallel to $\hat{x}$.

Following Goodall *et al.* 2011, we first model the supernova as a sphere of high density and velocity, against a static interstellar medium (ISM). During the free expansion phase, the velocity is self-consistently determined if we assume an energy of $10^{51} ergs$ for the SN blast and a mass of $6M_{\odot}$ for the ejected material; this results in a velocity of $v_{SN} \sim 10^4 km/s$. The jet is injected with a velocity of $v_j \simeq 0.25c$ (or Lorentz factor of $\gamma \sim 1.03$).

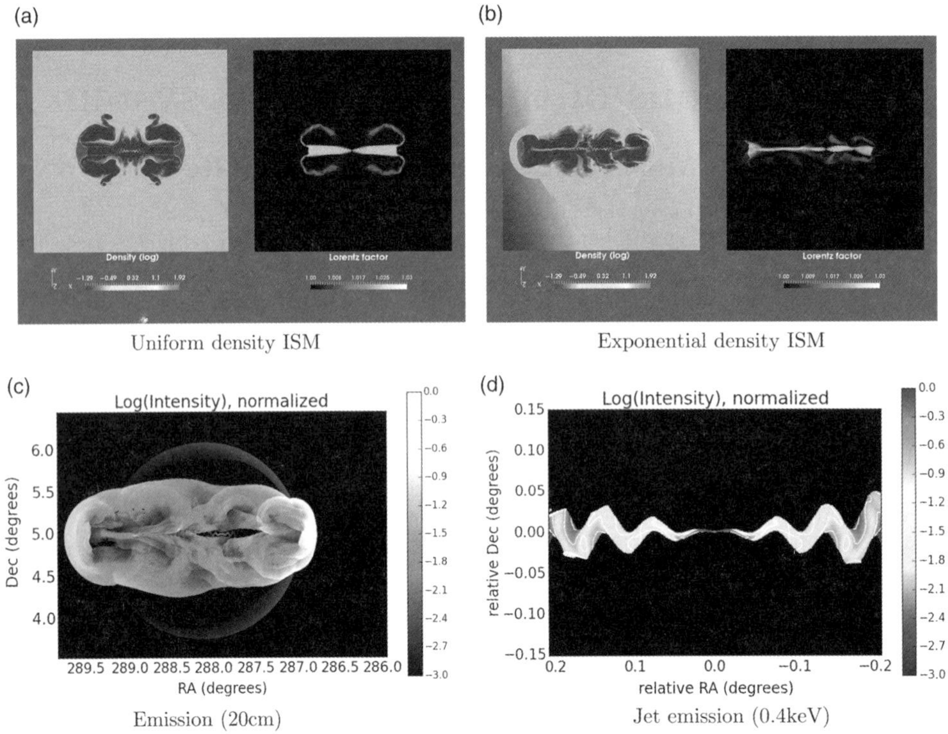

Figure 1. *Top:* Slice of a 3D run: uniform (left) vs exponential (right) density profile for the ISM. For each case we present the density (in log scale) and the Lorentz factor. *Bottom:* Emission map of the full system in radio, at the approximate location of W50-SS433 ($\sim 5.5kpc$) (left) and a zoomed image of the jet in X-rays (right).

The density is calculated via the kinetic luminosity of the jet, $L_{kin} = 10^{39} erg/s$ and the injection radius.

For the ISM we examine two different scenarios: (i) a uniform and (ii) an exponential density profile. In both scenarios, the number density in the center of the domain is fixed to n=1 particle/cm^3. The exponential profile for the ISM can be described as follows:

$$\rho(r, z) = \rho_o \exp(-R_m/R_d - r/R_d - z/Z_d) \tag{2.1}$$

where r is the distance from the galactic centre, z is the distance from the galactic disk and $R_m = 4kpc$, $R_d = 5.4kpc$, $Z_d = 40pc$ are constants determining the relevant scalelengths.

In Figs. 1a,1b we present a slice from the 3D output, perpendicular to the $\hat{z}$ axis, with the density and the Lorentz factor for the two cases. We verify that in order to obtain the observed asymmetry of the SNR, the density gradient is necessary. Synchrotron emission assuming equipartition, i.e. substituting the magnetic field energy density with a fraction of the internal energy density of the fluid (Fromm *et al.* (2016)), has been calculated by post-processing the output. For the exponential case, the emission map from the full simulation is shown in Fig. 1c and a zoomed image of the jet in Fig. 1d.

References

Dubner, G. M., Holdaway, M., Goss, W. M. & Mirabel, I. F. 1998, *AJ*, 116, 1842
Monceau-Baroux, R., Porth, O., Meliani, Z., & Keppens, R. 2014, *A&A*, 561, A30

Monceau-Baroux, R., Porth, O., Meliani, Z., & Keppens, R. 2015, *A&A*, 574, A143
Goodall, P. T., Alouani-Bibi, F., & Blundell, K. M. 2011, *MNRAS*, 414, 2838
Porth, O., Xia, C., Hendrix, T., Moschou, S. P., & Keppens, R. 2014, *ApJS*, 214, 4
Keppens, R., Meliani, Z., van Marle, A. J., Delmont, P., Vlasis, A., & van der Holst, B. 2012, *JCoPh*, 231, 718
Fromm, C. M., Perucho, M., Mimica, P., & Ros, E. 2016, *A&A*, 588, A101

Perseus in Sicily: From Black Hole to Cluster Outskirts
Proceedings IAU Symposium No. 342, 2018
K. Asada, E. de Gouveia Dal Pino, M. Giroletti,
H. Nagai & R. Nemmen, eds.

doi:10.1017/S1743921318007305

Black hole demographics from TDE modeling

T. Mageshwaran[1] and A. Mangalam[2]

[1]Tata Institute of Fundamental Research, Mumbai, India
email: t.mageshwaran@tifr.res.in

[2]Indian Institute of Astrophysics, Bangalore, India
email: mangalam@iiap.res.in

Abstract. The occurence rate of tidal disruption events (TDEs) by survey missions depend on the black hole mass function of the galaxies, properties of the stellar cusp and mass of the central black hole. Using a power law density profile with Kroupa mass function, we solve the steady state Fokker-Planck to calculate the theoretical capture rate of stars by the black hole. Using a steady accretion model, the Schechter black hole mass function (BHMF) and the cosmological parameters, we calculate the detection rate of TDEs for various surveys which is then fit with the observed TDE rates to extract the Schechter parameters. The rate tension between observation ($\sim 10^{-5}$ yr^{-1}) and theory ($\sim 10^{-4}$ yr^{-1}) for individual galaxies is explained by the statistical average over the BHMF.

Keywords. galaxies: nuclei, galaxies: luminosity function, mass function, Galaxy: kinematics and dynamics, galaxies: statistics

---

1. Introduction

A star orbiting close to the black hole such that the black hole's tidal gravity exceeds the star's self-gravity is tidally disrupted at the pericenter $r_p \leq r_t$, where $r_t = (M_\bullet/M_\star)^{1/3} R_\star$ is the tidal radius and is called as tidal disruption events (TDEs) [Rees 1998]. The physical parameters crucial for the study of TDEs are the black hole (BH) mass $M_\bullet$, specific orbital energy E and angular momentum J, star mass $M_\star$ and radius $R_\star$. For a stellar density $\rho(r) \propto r^{-\gamma}$ in the galactic center and the mass function $\xi(m)$ given by Kroupa (2001), Mageshwaran & Mangalam (2015) (hereafter MM15) solved the steady state Fokker-Planck equation (Merritt 2013) to obtain the capture rate $\dot{N}_t \propto M_\bullet^{-0.3}$. We use TDEs as a probe to derive the BHMF from observed detection rate for various surveys. Milosavljević *et al.* (2006) have showed that TDEs makes a negligible contribution at the higher end of luminosity functions. Stone & Metzger (2016) using the Schechter luminosity function showed that the volumetric rate of TDE detection is sensitive to the occupation fraction of low-mass black holes. Van Velzen (2018) using optical/UV selected TDEs and forward modeling showed that luminosity function $\propto L^{-2.5}$.

2. Black hole mass function

We consider a separable form of BHMF $\Phi(M_\bullet,\ z) = \mu(M_\bullet) X(z)$, where $X(z) = 1 - \delta(z)$ with duty cycle $\delta(z) = 10^{-3}(z/0.1)^{2.5}$ [Chen *et al.* 2007] and $\mu(M_\bullet)$ obtained using Schechter luminosity function $\Phi(L_R) = \Phi_\star (L_R/L_\star)^{-\alpha} e^{-L_R/L_\star}$, where L_R is the R-band luminosity (Schechter 1976). Combining the Faber-Jackson law given by $\sigma \propto L_R^{1/n}$ and $M_\bullet - \sigma$ relation given by $\sigma \propto M_\bullet^{\frac{1}{\lambda}}$, we have $\mu(M_\bullet) = (\Phi_\star \epsilon / M_s)\,(M_\bullet/M_s)^\beta \exp(-(M_\bullet/M_s)^\epsilon)$, where $M_s \propto L_\star^{\lambda/n}$, $\beta = \epsilon(1-\alpha) - 1$ and $\epsilon = n/\lambda$. We take $n = 4$ (Stone & Metzger 2016) and $\lambda = 4.86$ (Ferrarese & Ford 2005).

Table 1. The TDE surveys with the duration of survey t_s and the number of TDEs detected.

Survey	t_s **(yr)**	$N_{D,obs}$
ASAS-SN	4	3
PTF	4	4
iPTF	4	2
PS-MDS	3.88	2
GALEX	10.16	3

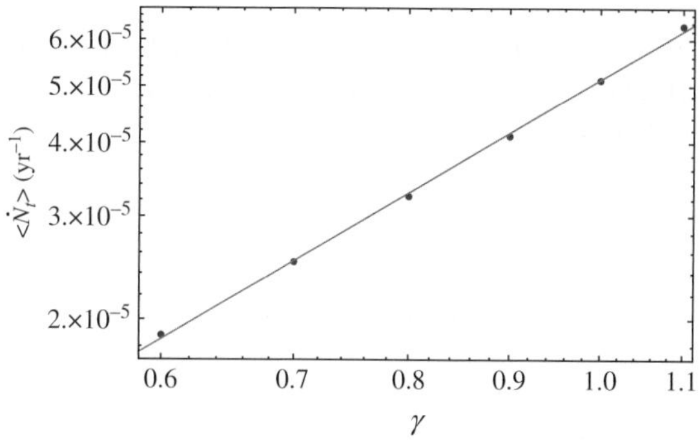

Figure 1. The galaxy averaged theoretical capture rate is shown in blue for various γ and red line shows best fit $< \dot{N}_t > \sim 2 \times 10^{-5}\gamma^2$ yr^{-1}.

The peak accretion rate and time of peak accretion for a star disrupted at r_t using MM15, are given by $\dot{M}_p = 7.9 \times 10^{25}$ g sec^{-1} $k^{\frac{3}{2}} m^{0.8} M_6^{-\frac{1}{2}}$ and $t_p =$ 38.55 days $k^{-\frac{3}{2}} m^{0.2} M_6^{\frac{1}{2}}$, where k is spin-up factor taken to be 3, $M_6 = M_\bullet/(10^6 M_\odot)$ and $m = M_\star/M_\odot$. Assuming that the accretion rate follows the fall back rate, the peak luminosity is given by $L_p = GM_\bullet \dot{M}/r_{in} = 7.1 \times 10^{46}$ erg sec^{-1} $k^{\frac{3}{2}} m^{0.8} M_6^{-\frac{1}{2}} Z^{-1}(j)$, where $r_{in} = (GM_\bullet/c^2)Z(j)$, j is the black hole spin and $Z(j)$ is given by Bardeen *et al.* (1972). If we can calculate the L_p and t_p from observations, we can estimate $M_\bullet$ and $M_\star$. We consider the steady accretion disk model of MM15 for a star on a parabolic orbit with $r_p = r_t$, to calculate the peak luminosity $L_p(M_\bullet,\ M_\star,\ z)$ in the given spectral bands with the corresponding time $t_p(M_\bullet,\ M_\star,\ z)$ and assume the luminosity given by $L = L_p\left((t_p + \delta t)/t_p\right)^{-5/3}$ in the declining phase to calculate the duration of flare detection $\delta_f = t_p[\left(L_p/(4\pi f_l d_L^2(z))\right)^{3/5} - 1]$ where f_l is the sensitivity of the detector. Assuming the probability of detection to be $P = \mathrm{Min}[1,\ \delta_f/(t_{cad} + t_{int})]$, where t_{cad} and t_{int} are the cadence and the integration times of the survey instrument, we calculate the maximum redshift $z_m(M_\bullet,\ M_\star)$ of detection. The detection rate of TDEs for a given survey is given by $\dot{N}_D = 4\pi f_s d_H^3 \int_{10^6 M_\odot}^{10^8 M_\odot} \mathrm{d}M_\bullet \mu(M_\bullet) \int_{0.8M_\odot}^{150M_\odot} \mathrm{d}M_\star \left(\mathrm{d}\dot{N}_t(\gamma,\ M_\bullet,\ M_\star)/\mathrm{d}M_\star\right) \int_0^{z_m} \mathrm{d}z\ K(z)$, where f_s is fraction of sky observed, $d_H = c/H_0$, $\mathrm{d}\dot{N}_t/\mathrm{d}M_\star$ is obtained by integrating eqn (48) given in MM15 over $\bar{e}$ and ℓ, $K(z) = (X(z)/(1+z))(I^2(z)/W(z))$ and $I(z)$ and $W(z)$ are given by eqn (86) in MM15. We calculate the number of detections in survey time t_s for the surveys given in Table 1 and compare it with the observed rate to calculate the Schechter parameters.

3. Results

We found that $\gamma = 0.9$, is the most possible solution to match the observed Schechter parameters and the best inferred Schechter parameters from our model fit is $\alpha = 1.093$, $\phi_\star = 10^{-4}$ Mpc^{-3} and $L_\star = 1.13 \times 10^{42}$ erg sec^{-1}. The galaxy averaged capture rate calculated using derived BHMF is $< \dot{N}_t > \sim 2 \times 10^{-5} \gamma^2$ yr^{-1} as shown in Fig 1 which is comparable with observed TDE rates $\sim 10^{-5}$ yr^{-1} (Donley *et al.* 2002). The estimation of BHMF from TDEs will improve with larger sample sizes which will be possible in the near future with the upcoming surveys like LSST, eROSITA, and zPTF.

References

Bardeen, J. M., Press, W. H., & Teukolsky, S. A. 1972, *ApJ*, 178, 347

Chen, Y.-M., Wang, J.-M., & Zhang, F. 2007, *in Astronomical Society of the Pacific Conference Series*, Vol. 373, 667

Donley, J. L., Brandt, W. N., Eracleous, M., & Boller, T. 1995, *AJ*, 124, 1308

Ferrarese, L., & Ford, H. 2005, *Space Science Reviews*, 116, 523

Kroupa, P. 2001, *MNRAS*, 322, 231

Mageshwaran, T., & Mangalam, A. 2015, *ApJ*, 814, 141

Merritt, D 2013, *Dynamics and Evolution of Galactic Nuclei* (Princeton University Press)

Milosavljević, M., Merritt, D., & Ho, L. C. 2006, *ApJ*, 652, 120

Schechter, P. 1976, *ApJ*, 203, 297

Stone, N. C., & Metzger, B. D. 2016, *MNRAS*, 455, 859

van Velzen, S. 2018, *ApJ*, 852, 72

Perseus in Sicily: From Black Hole to Cluster Outskirts
Proceedings IAU Symposium No. 342, 2018
K. Asada, E. de Gouveia Dal Pino, M. Giroletti,
H. Nagai & R. Nemmen, eds.

doi:10.1017/S1743921318006026

Multi-frequency monitoring of S5 0716+714

Jee Won Lee and Bong Won Sohn

Korea Astronomy and Space Science Institute, Republic of Korea Republic of Korea
email: jwlee78@kasi.re.kr

Abstract. We report preliminary results of long-term multi-frequency monitoring observations of S5 0716+714. We conducted observations at 22 and 43 GHz using Korean VLBI Network (KVN) radio telescopes and combined 8 (UMRAO), 15 (OVRO), 95 (CARMA), and 230 GHz (SMA) data to investigate characteristics of radio flares. We identified six flares (P1-P6) from 2010 November to 2014 June. The magnetic field strengths by assuming a synchrotron self-absorption model are in the range of 9 mG - 11 G in P2, 26 mG - 3G in P3, 3 mG - 38 G in P4, and 1 mG - 8 G in P6.

Keywords. BL Lacertae objects: individual (0716+714), galaxies: active, galaxies: jets

1. Introduction

S5 0716+714 is known as an extremely variable and a bright BL Lacertae-type blazar which has a relativistic jet pointing to the line of sight. Due to the orientation of jet, radio emission is Doppler boosted. The source has been observed complex variability in flux density and polarization emission on time scales of days to months across the entire electromagnetic spectrum from radio to gamma-ray bands (Rani *et al.* 2013; Lee *et al.* 2016; Lee *et al.* 2017a; Lee *et al.* 2017b). In our multi-frequency observations over three years and seven months, six local flares on time scales of weeks to months were detected at all observing frequencies. We investigate the individual flares showing different characteristics of S5 0716+714 to improve our understanding of the emission mechanism.

2. Observations and Data

We performed simultaneous multi-frequency single-dish observations of S5 0716+714 at 22 and 43 GHz using 21-m KVN radio telescopes from 2010 November to 2014 June (MJD 55509 to 56814) with a mean cadence of ~5 days. We collected multi-frequency data at 8 GHz (UMRAO: University of Michigan Radio Astronomy Observatory) from MJD 55504 to 56060, 15 GHz (OVRO: Owens Valley Radio Observatory), 95 GHz (CARMA:Combined Array for Research in Millimeter-wave Astronomy) from MJD 56157 to 56821, and 230 GHz (SMA:Submillimeter Array).

3. Results

S5 0716+714 shows active flux variability over observing period. Six local flares were detected (see Fig. 1). In order to analyze spectra, we used linear interpolation to the data then we fitted the spectra using synchrotron self-absorption model. We found that the individual flares have different ranges of turnover frequencies (see (c) in Fig. 2). The synchrotron self-absorption magnetic field strengths were estimated using a function in Marscher (1983). We adopted angular size θ of the source as 0.04 mas, Doppler factor δ as 7, following Rani *et al.* (2013), redshift z as 0.127, and b(α) as 2.92 (Lee *et al.* 2017a). The preliminarily estimated magnetic field strengths are in the range of 9 mG - 11 G in

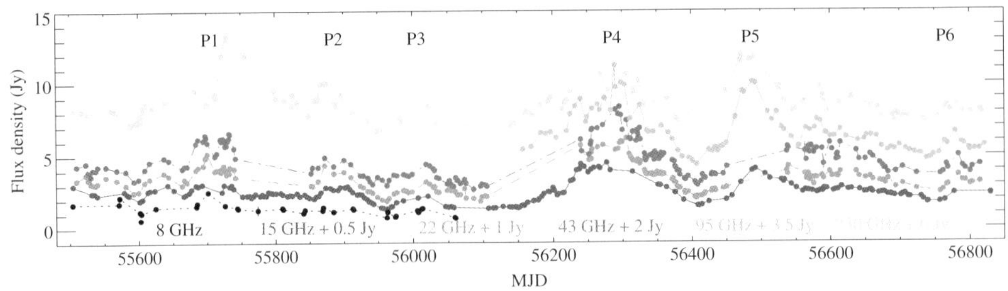

Figure 1. Light curves of S5 0716+714 observed 8 to 230 GHz over 3 years and 7 months from November 2010 to June 2014.

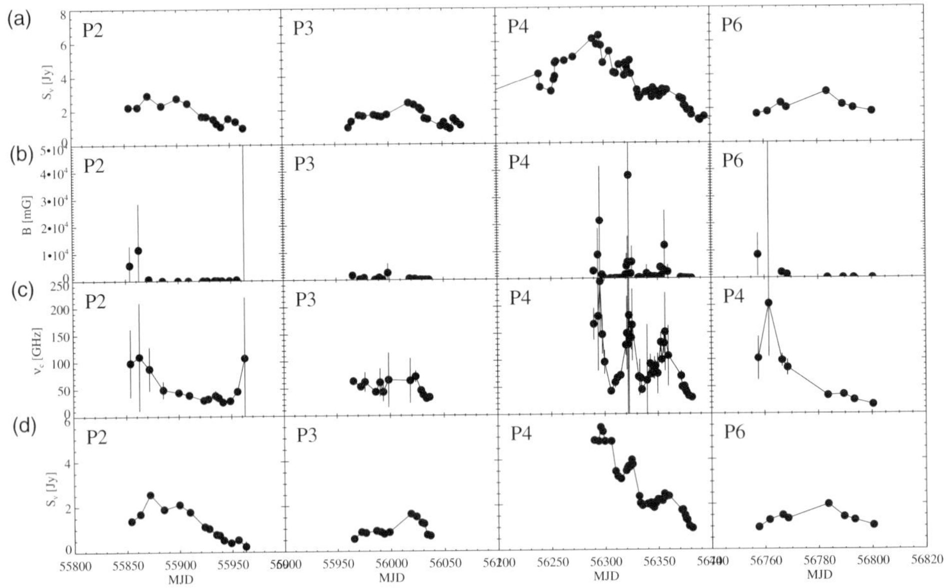

Figure 2. Plots of flux density light curves at 43 GHz (a), B-field (b), turnover frequency ν_c (c), and peak flux density S_m (d) in time domain for P2 to P6

P2, 26 mG - 3G in P3, 3 mG - 38 G in P4, and 1 mG - 8 G in P6. From top to bottom in Fig. 2, flux density light curves at 43 GHz for individual flares, magnetic field strength, turnover frequency, and peak flux density in the time domain are displayed.

Acknowledgement

We are grateful for the support of the National Research Council of Science and Technology, Korea(Project Number EU-16-001). This research has made use of data from the OVRO 40-m monitoring program (Richards, J. L. *et al.* 2011, ApJS, 194, 29) which is supported in part by NASA grants NNX08AW31G, NNX11A043G, and NNX14AQ89G and NSF grants AST-0808050 and AST-1109911. We thank Talvikki Hovatta for providing us the 95 GHz CARMA data obtained through the MARMOT program. The Submillimeter Array is a joint project between the Smithsonian Astrophysical Observatory and the Academia Sinica Institute of Astronomy and Astrophysics and is funded by the Smithsonian Institution and the Academia Sinica.

References

Lee, J. W., Lee, S.-S., Kang, S., Byun, D.-Y., & Kim, S. S. 2016, *A&A*, 592, L10
Lee, J. W., Lee, S.-S., Hodgson, J. A., *et al.* 2017a, *ApJ*, 841, 119
Lee, J. W., Sohn, B. W., Byun, D.-Y., Lee, J. A., & Kim, S. S. 2017b, *A&A*, 601, A12
Marscher, A. P. 1983, *ApJ*, 264, 296 Marscher, A. P., & Gear, W. K. 1985, *ApJ*, 298, 114
Rani, B., Krichbaum, T. P., Fuhrmann, L., *et al.* 2013, *A&A*, 552, A11

Perseus in Sicily: From Black Hole to Cluster Outskirts
Proceedings IAU Symposium No. 342, 2018
K. Asada, E. de Gouveia Dal Pino, M. Giroletti,
H. Nagai & R. Nemmen, eds.

doi:10.1017/S1743921318007317

Survival of Population III stars

Jayanta Dutta[1,2], Sharanya Sur[3], Athena Stacy[4] and Jasjeet Singh Bagla[1]

[1]Department of Physical Sciences, IISER, Mohali, 140306, India

[2]Department of Physics, University of *Évora*, 7002-554 *É*vora, Portugal

[3]Indian Institute of Astrophysics, 2nd Block, Koramangala, Bangalore 560034, India

[4]University of California, Berkeley, CA 94720, USA

`jd.astrop@gmail.com`

Abstract. In our earlier study Dutta (2016a), it has been shown that a number of primordial protostars (the 'first stars' in the Universe, also known as Population III or Pop III stars) are being ejected from the cluster of their origin with the velocity exceeding their escape velocity. Hence there is possibility that some of these protostars can enter main sequence and survive till present epoch, even in Milky Way. We ask the question if the protostars can avoid core collapse, and stop accreting before being ejected from the cluster, with the final mass of stars as 0.8 $M_{\odot}$.

Keywords. stars: Pop III – accretion – hydrodynamics – instabilities – methods: numerical

1. Motivation

The question of survival of the oldest stars in the Universe becomes interesting in the light of recent discoveries of extremely metal poor stars both in the halo of Milky Way as well as in the bulge. Present numerical simulations of collapse of primordial gas inside minihalos suffers from two major shortcomings (Hartwig *et al.* (2015), Dutta (2016b), Dutta *et al.* (2017)). Thus, the final fate of the protostars remains unclear. Here we use the typical orbital parameters of the protostars and the properties of the clusters, as determined from numerical simulations. We then determine a lower limit for Pop III stars that could have avoided the core collapse and survived for a significantly long time.

2. Numerical Method

We develop a semi-analytical model of Bondi-Hoyle accretion (Bondi & Hoyle (1944)) guided by results of cosmological simulations of Dutta *et al.* (2015a) and Dutta (2015b) to study the mass accretion by these protostars as a function of original stellar mass, and other parameters such as angular momentum and gravitational drag due to ambient gas. We also model the trajectories of protostars inside the gravitationally bound system.

3. Results

Mass accretion of the protostars that escape cluster is higher for a lower initial radial velocity. On the contrary protostars that remain in the cluster can have typical mass in the range $\sim 0.1 - 10\, M_{\odot}$ depending on their radial and rotational support. In addition, we have found that the Bondi-Hoyle accretion is a good approximation to understand the evolution of Pop III stars in an unstable self-gravitating disk. Protostars with speed greater than the escape speed can overcome the gravitational drag by the ambient gas and accrete only a small amount of mass before being ejected as low-mass Pop III stars

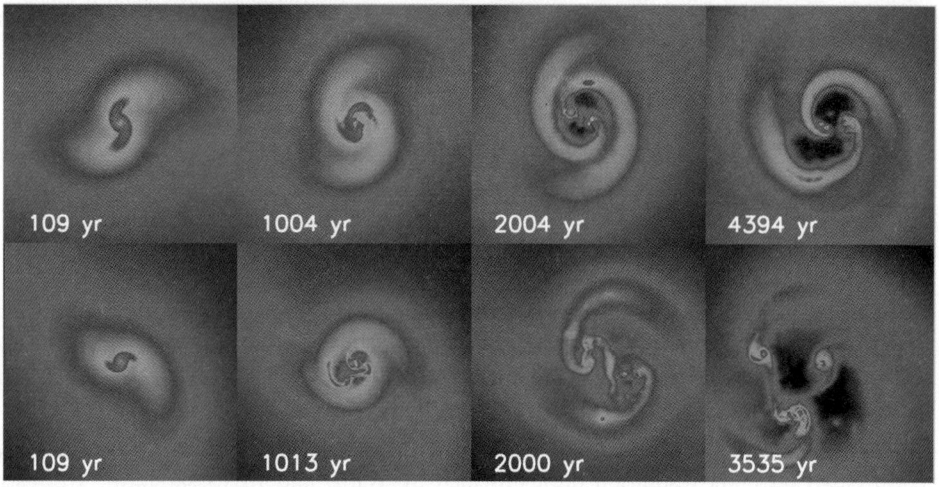

Figure 1. Evolution of gravitationally unstable disk that fragments to form Pop III cluster.

($M_* \sim 0.8\,M_\odot$). We also consider the evolution of a group of Pop III stars in a gas cloud and it is possible that many body interactions lead to a more nuanced understanding.

4. Conclusion

We expect that Pop III protostars that initially form within a certain range of mass and velocity larger than the escape velocity may survive with a mass lower than the cutoff. Hence, these may be found in the Milky Way or its satellites. Based on our theoretical as well as observational investigation, we conclude that the low-mass and low-metallicity stars are likely to exist in the halo or loosely bound structures like low-mass satellites.

5. Discussion

We have investigated in detail the gravitational collapse of primordial gas in minihalos and resulting fragmentation of circumstellar disk. We have ignored star-star interactions that may affect their radial and rotational velocities. During the interaction with ambient medium, stars gain or loose angular momentum that can change their velocity enough to escape the cluster. In future, we will investigate the N-body simulation using the spherical Bondi-Hoyle accretion to get more realistic evolution for their dynamical behaviour.

References

Bondi H. & Hoyle F., 1944, *MNRAS*, 104, 273B
Dutta, J, Nath, B, B., Clark, P. C., & Klessen, R. S., 2015, *MNRAS*, 450, 202D
Dutta, J., 2015, *ApJ*, 811, 98D
Dutta, J., 2016, *A&A*, 585A, 59D
Dutta, J., 2016, *Ap&SS*, 361, 35D
Dutta, J., Sur S., Stacy A., Bagla J. S., arXiv:1712.06912v2 (2017)
Hartwig, T., Bromm V., Klessen, R. S, & Glover, S. C. O., 2015, *MNRAS*, 447, 3892

Perseus in Sicily: From Black Hole to Cluster Outskirts
Proceedings IAU Symposium No. 342, 2018
K. Asada, E. de Gouveia Dal Pino, M. Giroletti,
H. Nagai & R. Nemmen, eds.

doi:10.1017/S174392131800738X

Infrared Diagnostics of the ISM in the Circumnuclear Environments of the Youngest Radio Galaxies

E. Kosmaczewski[1,2], **Ł. Stawarz**[2], **A. Wójtowicz**[2], ***et al.***

[1]Rensselaer Polytechnic Institute (RPI), 110 8th Street, Troy, NY 12180, USA
email: `emily.kosmaczewski@student.uj.edu.pl`

[2]Astronomical Observatory of Jagiellonian University, ul. Orla 171, 30-244 Kraków, Poland

Abstract. Here we present a systematic analysis of the mid-infrared properties of young radio galaxies, based on lower-resolution data provided by WISE and IRAS satellites. We restrict our analysis to sources in the earliest phase of radio galaxy evolution, with corresponding ages of the radio structures $\leq 3,000$ yrs. In our sample of 29 objects, we find a variety of WISE colors, which suggests that the mid-infrared continua of studied sources are not exclusively contributed to by the circumnuclear dust. A comparison of the total mid-infrared and absorption-corrected X-ray luminosities for our sample reveals a clear correlation between the two bands. This favors the scenario in which the observed X-ray emission of young radio galaxies — at least the high-luminosity ones — originates predominantly in accretion disk corona.

Keywords. ISM: jets and outflows— galaxies: active— galaxies: jets— infrared: galaxies— X-rays: galaxies

---

1. Introduction

Young radio sources constitute a particularly interesting yet diverse class of active galactic nuclei (AGN), with newly born radio structures (jets and lobes) fully confined within their host galaxies. We focus on a systematic investigation of the mid-infrared (MIR) properties of the youngest radio galaxies. Sources have been confirmed spectroscopically in the radio domain as 'GHz-peaked spectrum' (GPS) sources, and/or morphologically (using high-resolution radio interferometers) as 'compact symmetric objects' (CSOs), using the data provided by the Wide-field Infrared Survey Explorer (WISE), as well as the archival data from the Infrared Astronomical Satellite (IRAS) augmented in a few cases by the *Spitzer* Space Telescope observations.

2. Discussion & Conclusions

In Figure 1 (left), all objects in the sample confirmed as Compton-thick based on the X-ray spectroscopy (1404+286, 1511+0518 and 2021+614; see Guainazzi *et al.* 2004; Siemiginowska *et al.* 2016; Sobolewska *et al.* 2018, respectively), are characterized by the WISE colors consistent with the quasar identification, as in fact expected. The most striking feature of the diagram is, however, the fact that the youngest radio galaxies with colors W1–W2 ≤ 0.5 mag, i.e. those dominated in MIR by the ISM radiative output, populate almost the entire region occupied by galaxies (from elliptical to starbursts) with the W2–W3 colors between 1.0 and 4.0, meaning a wide range of star formation activity within their hosts.

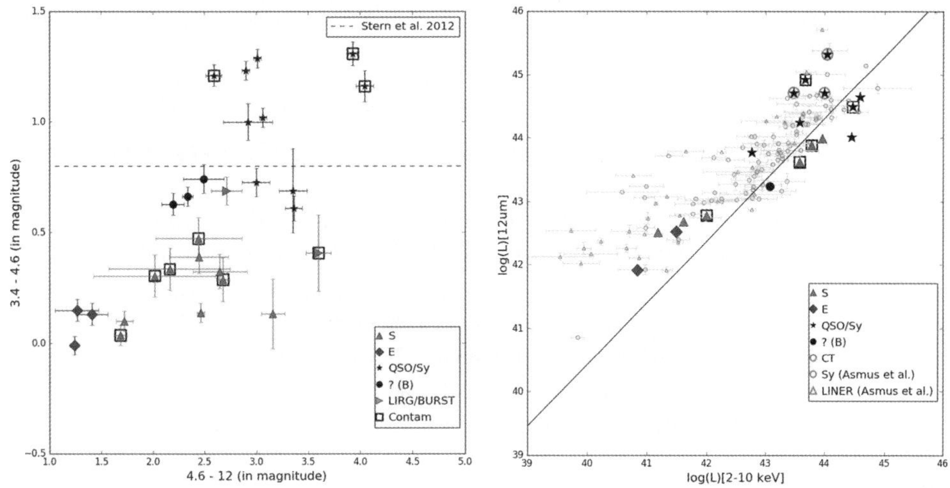

Figure 1. (Left) WISE color magnitudes for young radio galaxies included in our sample. Different symbols denote WISE color classification following Wright *et al.* (2010); sources marked with a black square may have significant contamination from nearby objects; red circles indicate objects confirmed as Compton-thick. The horizontal line delineates Stern *et al.* (2012) cut: the MIR emission of sources located above the line are dominated by an AGN component. (Right) Relation of the IRAS 12 μm and the intrinsic (absorption-corrected) 2–10 keV luminosities for our young radio galaxies denoted by large symbols (same as in the image on the left). The solid line shows the correlation established for nearby AGN by Asmus *et al.* (2015) based on the high-angular resolution MIR (ground-based) observations; small gray symbols denote the AGN from the Asmus *et al.* (2014) sample with IRAS 12 μm luminosities.

In Figure 1, right, we plot the IRAS 12 μm and the intrinsic (absorption-corrected) 2–10 keV luminosities for the sample of our young radio galaxies, along with a comparison sample of Seyfert galaxies and LINERs from Asmus *et al.* (2014), for which the 12 μm luminosities were obtained similarly with the IRAS satellite. In the cases of two young radio galaxies for which the IRAS 12 μm were not available, we used the *Spitzer*/MIPS 24 μm luminosities instead; note that the $\sim 6''$ MIPS resolution at 24 μm, is comparable to the IRAS angular resolution of $\sim 30''$ at 12 μm. The AGN from the comparison sample are located mostly above the best-fit correlation line: this is because with the relatively poor angular resolution of IRAS, the MIR luminosities derived for the sample are contributed not only by the circumnuclear dust emission, but also by the star formation activity within the host galaxies, while the best-fit correlation was established based on sub-arcsec MIR photometry, so without any significant ISM emission component.

References

Asmus, D., Hönig, S. F., Gandhi, P., Smette, A., & Duschl, W. J. 2014, *MNRAS*, 439, 1648
Asmus, D., Gandhi, P., Hönig, S. F., Smette, A., & Duschl, W. J. 2015, *MNRAS*, 454, 766
Guainazzi, M., Siemiginowska, A., Rodriguez-Pascual, P., & Stanghellini, C. 2004, *A&A*, 421, 461
Siemiginowska, A., Sobolewska, M., Migliori, G., *et al.* 2016, *ApJ*, 823, 57
Sobolewska, M., *et al.* 2018, *ApJ*, submitted
Stawarz, Ł., Ostorero, L., Begelman, M. C., *et al.* 2008, *ApJ*, 680, 911
Stern, D., Assef, R. J., Benford, D. J., *et al.* 2012, *ApJ*, 753, 30
Wójtowicz, A., *et al.* 2018, *ApJ*, submitted
Wright, E. L., Eisenhardt, P. R. M., Mainzer, A. K., *et al.* 2010, *AJ*, 140, 1868

Perseus in Sicily: From Black Hole to Cluster Outskirts
Proceedings IAU Symposium No. 342, 2018
K. Asada, E. de Gouveia Dal Pino, M. Giroletti,
H. Nagai & R. Nemmen, eds.

doi:10.1017/S1743921318007767

A Photoionization Method for Estimating Black Hole Masses in Quasars

C. Alenka Negrete[1], Deborah Dultzin[2], Paola Marziani[3], Jack W. Sulentic[4] and M. L. Martínez-Aldama[4]

[1]CONACyT Fellowship – Instituto de Astronomía, UNAM, Mexico
email: alenka@astro.unam.mx

[2]Instituto de Astronomía, UNAM, Mexico

[3]Osservatorio Astronomico di Padova, INAF, Italy

[4]Instituto de Astrofísica de Andaluca, CSIC, Spain

Abstract. We present a method that uses photoionization codes (CLOUDY) to estimate the supermassive black hole masses ($M_{\rm BH}$) for quasars at low and high redshift. This method is based on the determination of the physical conditions of the broad line region (BLR) using observational diagnostic diagrams from line ratios in the UV. We also considered that the density and metallicity of the BLR in quasars at high z could be different from those at the nearby Universe. The computed black hole masses obtained using this method are in agreement with those derived from the method of reverberation mapping.

Keywords. galaxies: active, quasars: general, quasars: emission lines

---

Our "photoionization method" (ϕ) estimates the size of the BLR ($r_{\rm BLR}$) based on the ionization parameter (U) definition. Solving for $r_{\rm BLR}$ we obtain:

$$r_{BLR} = \left[\frac{\int_{\nu_0}^{+\infty} \frac{L_\nu}{h\nu} d\nu}{4\pi n_{\rm H} U c} \right]^{1/2} \quad (1)$$

Where the integral represents the number of ionizing photons, and $n_{\rm H}$ is the hydrogen density.

Using diagnostic line ratios of the broad components (BC) in the UV spectra, we can estimate the product $n_{\rm H}$U (Negrete *et al.* 2012): AlIIIλ1860/SiIII]λ1892, sensitive to $n_{\rm H}$; CIVλ1549/SiIII]λ1892, a marker of ionization level; SiIVλ1397/SiIII]λ1892, sensitive to ionization, roughly independent of metallicity.

CLOUDY simulations (Ferland *et al.* 2013) were performed at fixed $n_{\rm H}$ and U values spanning in the ranges 7.00 $\leqslant$ log $n_{\rm H}$ $\leqslant$ 14.00 and -4.50 $\leqslant$ log U $\leqslant$ 00.00. We assumed plane-parallel geometry, column density $N_c = 10^{23} cm^{-2}$, a "standard" quasar continuum (Mathews & Ferland 1987), and three metallicities: 1$Z_\odot$, 5$Z_\odot$ and 5$Z_\odot$SiAl (an overabundance of Si and Al due to type II supernovae; Sani *et al.* 2010, Negrete *et al.* 2014).

To obtain the line fluxes; and then the line ratios, we used the IRAF task Specfit (Kriss 1994) to fit the spectral components and isolate the BCs. These line ratios allow us to draw isopleths i. e., curves representing measured values of these ratios in the $n_{\rm H}$ vs. U plane (Fig. 1).

Having the $r_{\rm BLR}$ we can compute the $M_{\rm BH} = \left(\frac{3}{4G}\right) f_{0.75} FWHM^2 r_{\rm BLR}$; where $f_{0.75} = 1.4$ is the geometry factor and G is the gravitational constant. Figure 2 shows that the $M_{\rm BH}$ computed following this method is in agreement with the ones derived from the reverberation mapping directly (RM; Bentz *et al.* 2009) and by the luminosity-$M_{\rm BH}$

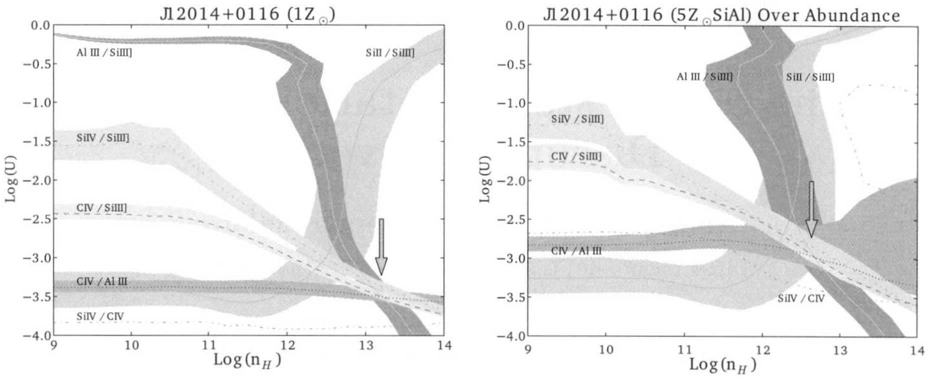

Figure 1. Isocontour maps considering $1Z_\odot$ and $5Z_\odot$SiAl metallicities. The intersection points give us the best estimates of $n_{\rm H}$ and U. Shaded areas are the uncertainties.

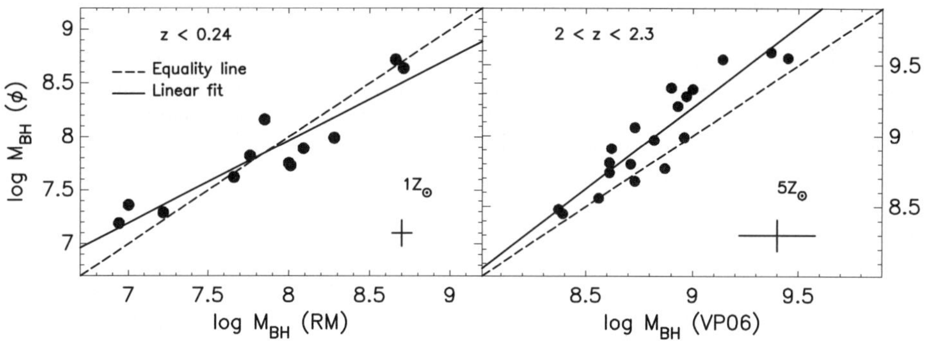

Figure 2. Left – $M_{\rm BH}$ comparison derived with our method and RM results for a low-z sample described in Negrete *et al.* (2013). Right – Preliminar results for a high-z sample of high accreting quasars (Martnez-Aldama *et al.* in prep).

relation (Vestergaard & Peterson 2006). It is worth to mention that VP06 computations are based on CIVλ1549 measurements, a line that systematically shows blue asymmetries (Sulentic *et al.* 2017).

Conclusion

We are able to estimate BLR distances and black hole masses using an independent photoionization method that yields results consistent with reverberation values and previous works in the UV. We suggest that the derived $r_{\rm BLR}$ values can significantly improve $M_{\rm BH}$ estimation especially at z $\geqslant$ 2 when the intermediate ionization lines are shifted into the wavelength range accessible to optical spectrometers.

References

Bentz, M. C., *et al.* 2009, *ApJL*, 694, L166
Ferland, G. J., *et al.* 2013, *RevMexA&Ap*, 49, 137
Kriss, G. 1994, A.S.P. Conference Series, 61, 437
Mathews, W. G., & Ferland, G. J. 1987, *ApJ*, 323, 456
Negrete, C. A., Dultzin, D., Marziani, P., & Sulentic, J. 2012, *ApJ*, 757, 62
Negrete, C. A., Dultzin, D., Marziani, P., & Sulentic, J. W. 2013, *ApJ*, 771, 31
Negrete, C. A., Dultzin, D., Marziani, P., & Sulentic, J. W. 2014, *ApJ*, 794, 95
Sani, E. *et al.* 2010, *MNRAS*, 403, 1246
Sulentic, J. W., del Olmo, A., Marziani, P., *et al.* 2017, *A&A*, 608, 122
Vestergaard, M., & Peterson, B. M. 2006, *ApJ*, 641, 689

Perseus in Sicily: From Black Hole to Cluster Outskirts
Proceedings IAU Symposium No. 342, 2018
K. Asada, E. de Gouveia Dal Pino, M. Giroletti,
H. Nagai & R. Nemmen, eds.

doi:10.1017/S1743921318003691

The role of stellar rotation in Tidal Disruption Events

Andrea Sacchi[1] **and Giuseppe Lodato**[2]

[1]Dipartimento di Fisica, Università degli Studi di Pavia,
Via Bassi 6, Pavia, Italy
email: andrea.sacchi05@universitadipavia.com

[2]Dipartimento di Fisica, Università degli Studi di Milano,
Via Celoria 16, Milan, Italy
email: giuseppe.lodato@unimi.it

Abstract. Tidal Disruption Events (TDEs) are highly variable high energy phenomena originating from Galactic Nuclei (Komossa & Bade 1999). TDEs are thus powerful tools to study quiescent Galactic Nuclei given their extreme brightness (several times super-Eddington) and the possibility of being seen in non-AGN galaxies. A TDE is the violent disruption of a star passing by a Super Massive Black Hole (SMBH); after the disruption, roughly half of the star mass gains enough energy to escape from the Black Hole, while the other half is bound to the Hole, falls back and eventually accretes onto it. Early works, (Rees 1988), pointed out a $t^{-5/3}$ behaviour for the light curves of this event and since then such a time dependency became the signature of these events. Strong deviations are however introduced when one considers the internal stellar structure or if one considers partial disruptions. One feature that has never been taken into account is the effect of stellar rotation in the resulting fallback rate, which is the aim of the present work. Firstly, we will show analytical estimates of the impact of stellar rotation on the TDE and we will then present a set of Smoothed Particle Hydrodynamic simulations of the tidal disruption of rotating stars, performed in order to test these analytical estimates.

Keywords. galaxies: nuclei, hydrodynamics, black hole physics

---

1. Analytical estimates

A Tidal Disruption Event occurs when a star gets so close to a SMBH that the BH tides overcame its self-gravity. After the disruption half of the debris are launched into highly eccentric bound orbits around the SMBH. TDEs are characterized by the mass return rate to the pericentre that is proportional to the event light curve. This quantity is determined by three parameters (in the simplest model): the mass return rate peak, the characteristic time and the time dependency. In our work we focused on understanding how stellar rotation would change these parameters with respect to the non rotating case. Analytically we followed the calculations in Rees (1988) assuming that the star is a rigidly rotating sphere with spin perpendicular to the orbital plane and with uniform density. That is because tidal forces induce a rotation to the star up to its break up velocity along an axis parallel to the orbital angular momentum.

We obtained an analytical relation that links the mass return rate peak of a TDE to the stellar rotation: we found that a TDE can be brightened or dimmed or even completely turned off depending on whether the stellar rotation is prograde or retrograde with respect to the tidal induced one. We show such analytical relation in Fig. 1 (solid curve). To better study this configuration and to investigate other stellar and orbital spin alignment, we performed numerical simulation.

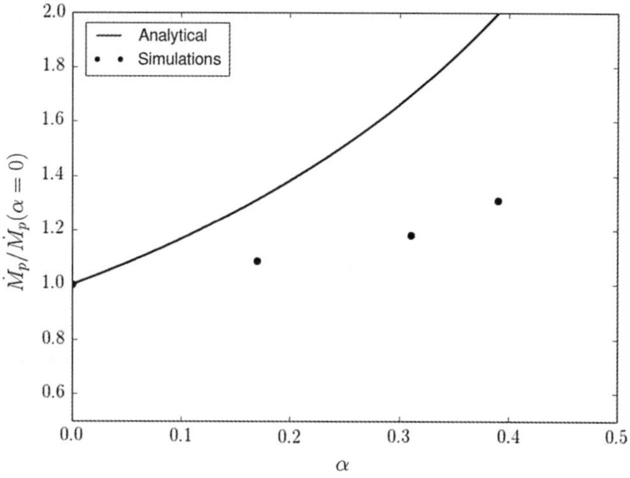

Figure 1. Mass return rate peak as a function of stellar rotation: analytical estimate and simulation outputs.

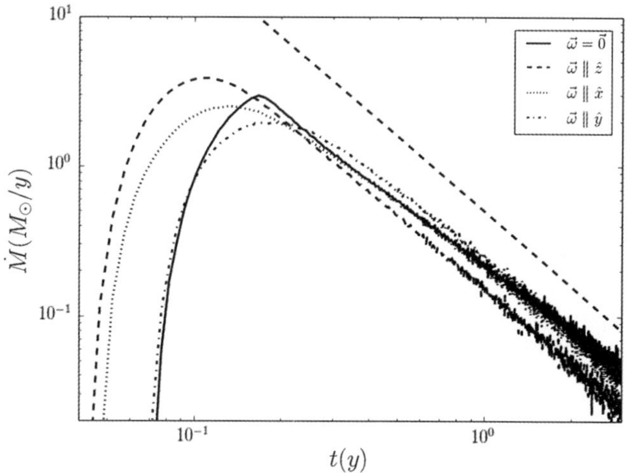

Figure 2. Mass return rate as a function of stellar spin orientation: the straight dashed line above every other curve is the $t^{-5/3}$ behaviour predicted.

2. Numerical set up

Our numerical simulations were performed with Smoothed Particle Hydrodynamic code PHANTOM (Price *et al.* 2017). Firstly, we set up a rotating star: we started from a polytropic sphere with polytropic index $\gamma = 5/3$ and we let the star relax in an empty space. Once relaxed we imposed a rigid rotation with magnitude $\alpha = \omega/\omega_b$ (where $\omega_b = \sqrt{2GM_\star/R_\star^2}$ is the break up angular velocity of the star) and then let the rotating star relax once more. Once the star is relaxed we check that the rotation is still rigid and then set the star on a parabolic orbit around a black hole of $10^6\ M_\odot$ with pericentre equal to the tidal radius (the distance at which tidal forces equal stellar self-gravity).

3. Results

Varying α, but keeping the star rotation prograde, we indeed observed brighter and shorter TDEs, however the enhancement is not as high as predicted analytically, this is probably due to the stellar polytropic structure (Fig. 1).

Varying stellar spin orientation with respect to the orbital plane we observed a distorted behaviour for TDEs light curves compared to the non rotating case (Fig. 2), however in both this and the scenario presented above the time dependency of $t^{-5/3}$ is reached asymptotically.

Very interesting is the case of a star rotating in retrograde way with respect to the orbital motion. In this case the simulations show that the debris are ripped off the star but stay bound to it and eventually fall back, forming a disc and getting accreted. Preliminary results indicate that this new kind of event will be shorter than a TDE (roughly some tens of days), but very bright: several times super-Eddington onto the star. It is also interesting to notice that the star (that initially was on a parabolic orbit) is bound to the black hole after the flyby and the tidal forces manage to completely reverse the spin of the star. We can thus predict that roughly one hundred years after the bright and short peak due to the material acrreting onto the star a normal TDE will take place.

References

Komossa S., Bade N., 1999, *A&A*, 343, 775
Price D. J., *et al.*, 2017, preprint, (arXiv:1702.03930)
Rees M. J., 1988, *Nature*, 333, 523

Perseus in Sicily: From Black Hole to Cluster Outskirts
Proceedings IAU Symposium No. 342, 2018
K. Asada, E. de Gouveia Dal Pino, M. Giroletti,
H. Nagai & R. Nemmen, eds.

doi:10.1017/S174392131800368X

On the likelihood of Gravitational Wave emission during the Tidal Disruption of stars by Super Massive Black Holes

Martina Toscani[1] **and Giuseppe Lodato**[2]

[1]Dipartimento di Fisica, Università degli Studi di Pavia,
Via Bassi 6, Pavia, Italy
email: mtoscani94@gmail.com

[2]Dipartimento di Fisica, Università degli Studi di Milano,
Via Celoria 16, Milan, Italy
email: giuseppe.lodato@unimi.it

Abstract. Tidal Disruption Events (TDEs) are a common feature between Active and Quiescent Galactic Nuclei; the study of these events is a very useful tool to probe phenomena that relate to the formation of an accretion disc or a jet. Also, the accretion rate at the beginning of the tidal flare is expected to be significantly super-Eddington and might result in high energy emission (in soft X-rays but sometimes up to the gamma regime, as in the the case of Swift J1644, see Komossa, 2015). These events may even play an important role in the newborn field of the Multimessenger Astronomy. This work is set within this context. Indeed, it is a study of generation of Gravitational Waves (GWs) from the hot accreting torus resulting after a TDE. Since the torus has only formed recently, magnetic fields are not expected to be strong enough, so that the torus is likely to be unstable to the Papaloizou-Pringle Instability (PPI), producing a strongly varying mass quadrupole. Here, the study of the evolution of such tori is developed, using both analytical calculation and a Smoothed Particle Hydrodynamics simulation (SPH). In particular the goal of this work is to determine the GW waveform and to compute the characteristic strain of these GWs in order to see if they are detectable by the Laser Interferometer Space Antenna (LISA).

Keywords. gravitational waves, black hole physics, accretion, accretion disks

---

1. Introduction

TDEs are known to be important sources of electromagnetic radiation; in addition, they could also be sources of gravitational radiation. As a matter of fact, during these events huge masses are involved, which undergo very high accelerations as they move close to the event horizon of a Supermassive Black Hole (SMBH). This work aims at studying the GW emission of a hot, thick accretion disc, resulting after a TDE, which is unstable with respect to the PPI. The work is divided in four parts: 1) an analytical study of the system, in order to see if these waves are detectable by LISA; 2) a numerical study of the system, starting from a simulation made by Nealon *et al.*, 2017, with PHANTOM, a Smoothed Particle Hydrodynamics and Magnetohydrodynamics code; 3) the study of the waveforms; 4) an extension of all the previous calculation in the case of rotating Black Holes.

2. Analytical estimates

In this first part we assume that the TDE occurs in the Virgo Cluster, at a distance of 16 Mpc. Results can be easily rescaled to other distances, since the strain scales

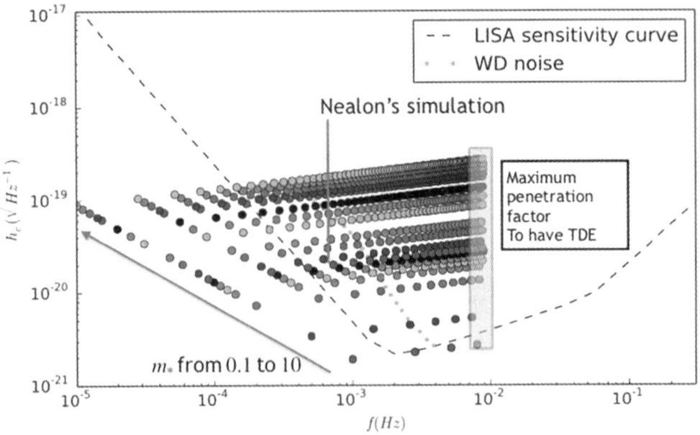

Figure 1. The plot shows the trend of h_c with respect to f, for Main Sequence Stars. The BH mass is $M_6 = 1$. The penetration factor β varies between $\beta_{\rm min} = 1$ and $\beta_{\rm max} = R_t/R_S$.

inversely with distance. In particular, we perform a study of the parameters of the system, considering different masses of the central Black Hole (BH), different types of stars that could be tidally disrupted and different closest approach between the star and the BH. We provide an analytical calculation of the magnitude of the gravitational wave strain tensor, using the approximated formula (e.g. see Thorne, 1998),

$$h \approx \frac{1}{r}\frac{4GE_{\rm kin}}{c^4}, \tag{2.1}$$

where h is the magnitude of the stress tensor which stands for the amplitude of the ripple in the fabric of space-time, G is the gravitational constant, c is the speed of light, $E_{\rm kin}$ is the kinetic energy and r is the distance of the source. The main goal of this study is to see if the characteristic strain, that is $h_c \doteq h\sqrt{f^{-1}}$, where f is the GW frequency, that we assume to be of the order of the Keplerian frequency at the circularization radius of the debris (see Bonnerot *et al.*, 2016), is above the LISA sensitivity curve.

3. Results

One important parameter for the study of this problem is the penetration factor $\beta = R_t/R_p$, where R_t is the tidal radius of the star, i.e. the maximum distance for the star to be disrupted by the BH and R_p is the pericentre of the stellar orbit before disruption. This factor varies between 1 (when $R_p = R_t$), and a maximum value, when R_p is equal to the Schwarzschild radius of the BH. We found that these waves may be detected by LISA, if discs resulting by Main Sequence (MS) and Red Giant (RG) stars are considered. In particular figure 1 shows the gravitational signal for MS stars; a BH of mass $M_6 = 1$ (that is $M_{\rm BH} = 10^6 M_\odot$) is considered. Each color represents a particular value of the stellar mass, which varies from $0.1M_\odot$ to $10M_\odot$. The factor β grows from left to right. It can be seen that, while β increasing, the gravitational signal is above the LISA sensitivity curve. We obtained a similar plot also for RG stars (not shown here). We expect that our analytical results overestimate the GW signal, because we suppose a 100% density perturbation. Thus, after our analytical study, we began a numerical study of the system based on the simulation by Nealon *et al.*, 2017, where they considered $M_{\rm disc} = 1M_\odot$, $\beta = 5$ and $M_6 = 1$; we rescaled their data to different values of the parameters involved. From this numerical study we estimate that the signal is two order of magnitude lower, but it is still in part above the LISA sensitivity curve, especially for the most penetrating events.

References

Komossa S., 2015, Journal of High Energy Astrophysics, Vol. 7, p 148-157
Thorne K. S., 1998, University of Chicago Press, p. 41
Nealon R., Price D. J., Bonnerot C., Lodato G., 2017, *MNRAS*, 1737-1745
Bonnerot C., Rossi E. M., Lodato G., Price D. J., 2016, *MNRAS*, 455, 2253

Perseus in Sicily: From Black Hole to Cluster Outskirts
Proceedings IAU Symposium No. 342, 2018
K. Asada, E. de Gouveia Dal Pino, M. Giroletti,
H. Nagai & R. Nemmen, eds.

doi:10.1017/S1743921318007883

Jet production efficiency in the sample of the youngest radio galaxies

Anna Wójtowicz, Łukasz Stawarz and Emily Kosmaczewski

Astronomical Observatory, Jagiellonian University, ul. Orla 171, 30-244 Kraków, Poland

Abstract. We investigate the sample of 16 the youngest radio galaxies with measured kinematic ages and available X-ray data from high-resolution Chandra or XMM-Newton observations. We characterize the accretion properties and derive the jet kinetic luminosities for our sources. We found high accretion rates ($> 1\%$ Eddington) and very high jet production efficiency for all the sources from our sample.This, along with the fact that the analyzed objects seem over-luminous in radio on the fundamental plane for the black hole activity, implies also that the radiative efficiency of the compact lobes is much higher than in the case of the evolved radio galaxies.

Keywords. techniques: spectroscopic, methods: data analysis, galaxies: jets, galaxies: active

1. Introduction

Compact radio galaxies, classified as Gigahertz-Peaked Spectrum (GPS) sources based on their radio spectra, are believed to represent the earliest phase in the evolution of radio-loud active galactic nuclei (AGN). As such they provide an excellent opportunity to study the jet formation in the very moment of triggering.

2. Method

In order to characterize the central engines for the objects in our sample, first we searched the SDSS database, and found spectra with sufficient quality for only four sources. We modeled these data with the STARLIGHT spectral synthesis code Cid Fernandes *et al.* (2005) and calculated (via velocity dispersion) the corresponding black hole masses. We also measured the $H\beta$ fluxes and derived the accretion disks bolometric luminosity using the scaling relation of Netzer (2009). For the remaining sources, we searched the literature for the measured $H\beta$ fluxes; if unavailable, we adopted the bolometric luminosity estimates following from different methods. We also adopted the remaining values for the black hole masses from Wu (2009), Willett (2010), Son (2012), and Trichas (2013).

We collected the 1.4 GHz fluxes for all the objects in our sample. With these, we estimated the jet total kinetic powers by means of the scaling relation introduced by Willott *et al.* (1999). We note that in some cases only the lower limit for the jet power could be derived due to pronounced absorption of the radio continua around GHz frequencies. However, since the scaling relation of Willott *et al.* (1999) has been established for the evolved sources, it may not be the most reliable one when applied to the youngest radio galaxies, for example due to the expected increased radiative efficiency of compact jets and lobes (see in this context Tadhunter (2016), Stawarz *et al.* (2008)). Therefore, we have also derived the *minimum* jet powers for all our sources by utilizing their measured kinematic ages, and assuming pressure equipartition within radio-emitting lobes. The jet kinetic luminosities derived in this way are about one order of magnitude lower than those following from the Willott *et al.* (1999) relation.

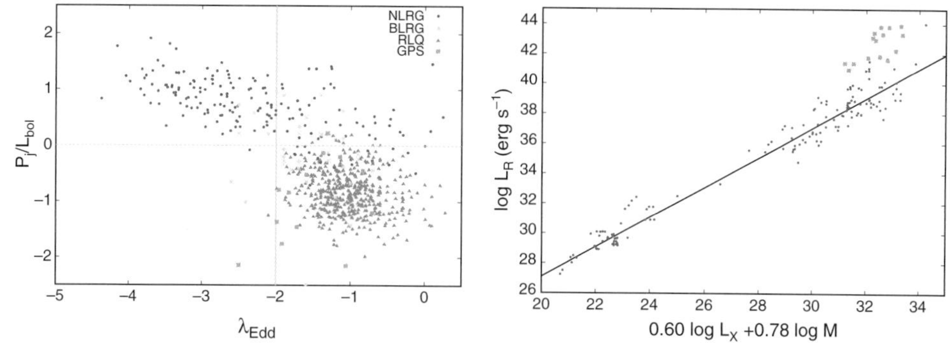

Figure 1. Left panel: Jet production efficiency vs. Eddington ratio estimated for our GPS/CSO sample, compared with evolved radio sources; jet powers estimated from measured kinematic ages, assuming pressure equipartition within the radio-emitting lobes. Right panel: fundamental Plane of the Black Hole Activity for Siemiginowska *et al.* (2016) sample of GPS compared with Merloni *et al.* (2003) sample. GPS are marked with red squares.

With such derived accretion rates, black hole masses, and jet kinetic powers, we were able to compare the jet production efficiencies in the youngest radio galaxies with those found in the evolved sources, including narrow-line radio galaxies (NLRGs), broad-line radio galaxies (BLRGs), and radio-loud quasars (RLQS), as listed in Rusinek *et al.* (2017) – see left panel of figure 1. With the addition of absorption-corrected (i.e., intrinsic) X-ray luminosities and 5 GHz radio luminosities available for our GPS/CSO sample, we also investigate the location of the youngest radio galaxies on the Fundamental Plane of the black hole activity (Merloni *et al.* (2003)) – see right panel figure 1.

3. Results & Conclusion

We found out that all but one (1718-649) the youngest radio galaxies studied in this work, accrete at high accretion rates (Eddington ratios $> 1\%$), and produce the jets with high efficiencies ($P_{\rm j}/L_{\rm bol} \sim 1\% - 100\%$), consistent with the prediction of the Blandford-Znajek model involving magnetically arrested disk (MADs). As such, they resemble more radio quasars and broad-line radio galaxies, rather than other types of radio-loud AGN. This, along with the fact that the objects from our sample seem over-luminous in radio on the fundamental plane for the black hole activity, implies also that the radiative efficiency of the compact lobes, i.e. a fraction of the jet kinetic luminosity converted to radio fluxes, is much higher than in the case of the evolved radio galaxies.

References

Cid Fernandes, R., Mateus, A., Sodré, L., Stasińska, G., & Gomes, J. M. 2005, *MNRAS*, 358, 363
Merloni, A., Heinz, S., & di Matteo, T. 2003, *MNRAS*, 345, 1057
Netzer, H. 2009, *MNRAS*, 399, 1907
Rusinek, K., Sikora, M., Kozie l-Wierzbowska, D., & Godfrey, L. 2017, *MNRAS*, 466, 2294
Siemiginowska, A., Sobolewska, M., Migliori, G., *et al.* 2016, *ApJ*, 823, 57
Son, D., Woo, J.-H., Kim, S. C., *et al.* 2012, *ApJ*, 757, 140
Stawarz, L., Ostorero, L., Begelman, M. C., *et al.* 2008, *ApJ*, 680, 911-925
Willett, K. W., Stocke, J. T., Darling, J., & Perlman, E. S. 2010, *ApJ*, 713, 1393
Willott, C. J., Rawlings, S., Blundell, K. M., & Lacy, M. 1999, *MNRAS*, 309, 1017
Wu, Q. 2009, *ApJL*, 701, L95
Trichas, M., Green, P. J., Constantin, A., *et al.* 2013, *ApJ*, 778, 188
Tadhunter, C. 2016, Astronomische Nachrichten, 337, 159

Author index